Handbook of Flood Risk Management in Developing Countries

This new Handbook brings together various views and experiences of the impacts of flooding and its management in Africa, Asia and Latin America by drawing from traditional and modern approaches adopted by communities, homeowners, academics, project managers, institutions and policy makers. Key stakeholders provide insights and perspectives on flood hazards, flood impacts, flood control and adaptation strategies across these regions. The inclusion of policy makers, emergency responders, leaders of key organizations and managers of flood defence projects makes this volume a unique addition to the flood management literature.

The chapters are organized to reveal various impacts and challenges associated with the management of flooding, including response and recovery. The chapter contributions bring together the different impacts of flooding and propose various mitigation approaches. They describe procedures for managing flooding and reducing the impacts from the perspectives of policy makers, environmental planners and restorers of flood-affected communities. Also, the book considers some of the related aspects including land use, waste management, drainage systems, security challenges, urban planning and development and their contributions to flooding.

The book's primary target is experienced researchers and practitioners in flood risk management. It would also serve as a key text for postgraduate students studying related programmes. Inhabitants of flood-prone communities in such developing countries will also find the text an important resource for guidance and understanding. This multi-disciplinary book represents a valuable contribution for a wide range of professionals (e.g. in engineering, built environment, health, retail, etc.) who are interested in flood control and management and/or faced with flood-related challenges in the course of their work.

Victor Oladokun, Ph.D., a Professor of Industrial and Production Engineering at the University of Ibadan, Nigeria, is a senior Fulbright scholar and a Commonwealth academic fellow. Professor Oladokun, a certified SAP trainer/consultant, is the Deputy Dean, University of Ibadan School of Business. He is a member of the Academic Board of SAP University Alliances Africa (ESEFA) and a member of the Commonwealth Scholarship Commission Alumni Advisory Panel. Engr. Oladokun, a member of the Nigerian Society of Engineers, worked in the heavy equipment service sector before joining academics. He has extensive experience in teaching, research, mentoring, leadership and curricula development and has been involved in several multidisciplinary and international collaborative initiatives. As the Chair of the Department of Industrial and Production Engineering, he led the successful development and deployment of a new professional master's program in engineering management to create a vital university–industry link. Professor Oladokun has served as a visiting research fellow at universities in the United Kingdom, United States and Nigeria. His research interests include disaster risk management and resilience modelling. His ongoing research includes the application of digital twin and IoT systems for disaster risk management and supply chain resilience improvement. Prof Oladokun teaches undergraduate and postgraduate courses in applied optimization, operation management, project management, scheduling, reliability engineering, soft computing, entrepreneurship, supply chain management and enterprise systems.

Professor David Proverbs is Dean of the Faculty of Science and Engineering at the University of Wolverhampton. He has over 25 years of experience in higher education and has held strategic leadership roles in three modern universities, where he has championed student learning as a research-driven, curriculum-active and enterprising academic. David has developed significant regional, national and international research and enterprise collaborations, drawing on a range of funding sources to pioneer innovative solutions towards improving resilience to flooding. He has pioneered the development of flood recovery approaches to the benefit of many governments, agencies, charities, companies and institutions worldwide. His research has had significant national and international impact in the development of UK climate change policy; as a trustee advising on global research grant awards; and through supporting the development of national flood risk strategies, for example, in Brazil, China, Nigeria, Peru and the United Kingdom. He is a member and lead innovation adviser to the Environment Agency's Regional Flood and Coastal Committee. He has published extensively on a range of flood risk management topics, including adaptation, resilience and recovery. David is an experienced editor and is Chief Editor of the *International Journal of Building Pathology and Adaptation* (Emerald), Guest Editor of two recent special issues of the *Water* journal on flood risk and Editor of a number of books related to flooding and climate change. He is Co-Chair to the bi-annual International Conference on Flood and Urban Water Management (FRIAR).

Oluseye Adebimpe is a young academic and doctoral student at the Department of Industrial and Production Engineering, University of Ibadan, Nigeria. He is a certified SAP trainer, SAP ERP consultant, Fulbright scholar and Erasmus scholar. Oluseye is registered with the apex engineering body in Nigeria (Council for the Regulation of Engineering in Nigeria) and is a member of the Nigerian Society of Engineers. Oluseye had vast experience in the manufacturing sector which spans production and management of engineering systems before joining academics. He is a teacher, researcher and mentor with publications in reputable journals. He has served as a research student in the United Kingdom and United States. His research interests include flood risk management and renewable energy storage. His ongoing research includes developing models for flood resilience measurement, measuring flood vulnerability and developing a model for evaluating integrated solar photovoltaic and pumped-hydro storage systems. Oluseye teaches undergraduate and professional courses in energy systems modelling, terotechnology, industrial quality control, manufacturing systems and supply chain management.

Dr. Taiwo Adedeji is a lecturer at the Department of Design, Manufacturing and Engineering Management at the University of Strathclyde, Glasgow. He has over eight years of teaching experience in higher education, with a passion for student learning and ability to impart complex information to audiences of all levels. He has developed particular expertise in flood risk management and is committed to helping communities respond and adapt to climate change. He has published his research findings in a range of outputs, including Q1 journals, book chapters and research reports. Taiwo is currently a reviewer for the *International Journal of Building Pathology and Adaptation* (Emerald).

Handbook of Flood Risk Management in Developing Countries

Edited by Victor Oladokun, David Proverbs, Oluseye Adebimpe and Taiwo Adedeji

LONDON AND NEW YORK

Designed cover image: Getty Images

First published 2023
by Routledge
4 Park Square, Milton Park, Abingdon, Oxon OX14 4RN

and by Routledge
605 Third Avenue, New York, NY 10158

Routledge is an imprint of the Taylor & Francis Group, an informa business

© 2023 selection and editorial matter, Victor Oladokun, David Proverbs, Oluseye Adebimpe and Taiwo Adedeji; individual chapters, the contributors

The right of Victor Oladokun, David Proverbs, Oluseye Adebimpe and Taiwo Adedeji to be identified as the authors of the editorial material, and of the authors for their individual chapters, has been asserted in accordance with sections 77 and 78 of the Copyright, Designs and Patents Act 1988.

All rights reserved. No part of this book may be reprinted or reproduced or utilised in any form or by any electronic, mechanical, or other means, now known or hereafter invented, including photocopying and recording, or in any information storage or retrieval system, without permission in writing from the publishers.

Trademark notice: Product or corporate names may be trademarks or registered trademarks, and are used only for identification and explanation without intent to infringe.

British Library Cataloguing-in-Publication Data
A catalogue record for this book is available from the British Library

Library of Congress Cataloging-in-Publication Data
Names: Oladokun, Victor Oluwasina, editor.
Title: Handbook of flood risk management in developing countries / edited by Victor Oluwasina Oladokun, David Proverbs, Oluseye Adewale Adebimpe, and Taiwo Adedeji.
Description: Abingdon, Oxon ; New York, NY : Routledge, 2023. | Includes bibliographical references and index.
Identifiers: LCCN 2022044515 (print) | LCCN 2022044516 (ebook) | ISBN 9780367749743 (hbk) | ISBN 9780367750404 (pbk) | ISBN 9781003160823 (ebk)
Subjects: LCSH: Flood control—Developing countries. | Floods—Risk assessment—Developing countries.
Classification: LCC TC527 .H36 2023 (print) | LCC TC527 (ebook) | DDC 627/.4091724—dc23/eng/20221205
LC record available at https://lccn.loc.gov/2022044515
LC ebook record available at https://lccn.loc.gov/2022044516

ISBN: 978-0-367-74974-3 (hbk)
ISBN: 978-0-367-75040-4 (pbk)
ISBN: 978-1-003-16082-3 (ebk)

DOI: 10.1201/9781003160823

Typeset in Bembo
by Apex CoVantage, LLC

Contents

Foreword ix
List of contributors xi

1 Handbook of flood risk management in developing countries 1
  Victor Oladokun, David Proverbs, Oluseye Adebimpe and Taiwo Adedeji

SECTION I
Impacts, challenges and particularities 7

2 Impacts of floods on infrastructures in developing countries: focus on Bangladesh 9
  Md. Humayain Kabir, Shahpara Nawaz, Md. Nazmul Hossen, Md. Lokman Hossain and Sayeda Umme Habiba

3 Impacts of flooding on agriculture and food security in developing countries: evidence from Southeastern Nigeria 24
  Thecla Iheoma Akukwe, Lilian Chinedu Mba, Onyinyechi Gift Ossai and Alice Atieno Oluoko-Odingo

4 Adoption pathway for flood-resilient construction and adaptation in Ghana 41
  Eric Kwame Simpeh, Henry Mensah and Divine Kwaku Ahadzie

5 Resettlement as a flood-preventive measure in Sri Lanka: investigation into the socio-economic impacts 55
  Senuri Siriwardhana, Udayangani Kulatunga and Bingunath Ingirige

6 A survey-based approach to estimating the downtime of buildings damaged by flood in developing countries 70
  Rayehe Khaghanpour-Shahrezaee, Melissa De Iuliis and Gian Paolo Cimellaro

SECTION II
Preparedness, prevention, responses and recovery 91

7 Application of spatial planning and other tools to support preparedness for flooding in developing countries; example from Nigeria 93
Oluseyi O. Fabiyi

8 Understanding engineering approaches to urban flood management in Indonesia 109
Rian Mantasa Salve Prastica, Wakhidatik Nurfaida, Amalia Wijayanti, Abdunnavi Alfath and Muhammad Sulaiman

9 Restoration and recovery of flood-affected communities 123
Anoradha Chacowry

10 Preparedness and management of (flood) disaster amid a pandemic in a developing country: lessons from Cyclone Amphan in southwestern Bangladesh 137
Sabiha Lageard and Namrata Bhattacharya-Mis

11 A review of the flooding events in southern Brazil: challenges and opportunities 161
Francisco Henrique de Oliveira, Frederico Rudorff, Guilherme Braghirolli, Guilherme Linheira, Maria Carolina Soares, Raidel Baez Prieto, Regina Panceri, Renan Furlan de Oliveira and Victor Luís Padilha

SECTION III
Risk assessments, flood mitigation and project management 181

12 Flood vulnerability in developing countries, international collaboration, and network visualization: a bibliometric analysis 183
Ahmed Karmaoui

13 Flood risk assessment in developing countries: dealing with data quality and availability 197
Srijon Datta, Shahpara Nawaz, Md. Nazmul Hossen, Mir Enamul Karim, Nure Tasnim Juthy, Md. Lokman Hossain and Md. Humayain Kabir

14 Flood risk management projects: financing and project implementation, the case of Ibadan Urban Flood Management Project 217
Adedayo Ayodele Ayorinde, Abiodun Adefioye, Olakunle Oladipo and Oluseyi O. Fabiyi

15 Current knowledge, uncertainties, and aspirations of flood risk
    management policy for developing countries   229
    Ugonna C. Nkwunonwo

SECTION IV
Infrastructure systems, urban systems and their management   **247**

16 Integrated water resources management and flood risk management:
    opportunities and challenges in developing countries   249
    *Rudresh Kumar Sugam, Md. Humayain Kabir, Sherin Shiny George and
    Mayuri Phukan*

17 Developing resilient cities in developing countries   266
    Bolanle Wahab and Oluwasinaayomi Kasim

18 Flood risk and urban infrastructure sustainability in a developing
    country: a case study of Central Java Province, Indonesia   282
    Purwanti Sri Pudyastuti and Isnugroho

19 A review of flood management in South Asia: approaches, challenges,
    and opportunities   302
    *Md. Arif Chowdhury, Shahpara Nawaz, Md. Nazmul Hossen, Syed Labib
    Ul Islam, Sayeda Umme Habiba, Mir Enamul Karim, Md. Lokman
    Hossain and Md. Humayain Kabir*

SECTION V
Community perspectives, resilience and adaptation   **319**

20 The role of education in flood risk management: building a resilient
    generation in developing countries   321
    Edson Munsaka

21 Emerging resilience to urban flooding in low-income communities:
    a socio-cultural perspective from Ghana   333
    Clifford Amoako and Irene-Nora Dinye

22 Towards a socially just flood risk management in developing
    countries: lessons from serving the last mile in Malawi   350
    Marc van den Homberg and Robert Šakić Trogrlić

23 Exploring the perspective of school children on flood risk
   management in developing countries: lessons from Ghana   375
   *Henry Mensah, Grace Wanma, Divine Kwaku Ahadzie and
   Eric Kwame Simpeh*

24 Conclusions and final remarks   390
   Victor Oladokun, David Proverbs, Oluseye Adebimpe and Taiwo Adedeji

Index   *398*

Foreword

Handbook of FRM in Developing Countries

Flooding is one of the most common and severe hazards disrupting people's lives and livelihoods around the world. Floods often cause unmitigated damage and suffering, especially in lower-income countries where infrastructure systems, including drainage and flood protection, tend to be less developed. While countries at all levels of development face flood risk, the vast majority of the world's flood-exposed people (89%) live in low- and middle-income countries. Critically, it is not only major, more infrequent floods, but also smaller, frequent events that can reverse years of progress in poverty reduction and development. The World Bank has estimated that 1.47 billion people globally are directly exposed to the risk of intense flooding – over a third of them, almost 600 million, are poor.

Furthermore, the most devastating long-term consequences of floods are often experienced by the poorest households – those who have next to no savings and limited access to support systems. Countries in sub-Saharan Africa face the greatest threat: We estimate that of the 171 million flood-exposed people in this region, at least 71 million people live in extreme poverty (i.e. living on less than $1.90 a day). Globally, 587 million poor people are exposed to flood risk, 132 million of whom live in extreme poverty.

The World Bank, through its Climate Change Action Plan 2021–2025, aims to support Green, Resilient, and Inclusive Development, in pursuing poverty eradication and shared prosperity with a sustainability lens. Our Action Plan is supporting countries and private sector clients to maximize the impact of climate finance, aiming for measurable improvements in adaptation and resilience and measurable reductions in greenhouse gas (GHG) emissions. Our Action Plan also considers the vital importance of natural capital, biodiversity and ecosystems services and will increase support for nature-based solutions, given their importance for both mitigation and adaptation. As part of our effort to drive climate action, the World Bank has a long-standing record of participating in key partnerships and high-level forums aimed at enhancing global efforts to address climate change. Our new plans represent a shift from efforts to "green" projects, to greening entire economies, and from focusing on inputs, to focusing on impacts. We are now focusing on (i) integrating climate and development, (ii) identifying and prioritizing action on the largest mitigation and adaptation opportunities and (iii) using those to drive our climate finance and leverage private capital in ways that deliver the most results. That means helping the largest emitters flatten the emissions curve and accelerate the downward trend and ramping up financing on adaptation to help countries and private sector clients prepare for and adapt to climate change while pursuing broader development objectives through this new approach.

Foreword

In light of these significant global changes and developments, I am delighted to see this new Handbook on Flood Risk Management in Developing Countries. This comprehensive collection represents an excellent contribution to the body of knowledge as well as providing a highly timely and relevant collection. There are not only contributions from a truly international group of authors, highlighting some of the key challenges, but also many successful case studies of good practice. The handbook includes contributions from academics, researchers, practitioners and development partners, addressing modern developments in flood hazard management with respect to developing countries. As such, this handbook provides sound insights on key aspects associated with the effective management of flooding, while highlighting the unique challenges involved in developing regions. The book provides a distinctive collection of chapters, organized around key thematic areas towards improving resilience to flooding in some of our most vulnerable communities. Many essential topics are addressed such as infrastructure systems, agriculture and food security, planning, data management, finance and community-based approaches, to name a few. This encompasses the latest approaches to managing these hazards and addressing the full cycle of flood risk management. This handbook represents a vital resource for all practitioners involved in managing these flooding events in developing countries while also being essential reading to those involved in research or studying the impacts of flooding.

In highlighting the unique challenges that these events bring to communities and in identifying workable solutions for those responsible for mitigating and adapting to their impacts, I have no hesitation in commending this handbook.

Abhas Jha,
Practice Manager
Climate Change and Disaster Risk Management
South Asia Region
World Bank Group

Contributors

Oluseye Adebimpe	University of Ibadan, Nigeria
Taiwo Adedeji	University of Strathclyde, UK
Abiodun Adefioye	Ibadan Urban Flood Management Project, Nigeria
Divine Kwaku Ahadzie	Kwame Nkrumah University of Science and Technology, Ghana
Thecla Iheoma Akukwe	University of Nigeria, Nsukka
Abdunnavi Alfath	Universitas Gadjah Mada, Indonesia
Clifford Amoako	Kwame Nkrumah University of Science and Technology, Ghana
Adedayo Ayodele Ayorinde	Ibadan Urban Flood Management Project, Nigeria
Namrata Bhattacharya-Mis	University of Chester, UK
Guilherme Braghirolli	Santa Catarina State University, Brazil
Anoradha Chacowry	Association Pour Le Développement Durable (ADD), Mauritius
Md. Arif Chowdhury	Jashore University of Science and Technology, Bangladesh
Gian Paolo Cimellaro	Politecnico di Torino, Italy
Srijon Datta	University of Chittagong, Bangladesh
Melissa De Iuliis	Politecnico di Torino, Italy
Francisco Henrique de Oliveira	Santa Catarina State University, Brazil
Renan Furlan de Oliveira	Federal University of Santa Catarina – UFSC, Brazil
Irene-Nora Dinye	Kwame Nkrumah University of Science and Technology, Ghana
Oluseyi O. Fabiyi	Ibadan Urban Flood Management Project, Nigeria
Sherin Shiny George	Independent Consultant, Bengaluru, India
Sayeda Umme Habiba	University of Cadiz, Spain
Md. Lokman Hossain	Hong Kong Baptist University, Hong Kong, China
	German University Bangladesh, Gazipur, Bangladesh
Md. Nazmul Hossen	University of Chittagong, Bangladesh
Bingunath Ingirige	University of Salford, UK
Isnugroho	Ministry of Public Work and Public Housing, Republic of Indonesia
Nure Tasnim Juthy	University of Chittagong, Bangladesh
Md. Humayain Kabir	University of Chittagong, Bangladesh
	University of Graz, Austria
Mir Enamul Karim	University of Chittagong, Bangladesh
	Transparency International Bangladesh

Contributors

Ahmed Karmaoui	Moulay Ismail University, Morocco
	Southern Center for Culture and Science, Morocco
Oluwasinaayomi Kasim	University of Ibadan, Nigeria
	University of Guyana, Guyana
Rayehe Khaghanpour-Shahrezaee	University of Tehran, Iran
Udayangani Kulatunga	University of Moratuwa, Sri Lanka
Sabiha Lageard	University of Chester, UK
Guilherme Linheira	Santa Catarina State University, Brazil
Lilian Chinedu Mba	University of Nigeria, Nsukka
Henry Mensah	Kwame Nkrumah University of Science and Technology, Ghana
Edson Munsaka	National University of Science and Technology, Zimbabwe University of Johannesburg, South Africa
Shahpara Nawaz	University of Chittagong, Bangladesh
Ugonna C. Nkwunonwo	University of Nigeria, Nigeria
Wakhidatik Nurfaida	Universitas Gadjah Mada, Indonesia
Olakunle Oladipo	Ibadan Urban Flood Management Project, Nigeria
Victor Oladokun	University of Ibadan, Nigeria
Alice Atieno Oluoko-Odingo	University of Nairobi, Kenya
Onyinyechi Gift Ossai	University of Nigeria, Nsukka
Victor Luís Padilha	Santa Catarina State University, Brazil
Regina Panceri	Civil Defence of Santa Catarina, Brazil
Mayuri Phukan	Northeast Development Agency (NEDA), Guwahati, India
Rian Mantasa Salve Prastica	Polytechnic of Public Works, Ministry of Public Works and Housing, Indonesia
Raidel Baez Prieto	Santa Catarina State University, Brazil
David Proverbs	University of Wolverhampton, UK
Purwanti Sri Pudyastuti	Universitas Muhammadiyah Surakarta, Indonesia.
Frederico Rudorff	Civil Defence of Santa Catarina, Brazil
Eric Kwame Simpeh	Kwame Nkrumah University of Science and Technology, Ghana
Senuri Siriwardhana	University of Moratuwa, Sri Lanka
Maria Carolina Soares	Santa Catarina State University, Brazil
Rudresh Kumar Sugam	Centre for Urban Regional Excellence, New Delhi, India.
Muhammad Sulaiman	Universitas Gadjah Mada, Indonesia
Robert Šakić Trogrlić	International Institute for Applied Systems Analysis (IIASA), Austria
Syed Labib Ul Islam	JPZ Consulting (Bangladesh) Ltd., Dhaka, Bangladesh
Marc van den Homberg	510 An Initiative of the Netherlands Red Cross, The Netherlands
Bolanle Wahab	University of Ibadan, Nigeria
Grace Wanma	Kalpohin Senior High School, Tamale, Ghana
Amalia Wijayanti	Universitas Gadjah Mada, Indonesia

1

Handbook of flood risk management in developing countries

Victor Oladokun, David Proverbs, Oluseye Adebimpe and Taiwo Adedeji

Introduction

As the occurrence of flooding events has increased around the world, the impacts on people, infrastructure and urban systems have become a major burden on development and the general welfare of communities. This burden is more pronounced in developing countries due to a number of complex and interrelated economic, environmental and political circumstances. Experts have acknowledged the need to generate an improved understanding of these impacts towards more effective management of flood hazards in the context of communities and systems in these countries. There exists a clear need to collate informed views from both research and policy stakeholders on the impacts of flooding from a variety of different dimensions, disciplines and perspectives highlighting the challenges and particularities of flood hazards in the context of such developing countries. Hence, this book brings together a carefully selected variety of views and experiences of the impacts of flooding and its management in developing countries, drawing on both traditional and modern approaches adopted by communities, homeowners, researchers, project managers, institutions and policymakers. The collection of chapters is multi-disciplinary by design, reflecting the distinct nature of these challenges and highlighting opportunities for the emergence of innovative approaches. Key stakeholders drawn from regions across the globe provide insights and perspectives on flood hazards, impacts and control and adaptation strategies across these regions. The inclusion of contributions from policymakers, emergency responders, leaders of key organizations, researchers and managers of flood management projects makes this volume a unique addition to the flood risk management literature.

Flooding in developing countries

Flood management in developing countries also presents some specific challenges because of the complexities arising from the interplay of the rapid rate of urbanization, exploding population, fragile economies and weak infrastructural systems. These challenges can be viewed from various perspectives represented by communities, policy makers and flood risk managers responsible for dealing with flooding in developing countries. While the frequency and intensity of flooding has increased globally due to extreme weather events driven by climate change, the impacts of flood events have increased due to the interplay of population growth, increased urbanization and anthropogenic activities on flood plains. Developing countries are particularly vulnerable to flooding due to fragile economies, lack of risk awareness, inadequate preparedness, weak coping capacities, lack of well-organized urban development plans and poor building standards (De Risi et al., 2018). Flood disasters pose a greater threat to human life, health and well-being in developing countries than in developed countries (Dewan, 2015). Flood events across developing countries in Africa, Asia and Latin America have negatively impacted the built and natural environments and have become the most damaging natural hazards in terms of the number of people affected and their long-term socio-economic and health implications. According to Shrestha (2008), losses due to flooding in developing countries, per unit of GDP, are five times higher than those of richer countries. The observed higher destructive consequences and impacts of flood events in developing countries can be attributed to a combination of factors. The relatively weak level of economic development, poor infrastructural systems and weak governance institutions in many developing countries aggravate the vulnerability of their communities to flood hazards. There is also a lack of sufficient disaster management structures and poor economic resources which exacerbate the challenges of poverty and welfare degradation across affected communities, resulting in a flood–poverty cycle (Osti, 2004).

Meanwhile, many developing countries are experiencing rapid economic growth accompanied by a rapid rate of urbanization, transforming these regions from mostly agricultural to industrial communities. In many instances, the urban management systems lag behind their urban growth. The increased rate of land use with a rapidly growing population has increased the overall exposure to flood hazards. Hence, while flood events might have caused limited damage in the past, many developing countries are witnessing more intense economic losses in recent times due to the increased exposure (Jonkman & Dawson, 2012). The availability of flood data and information is another major issue of concern to flood management stakeholders, as there appears to be a paucity of reliable flood data needed for the deployment of flood countermeasures and mitigations (Ekeu-wei, 2018). Many flood-prone areas lack adequate gauging stations or warning systems for systematic data collection, thereby complicating flood management across these ungauged catchments that, in most cases, span several transboundary river basins (Lumbroso, 2020).

On the contrary, flood management in many developing countries presents some opportunities for the adoption of sustainable approaches. The concept of 'living with water' and developing improved flood resilience through integrated blue-green (BG) infrastructures would seem to have huge potential in many developing regions (Alves et al., 2019). For instance, the virgin nature of communal infrastructural systems and the relatively low level of investment in existing infrastructures makes the adoption of, or transition to, the blue-green systems approach more economically attractive in developing countries (Oladokun & Proverbs, 2016). This is because the emerging infrastructural systems create some flexibility for developing countries to retrofit existing systems and adapt to blue-green systems at a much lower cost compared to developed countries.

The effectiveness of the governance structures needed to support flood risk management varies widely across countries in the developing regions. While a few countries have somewhat effective systems, a majority are largely inadequate due to factors such as poor policy formulation and planning processes, lack of institutional coordination and poor synergies between flood management strategies and research support systems (Abbas et al., 2016).

Hence this book considers a wide spectrum of flood management concepts and approaches from various regions of the global south. Concepts such as integrated and sustainable flood risk management, land use, waste management, drainage systems, security challenges, urban planning and development and their contributions to improving our understanding of resilience and flood management are covered. Concepts and models on urban systems and the development of the built environment in the context of flood risk management are presented from multidisciplinary and multicultural perspectives. These include topics and themes on sewage, drainage, waste and water supply systems, urban master plans, land use, flood plain management and other 'soft' issues affecting flood management. Topics on economic, behavioural and socio-cultural systems in the context of resilience, adaptation and related dimensions of flood risk management are treated extensively.

Organization of the book

The chapters are organized to reveal various impacts and challenges associated with the management of flooding across the various phases of the flood risk management circle encompassing preparation, response and recovery and risk assessment as well as community perspectives. The approach adopted is to accommodate topics across all phases of the flood risk management cycle from both descriptive and normative perspectives. For instance, theories and case studies on mitigation and intervention projects are discussed and aligned with risk assessments, flood mitigation, project management and project financing from research and policy perspectives.

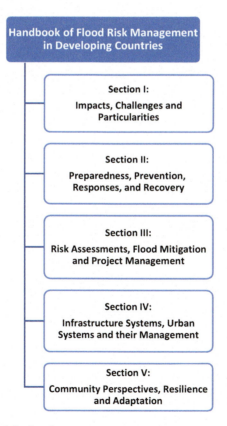

Figure 1.1 Organization of the book

The book thus brings together contributions on different impacts of flooding and proposes various mitigation approaches, drawing on case studies of good practice across various global regions. The authors describe procedures for managing flooding and reducing the impacts from the perspectives of policy makers, environmental planners and various stakeholders of flood-affected communities.

Section I: impacts, challenges and particularities

This section discusses the impacts of floods from various dimensions and perspectives of various stakeholders with themes that highlight the challenges and particularities of flood hazards in the context of developing countries. Topics include the impacts of flooding on agriculture and food security in developing countries, pathways for the development and construction of flood resilience and investigations of socio-economic impacts of flood management programmes.

Section II: preparedness, prevention, responses and recovery

Section II includes topics across all phases of the flood risk management cycle from both descriptive and normative perspectives. Topics include application of spatial planning and engineering tools and preparedness approaches for dealing with various flood scenarios in developing countries.

Section III: risk assessment, flood mitigation and project management

Section III covers methodologies, tools and approaches for evaluating level of flood risk. Theories and case studies on mitigation and intervention projects from research and policy perspectives are included in this section. Topics revolve around flood risk assessment, data quality and data availability as well as issues that border on financing and implementation of flood management projects.

Section IV: infrastructure systems, urban systems and their management

This section deals with topics on civil infrastructure and urban systems that support the built environment in the context of flood risk management. This includes topics on sewage, drainage, waste, water supply systems, urban master plans, land use and flood plain management. Challenges and opportunities for integrated water resources management and flood risk management in the context of resilient urban cities and sustainability are considered from multiple perspectives.

Section V: community perspectives, resilience and adaptation

Several 'soft' issues that centre on flood management relating to the influence of behavioural and socio-cultural systems on building resilience in communities are considered in this section. Emerging trends in adaptation, achieving socially just flood risk management practices and the role of education in support of more sustainable flood management are covered.

Summary

While each chapter focuses on the current state of affairs and the research in a particular area, each contribution is not overly descriptive but instead gives a critical analysis of the matter and contains suggestions for improvement and further developments. Given the book's envisaged interdisciplinary audience, authors have been encouraged to avoid unnecessary technical jargon (where necessary terminology needs were defined) so that the book can present an enjoyable and insightful reading experience. The scope covers various views and experiences of the impacts of flooding and its management in developing countries, drawing on examples and cases from a wide range of regions. Many of the contributions provide a useful historical context and evolution of ideas as well as highlights of major controversies, debates and research questions that characterize a research topic or case study area. Many chapters include a systematic review of current research in an area pointing to future research directions. Alternatively, some of the contributions draw on the authors' own research or present a case study from their own experience as examples of good practice in regard to a specific aspect of flood risk management. As such, many of the chapters draw on real-life projects and experiences or findings from empirical, applied research.

References

Abbas, A., Amjath-Babu, T. S., Kächele, H., Usman, M., & Müller, K. (2016). An overview of flood mitigation strategy and research support in South Asia: Implications for sustainable flood risk management. *International Journal of Sustainable Development & World Ecology, 23*(1), 98–111. https://doi.org/10.1080/13504509.2015.1111954

Alves, A., Gersonius, B., Kapelan, Z., Vojinovic, Z., & Sanchez, A. (2019). Assessing the co-benefits of green-blue-grey infrastructure for sustainable urban flood risk management. *Journal of Environmental Management, 239*, 244–254. https://doi.org/10.1016/j.jenvman.2019.03.036

De Risi, R., De Paola, F., Turpie, J., & Kroeger, T. (2018). Life cycle cost and return on investment as complementary decision variables for urban flood risk management in developing countries. *International Journal of Disaster Risk Reduction, 28*, 88–106. https://doi.org/10.1016/j.ijdrr.2018.02.026

Dewan, T. H. (2015). Societal impacts and vulnerability to floods in Bangladesh and Nepal. *Weather and Climate Extremes, 7*, 36–42. https://doi.org/10.1016/j.wace.2014.11.001

Ekeu-wei, I. T. (2018). Evaluation of hydrological data collection challenges and flood estimation uncertainties in Nigeria. *Environment and Natural Resources Research, 8*(2), 44. https://doi.org/10.5539/enrr.v8n2p44

Jonkman, S., & Dawson, R. (2012). Issues and challenges in flood risk management – editorial for the special issue on flood risk management. *Water, 4*(4), 785–792. https://doi.org/10.3390/w4040785

Lumbroso, D. (2020). Flood risk management in Africa. *Journal of Flood Risk Management, 13*(3). https://doi.org/10.1111/jfr3.12612

Oladokun, V. O., & Proverbs, D. (2016). Flood risk management in Nigeria: A review of the challenges and opportunities. *International Journal of Safety and Security Engineering, 6*(3), 485–497. https://doi.org/10.2495/SAFE-V6-N3-485-497

Osti, R., 2004. Forms of community participation and agencies' role for the implementation of water-induced disaster management: Protecting and enhancing the poor. *Disaster Prevention and Management: An International Journal, 13*(1), 6–12.

Shrestha, M. S. (2008). *Impacts of flood in South Asia.* www.researchgate.net/publication/259484329

Section I
Impacts, challenges and particularities

2
Impacts of floods on infrastructures in developing countries
Focus on Bangladesh

Md. Humayain Kabir, Shahpara Nawaz, Md. Nazmul Hossen, Md. Lokman Hossain and Sayeda Umme Habiba

Introduction

Bangladesh, one of the most flood-prone countries in the world, came in seventh place on the Global Climate Risk Index 2020 when the effects of weather-related events (storms, floods, and heat waves) between 1999 and 2018 were taken into account. About 191 climate-related extreme weather events occurred in Bangladesh during this time, ranking it 9th among all nations in terms of the number of fatalities, 37th in terms of fatalities per 100,000 people, 17th in terms of losses, and 40th in terms of losses per unit of GDP (Eckstein et al., 2019). Bangladesh is among the nations that are most vulnerable to the effects of climate change. It is anticipated that the increase in sea level rise brought on by climate change driven by greenhouse gases in the earth's atmosphere will continue to fuel this trend. In Bangladesh, this will have a significant impact on all socioeconomic sectors. This country is particularly at risk because of its physical setting, geology, and huge delta plain, as well as its approximately 230 rivers (FAO, 2012). Due to the combined impact of these rivers, Bangladesh experiences floods frequently, which often result in fatalities and significant economic damage. The worst of these floods, the one that struck Bangladesh in 1998, damaged 53 of the nation's 64 districts (BWDB, 2014).

In addition, the presence of the Bay of Bengal in the south and river runoff from melting Himalayan glaciers in the north make the nation vulnerable to dangerous level of flooding. About 80 percent of the country is covered by the Ganges, Brahmaputra, Meghna, and numerous other rivers' floodplains (Brouwer et al., 2007). Every ten years, one-third of the country experiences severe flooding, with the terrible floods of 1988, 1998, and 2004 submerging more than 60 percent of the country (CEGIS, 2002). The incidence of severe storms brought on by climatic stress is making the issue worse (Dastagir, 2015). The entire Bangladeshi coast is at risk of cyclones and storm surges. By 2050, the coast is likely to be inundated by an additional 15 percent cyclone and storm surge intensity, experience monsoon flooding that is more intense and widespread due to climate change, which may result in an additional 4 percent of the country being flooded and a 15 cm increase in the depth of the water (World Bank, 2010). Considering this background, this chapter aims to review the impacts of flood on infrastructure in Bangladesh and some selected developing countries and do a geospatial analysis of Bangladesh with respect to flood impact.

Impacts of flood on infrastructure: developing countries' perspective

Numerous studies have been done to account for experiences of floods' effects on infrastructure in various nations. For instance, Shah (1999) referred to the floods that ravaged Bangladesh in 1998 for 65 days as the worst to ever occur there, with millions of people affected and over two-thirds of the nation submerged. The floods ruined homes, businesses, crops, and livestock in addition to damaging key infrastructures like roads and bridges. Similarly, Mustafa (2002) claims that Pakistan is still at risk of flooding despite making significant investments in its water industry. Major floods struck Pakistan in 1950, 1956, 1973, 1976, 1988, and 1992, each of which claimed more than 10,000 lives. Floods also struck Pakistan in 1987, causing significant damage to the buildings, bridges, roads, trains, telephone connections, and dams in riverside cities.

Flooding, particularly as a result of heavy precipitation, is the main reason for weather-related disruption in many sectors and nations. For instance, flood events have a significant impact on the transportation industry. Transport networks support economic activity by facilitating the mobility of people and goods, according to Pregnolato et al. (2017). Extreme weather events can damage transportation infrastructure either directly or indirectly, endangering human safety and resulting in major disruptions, as well as accompanying economic and social effects. Heavy rain typically results in overland flow, which pushes drainage systems beyond their capacity and increases the risk of debris blockage. The effects of floods on infrastructure in various nations are further illustrated by studies like Sultana et al. (2016), Kenley et al. (2014), Winter et al. (2016), and Habiba et al. (2013), among others. Because public buildings and infrastructure provide services that the entire population relies on, flood damage to infrastructure has varying ramifications (EOD Resilience Resources, 2016).

Bangladesh

Flood of 2017 in Bangladesh

Regarding duration and size, the 2017 flood's characteristics serve as an example of a severe one. While the 2017 monsoon floods were strong, they only lasted for a short time in the north. In August, a number of rivers in the Brahmaputra basin broke records. The northeast region saw a catastrophic flash flood very early in April, and the southeast steep region experienced numerous landslides in June. The centre region (near the Padma River) saw considerable long-lasting flooding. The duration of some rivers' flooding, notably Kushiyara, was lengthier during the monsoon season. Overall, Bangladesh had moderate to severe flooding during the 2017 monsoon season.

During the 2017 monsoon (May to October), the nation as a whole received just 1.20 percent less precipitation than average, which is within the range of what is considered normal for the monsoon. The Ganges, Meghna, and South Eastern Hill basins received 1.79 percent, 0.78 percent, and 4.25 percent more rainfall than usual, respectively, while the Brahmaputra basin had 17.90 percent less rainfall. In August 2017, all basins – with the exception of the South Eastern Hill basin – recorded rainfall above their corresponding monthly averages. Maximum area inundated during the 2017 monsoon was 42 percent of the entire country (62,000 sq-km approximately) (FFWC, 2017).

The Brahmaputra and Ganges basins were impacted by the 2017 flood throughout the months of July and August. The flood's duration was short to medium in most areas of the basins, with a few exceptions, and its size ranged from mild to severe. The 2017 flooding in the Meghna basin was erratic because, in addition to intense flash flooding during the pre-monsoon, there was a lengthy period of flooding (up to 92 days) during the monsoon, particularly in the low-lying areas of Sylhet district, which were affected by moderate to severe intensity. However, a few brief flood events with moderate to severe intensity occurred in the southeastern hill basin in 2017: there were 35 flood-affected districts, or 42 percent of the country as a whole.

Water level (WL) crossed and remained above respective danger levels (DLs) at 22 of the 30 WL monitoring sites in the Brahmaputra basin in 2017. The year's flood wave struck the basin twice, once in the first week of July and again in the second week of August, separated by nearly one month. Both episodes lasted for around two weeks, but the second wave was far more intense and resulted in more flooding. The following stations crossed and stayed over DLs during the months of July and August: Dharala at Kurigram for 12 days, Teesta at Dalia for 6 days, Jamuneswari at Badarganj for 8 days, Ghagot at Gaibandha for 15 days, Karatoa at Chakrahimpur for 10 days, Chilmari for 14 days, Jamuna at Bahadurabad for 25 days, Sariakandi for 24 days, and Kazipur for 29 days (BWDB, 2017).

Flood of 2019 in Bangladesh

The characteristics of the flood in 2019 as a whole are severe in terms of their scale. All of the significant nationwide flood incidents that occurred during the 2019 monsoon occurred in July. In the Northern, North-Western, and North-Central regions, the flood lasted up to slightly longer than medium duration and was of severe size. The year's flood, which exceeded the greatest water level ever recorded in Bahadurabad and Fulchari, was mostly caused by the Brahmaputra-Jamuna River. As a result, one of the worst floods in recorded history occurred in the low-lying areas of Jamalpur and Gaibandha. Flooding of moderate to above-moderate magnitude and duration occurred along the Padma River in the country's central area. During that time, the North-Eastern and South-Eastern regions had moderate to severe flash floods, all of which persisted only briefly. However, the Ganges River at Hardinge Bridge ran over danger level during the late monsoon period from October 1 to October 5, causing a typical flood of brief duration in the West-Central region of the country. The Ganges at Hardinge Bridge flowed above the danger threshold for the first time in 16 years. Due to rainfall brought on by cyclonic storm Fani, some low-lying areas of the Haor basins were slightly, sooner than usual, submerged during the first week of May, but no serious agricultural damage occurred. In the monsoon of 2019, evaluation found that the flood forecasting and warning centre's (FFWC) five-day deterministic flood forecasts had an average accuracy of 95, 89, 82, 76, and 75 percent for lead times of 24, 48, 72, 96, and 120 hours, respectively (BWDB, 2019).

Flood of 2020 in Bangladesh

The Northern, North-Eastern, and South-Eastern regions of Bangladesh were all affected by the monsoon floods of 2020. The floods affected 21 districts, with 16 districts suffering moderate to severe effects. With a 71 percent likelihood of heavy flooding, the Bahadurabad location was expected to experience the flooding's highest peak. On July 18, 2020, flood waters were expected to be at their highest. As of July 22, 2020, floods had affected 3.3 million people and had waterlogged 7,31,958 people throughout 102 upazila and 654 unions. The majority of the 93 fatalities died due to drowning, which has claimed the lives of 41 children since June 30, 2020. The COVID-19 epidemic, persistent flooding, and monsoon floods all have an aggravating effect on the flood victims. Due to the fact that people impacted by the flood were relocated and moved to shelters that were crowded with inadequate water, sanitation and hygiene facilities, it was very difficult to maintain critical practices like social distance and handwashing during the 2020 monsoon flood. For different unions – the smallest unit of local administration – of coastal Bangladesh with a high rate of internal displacement, the likelihood of economic and social disruptions is significant. According to primary data, 93 percent of unions experienced disruptions in social and income-generating activities, and 24 percent of unions had more than 40 percent of their members displaced or living in other locations. Many people were living together as a result of shelter damage, thereby raising the risk of COVID-19 spreading.

In addition to damages caused by the 2020 floods, many flood prevention infrastructures like dikes and embankments had already been compromised by earlier monsoon floods. There were already 220 reports of embankment damage. The recovery period following a disaster typically lasts three to five years, but in recent disasters, this period has been shortened (Needs Assessment Working Group [NAWG] Bangladesh, 2020). Additionally, this flood jeopardized the repair and reconstruction activities of damaged infrastructures hindered by the COVID-19 pandemic. Infrastructures that are not fixed and maintained are further exposed to impending disaster. Consequently, rehabilitation becomes a more difficult yet essential stage.

Impacts on buildings and infrastructures

Research reported that floods can have a significant impact on housing and households. While slowly rising floodwater ruins structures, fast-moving floodwaters have the power to wipe out entire slums. Homes with mud walls, coconut leaf walls, and tin walls fell in rural Bangladesh, leaving people and property exposed and unsafe. Since around 32 percent of Bangladesh's population live in slums (Miyan, 2012; Rahman, 2011), flooding frequently leaves a huge number of people trapped and destitute. According to Jha et al. (2012), damage to important cultural monuments like mosques and temples as well as public structures like hospitals, clinics, and educational facilities have a ripple effect. For instance, disruptions in schooling affect academic sessions and result in poor literacy rates that continue to exist in both countries. The ability to provide both emergency and long-term healthcare and support has also been significantly reduced. Bangladesh frequently experiences floods that render roads impassable or partially destroyed, making travel difficult. When there is flooding, the city's waste management system suffers because garbage is left lying around, obstructing the drainage system and harming the environment (Jha et al., 2012; NAPA, 2005; Dasgupta et al., 2010).

Climate change-induced flooding impacts on urban infrastructure

Numerous studies have shown that climate change will exacerbate the flood scenario in Bangladesh (Hossain et al., 2018; Whitehead et al., 2018). In the recent past, 2014, Bangladesh had published municipal-level master plans for 222 municipalities in order to ensure sustainable urban development. However, this master plan did not consider flood intensification in the context of changing climate, which shows the incompatibility of the plan in the long term. Using a geographic information system (GIS), Rahman and Islam (2020) analysed the risk of flooding to specific infrastructures in Ullapara Municipality. These infrastructures include those related to transportation, education, health care, and other urban facilities. The analysis of flood exposure was done using floods caused by climate change for the year 2040. According to the flood exposure analysis, due to future climate change, approximately 33.99 percent of roads would be exposed to inundation levels between 1.5 and 2 metres; seven primary schools, six secondary schools, and four colleges would be highly exposed to inundation levels between 2.0

Impacts of floods on infrastructures in developing countries

and 2.50 metres; and four health facilities would be exposed to inundation levels between 1.0 and 2.0 metres. This situation of prolonged flooding will cause the affected infrastructure to malfunction, which will then jeopardize the integrity of Ullapara Municipality's socioeconomic system (Rahman and Islam, 2020).

Future damage by floods to infrastructure

Two-thirds of the country are located at sea level or lower. According to historical data on monsoon flooding, 42 percent of the country is at risk of floods of varying intensities, while another 21 percent is exposed to annual flooding. Even though regular yearly flooding has historically been advantageous, producing nutrient-rich sediments and replenishing groundwater aquifers, the nation frequently sees severe floods during a monsoon with negative impact on rural livelihoods and production. Two-thirds of the land area was submerged by the 1998 flood, which also caused losses and damages of approximately US$2 billion, or 4.8 percent of GDP. Increased precipitation, greater transboundary water flows, and sea level rise, according to climate models, will all increase the destructive impact of monsoon floods. Dasgupta et al. (2011) estimate an incremental cost of US$2.671 billion initially and US$54 million in annual recurrent costs to climate-proof roads and railways, river embankments protecting productive agricultural lands, and drainage systems and erosion control measures for major towns. These costs are based on climate change scenarios out to 2050 and hydrological and hydrodynamic models (Dasgupta et al., 2011).

Geospatial mapping of flood impacts: case of Bangladesh

We conducted a geospatial analysis of two selected major floods in Bangladesh. To do so, we followed a number of steps (Figure 2.1) to find the potential flood hazards in Bangladesh. In our analysis, we considered pre-, during, and post-flood events (Table 2.1). We found that flooding in 2017 and 2019 affected road systems and vulnerable people across the country. Particularly, in 2019, the northern part of the country was more exposed to flooding (Figure 2.2).

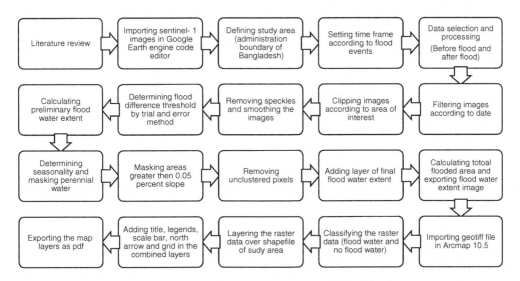

Figure 2.1 Steps of geospatial mapping of flood impacts on infrastructure in Bangladesh

Table 2.1 Months considered for flood mapping

Flood year	Pre-flood	During flood (from EM-DAT data)	Post-flood
2017	January–February	July–August	November–December
2019	January–February	June–July	November–December

Case study in Pakistan

In 2011–2012, a catastrophic flood hit Pakistan. Unexpected significant rainfall during the 2011–2012 monsoon season led to a flood in the country's south. The flood's extreme vast scope resulted in the flooding of numerous locations and the breaching of dams. Three months of heavy rain were experienced. It quickly spread to more than 20 districts in Sindh Province and the northern regions of Baluchistan Province. As a result, this unanticipated rain began harming the economy, infrastructure, crops, and people (Khalid and Ali, 2019). Later, it caused canal damage, flash floods, drainage catastrophes, and river overflows. The consequences were particularly severe because there were two floods within a space of two years. Another flood struck in 2011, as many were still recovering from the 2010 catastrophe. Finance (2012) reported that there were 9.6 million individuals affected, with close to 520 fatalities and tens of thousands of injuries. This tragic occurrence seriously disrupted economic activity. Demand and overall output both fell sharply; 15.06 percent was lost by the agriculture industry, 9.7 percent by the service sector, and 2.23 percent by the transportation sector (Finance, 2012).

Infrastructure and economic sectors are linked and are susceptible to disasters like floods. The key sector's resilience, or its capacity to bounce back swiftly from a setback, might also lessen the disaster's effects. Khalid and Ali (2019) used a resilience and recovery time dynamic inoperability input-output model (DIIM) to estimate Pakistan's economic system, and a case study on the flooding in Pakistan in 2011–2012 was conducted. More specifically, this study's goal was to provide a rough estimate of the disruption's overall impact and its knock-on effects on the sectors that persisted for several days following the flood. They found that the majority of the crucial industries are related to the agricultural and service sectors in terms of economic loss and operational inefficiency, respectively. In order for policymakers, disaster management authorities, and health agencies to appropriately respond, the study's findings provided useful information (Khalid and Ali, 2019).

Flood impacts on infrastructure in Kenya

Floods and other climate change–related events make infrastructure services more frequently unavailable. When it comes to controlling the consequences of floods on infrastructure, developing countries tend to be reactive rather than proactive. This is well proven by the insufficient degree of preparedness encountered prior to, during, and following flood occurrences. Kenya's infrastructure is vulnerable to flooding, which is a source for concern (Njogu, 2021). Kenya has recently experienced significant floods, which have resulted in numerous instances of infrastructure failure and consequent socioeconomic damage (Opondo, 2013). Between 1997 and 1998, El Nino–related floods were severe, broad, and intense, affecting over 1.5 million people and causing property damage in Kenya worth US$151.4 million. In 46 of Kenya's 61 districts, the floods destroyed and disrupted important social infrastructure, including housing, transportation, hydropower dam silting, and communication. As a result of the 2010 floods, 40 bridges,

Impacts of floods on infrastructures in developing countries

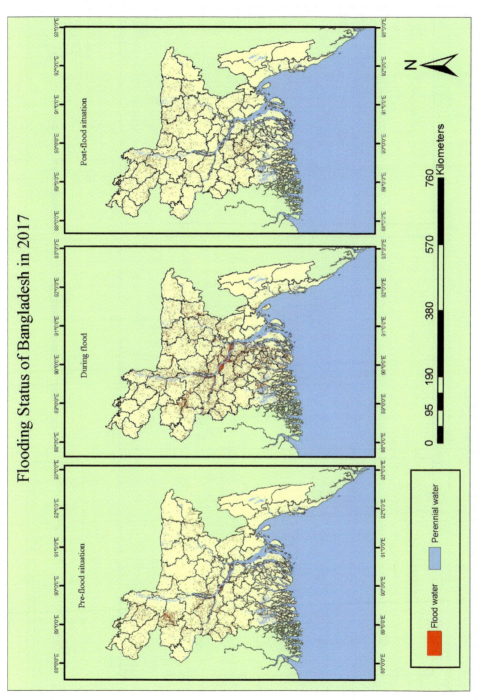

Figure 2.2 Flooded area, inundated road networks, and exposed population due to flooding in 2017 (left panel) and 2019 (right panel)

Note: Maps throughout this chapter were created using ArcGIS software by Esri. ArcGIS and ArcMap are the intellectual property of Esri and are used herein under license. Copyright © Esri. All rights reserved. For more information about Esri software, please visit www.esri.com.

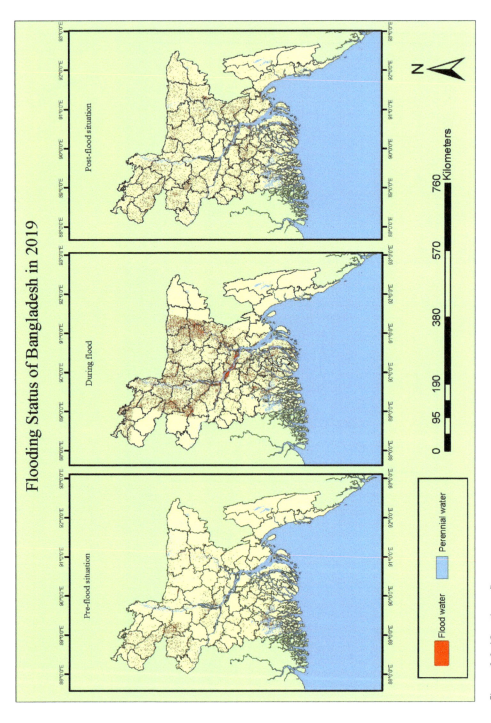

Figure 2.2 (Continued)

Impacts of floods on infrastructures in developing countries

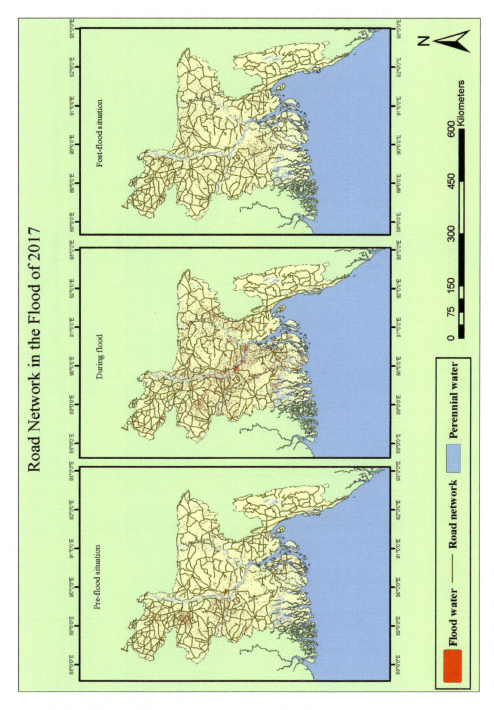

Figure 2.2 (Continued)

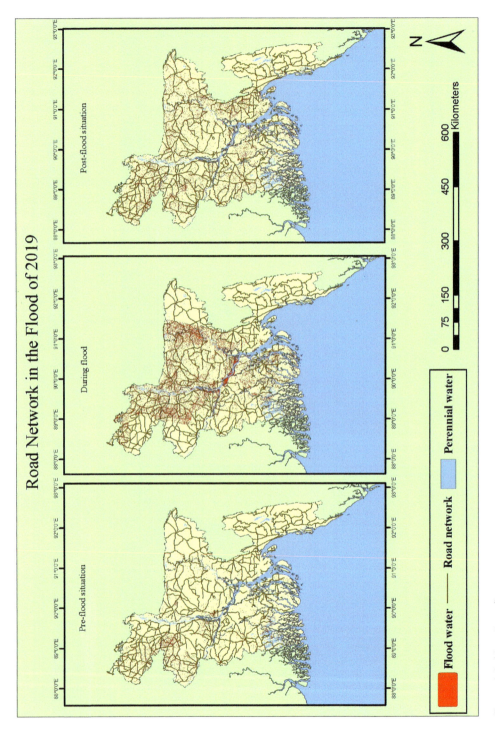

Figure 2.2 (Continued)

Impacts of floods on infrastructures in developing countries

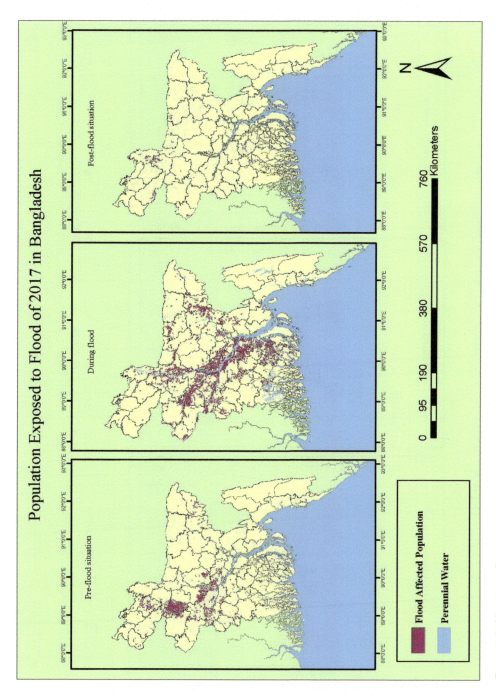

Figure 2.2 (Continued)

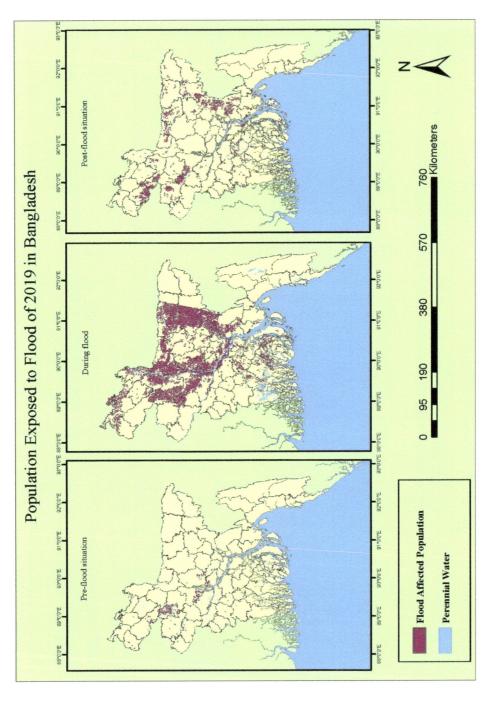

Figure 2.2 (Continued)

roads, and pieces of the water infrastructure were destroyed in the North Rift, South Rift, Upper Eastern, North Eastern, and South Rift regions (Opondo, 2013). According to research by the Kenya Institute for Public Policy Research and Analysis (KIPPRA), the recent floods that occurred in 27 of Kenya's counties in 2017 and 2018 had a significant impact on infrastructure users in all of the counties.

Based on a primary study conducted in 27 of Kenya's 47 counties, Njogu (2021) examined the consequences of floods on infrastructure users in Kenya. The results of the study demonstrated how flooding in Kenya causes disruptions in the delivery of infrastructure services by destroying buildings, social facilities, and road networks. According to the analysis, Kenya expended a significant amount of resources throughout the flood recovery period. This study found that local communities are crucial to flood response and recovery. Kenya lacks climate-sensitive policies, laws, and development plans that mention resilience in order to safeguard infrastructure against known and anticipated climate hazards. According to the report, constructing resilient infrastructure to react to and survive the effects of floods necessitates shared vision across all stakeholders (Njogu, 2021).

Conclusion

Flooding is a common phenomenon in both in developed and developing countries. However, the least-developed and developing countries like Bangladesh are more vulnerable to floods. Infrastructures are affected severely both during and after flood events. From our analysis in Bangladesh, we conclude that major recent floods of 2017 and 2019 had negative impacts on critical roads and huge populations. We reviewed major impacts of flooding in other developing countries. In Pakistan and Kenya, flooding events also affected railways, roads, and other infrastructure. Several studies claimed that these impacts will worsen in the near future due to global climate change. Based on our findings, we believe that relevant institutions on flood risk management will take necessary actions to mitigate the negative impacts of flooding events.

References

Brouwer, R., Akter, S., Brander, L., & Haque, E. (2007). Socioeconomic vulnerability and adaptation to environmental risk: A case study of climate change and flooding in Bangladesh. *Risk Analysis*, 27(2), 313–326.

BWDB (2014). *Bangladesh water development board: Observed river data and reports of 1988, 1998, and 2004 floods.* www. bwdb.gov.bd

BWDB (2017). *Annual flood report 2017.* Dhaka, Bangladesh: Flood Forecasting and Warning Centre (FFWC) Bangladesh Water Development Board (BWDB) WAPDA Building (8th Floor), Motijheel C/A. http://www.ffwc.gov.bd/

BWDB (2019). *Annual flood report 2019.* Dhaka, Bangladesh: Flood Forecasting and Warning Centre (FFWC) Bangladesh Water Development Board (BWDB,) WAPDA Building (8th Floor), Motijheel C/A. http://www.ffwc.gov.bd/images/annual19.pdf

CEGIS (2002). *Analytical framework for the planning of integrated water resources management.* Dhaka, Bangladesh: Center for Environmental and Geographic Information Systems.

Dasgupta, S., Huq, V., Khan, Z.H., Masud, M.S., Ahmed, M.M.Z., Mukherjee, N., & Pandey, K. (2010). *Climate proofing infrastructure in Bangladesh, the incremental cost of limiting future inland monsoon flood damage.* http://elibrary.worldbank.org/doi/pdf/10.1596/1813-9450-5469

Dasgupta, S. et al. (2011). Climate proofing infrastructure in bangladesh: The incremental cost of limiting future flood damage. *Journal of Environment and Development*, 20(2), 167–190. https://doi.org/10.1177/1070496511408401

Dastagir, M.R. (2015). Modeling recent climate change induced extreme events in Bangladesh: A review. *Weather and Climate Extremes*, 7, 49–60. https://doi.org/10.1016/j.wace.2014.10.003

Eckstein, D., Künzel, V., Schäfer, L., & Winges, M. (2019). *Global climate risk index 2020 who suffers most from extreme weather events?* https://germanwatch.org/sites/germanwatch.org/files/20-2-01eGlobalClimateRiskIndex2020_14.pdf.

EOD Resilience Resources (2016). *Understanding risk and resilient infrastructure investment*. Crown. https://assets.publishing.service.gov.uk/media/57d6bc78e5274a34de000042/Understanding_Risk_Resilient_Infrastructure_Investment_27May_16_rev_external.pdf

FAO – Food and Agriculture Organization of the United Nations (2012). *Irrigation in Southern and Eastern Asia in figures, AQUASTAT survey-2011*. FAO Water Report #37. Rome: FAO.

FFWC (2017). *Annual flood report 2016 processing & flood forecasting circle Bangladesh water development board*. www.ffwc.gov.bd/images/annual17.pdf

Finance, M.O. (2012). *Pakistan economic survey*. Islamabad, Pakistan: Flood Impact Assessment.

Habiba, U., Shaw, R., & Hassan, W.R. (2013). Drought risk and reduction approaches in Bangladesh. *Environmental Hazards*, 10, 121–138. Springer.

Hossain, M.A.R., Ahmed, M., Ojea, E., & Fernandes, J.A. (2018). Impacts and responses to environmental change in coastal livelihoods of South-West Bangladesh. *Science of the Total Environment*, 637–638. https://doi.org/10.1016/j.scitotenv.2018.04.328

Jha, A.K., Bloch, R., & Lamond, J. (2012). *A guide to integrated urban flood risk management for the 21st century*. https://openknowledge.worldbank.org/handle/10986/2241

Kenley, R., Harfielda, T., & Bedggood, J. (2014). *Road asset management: The role of location in mitigating extreme flood maintenance*. Fourth International Conference on Building Resilience, Building Resilience, Elsevier, Amsterdam, 8–10 September, pp. 198–205.

Khalid, M.A., & Ali, Y. (2019). Analysing economic impact on interdependent infrastructure after flood: Pakistan a case in point. *Environmental Hazards*, 18(2), 111–126. https://doi.org/10.1080/17477891.2018.1496899

Miyan, M.A. (2012). *Vulnerabilities of the people of Bangladesh to disaster*. Proceedings of the Regional Consultative Meeting on Engaging SAARC for Disaster Resilience, Islamabad, Pakistan, 1–2 November.

Mustafa, D. (2002). Linking access and vulnerability: Perceptions of irrigation and flood management in Pakistan. *The Professional Geographer*, 54(1), 94–105. https://onlinelibrary.wiley.com/doi/abs/10.1111/0033-0124.00318

NAPA (2005). *National adaptation programme of action (NAPA), ministry of environment and forest, government of peoples Republic of Bangladesh*. https://unfccc.int/resource/docs/napa/ban01.pdf

Needs Assessment Workig Group (NAWG) Bangladesh (2020). *Monsoon floods 2020: Coordinated preliminary impact and needs assessment Bangladesh needs assessment working group*, July. https://reliefweb.int/report/bangladesh/bangladesh-monsoon-floods-2020-coordinated-preliminary-impact-and-needs-assessment

Njogu, H.W. (2021). Effects of floods on infrastructure users in Kenya. *Journal of Flood Risk Management*, 14(4), 1–10. https://doi.org/10.1111/jfr3.12746

Opondo, D.O. (2013). *Loss and damage from flooding in Budalangi District, Western Kenya*. Loss and damage in vulnerable countries initiative: Case study report. Tokyo: United Nations University Institute for Environment and Human Security.

Pregnolato, M., Ford, A., Wilkinson, S., & Dawson, R. (2017). The impact of flooding on road transport: A depth-disruption function. *Transportation Research Part D: Transport and Environment*, 55, 67–81.

Rahman, M.A. (2011). *Study on the changes of coastal zone: Chittagong to Cox's Bazar along the Bay of Bengal*. Baltimore, MD: Global Summit on Coastal Seas, EMECS 9, 28–31 August.

Rahman, M.M., & Islam, I. (2020). Exposure of urban infrastructure because of climate change-induced flood: Lesson from municipal level planning in Bangladesh. *Ecofeminism and Climate Change*, 1(3), 107–125. https://doi.org/10.1108/efcc-05-2020-0011

Shah, S. (1999). *Coping with natural disaster: The 1998 floods in Bangladesh*. Washington, DC: World Bank.

Sultana, M., Chai, G., Chowdhury, S., & Martin, T. (2016). Deterioration of flood affected Queensland roads: An investigative study. *International Journal of Pavement Research and Technology*, 9, 424–435.

Whitehead, P.G., Jin, L., Macadam, I., Janes, T., Sarkar, S., Rodda, H.J.E., & Nicholls, R.J. (2018). Modelling impacts of climate change and socio-economic change on the Ganga, Brahmaputra, Meghna, Hooghly and Mahanadi River systems in India and Bangladesh. *Science of the Total Environment*, 636. https://doi.org/10.1016/j.scitotenv.2018.04.362

Winter, M.G., Shearer, B., Palmer, D., Peeling, D., Harmer, C., & Sharpe, J. (2016). The economic impact of landslides and floods on the road network. *Advances in Transportation Geotechnics*, 143, 1425–1434.

World Bank (2010). *Economics of adaption to climate change: Bangladesh country study*. https://openknowledge.worldbank.org/handle/10986/12837

3

Impacts of flooding on agriculture and food security in developing countries

Evidence from Southeastern Nigeria

*Thecla Iheoma Akukwe, Lilian Chinedu Mba,
Onyinyechi Gift Ossai and Alice Atieno Oluoko-Odingo*

Introduction

Flooding occurs when a normally dry area is completely or partially submerged by runoff. Flooding is an intractable environmental problem that has negative impacts on agriculture and food security in developing countries. Developing countries have been noted to be predominantly affected by floods because of lack of resources, infrastructure and effective disaster management programs (D'Odorico & Rulli 2013). The international development community has recognized agriculture as being central to economic growth, food security and poverty alleviation, especially in countries where it is the main source of livelihood for the poor (World Bank 2007; FAO2011).

Agriculture enhances food security in two ways: food production and provision of livelihood sources. In Africa, about 70% of the population depends on agrarian sources of livelihood, 40% of export revenue is agriculture dependent and agriculture generates about one-third of their national income (Yaro 2004; McCusker & Carr 2006). The poorest countries in Africa have been observed to be those that depend on rain-fed subsistence agriculture for food, income and survival, and these countries are invariably the most vulnerable to any change in climate (Yaro 2004).

Climate change has been noted to represent abnormal climatic situations that influence temperature and water availability, thereby affecting agricultural production (Syaukat 2011), and would intensify the current food insecurity and hunger problems in developing countries which already contend with chronic food problems, since this has a multiplier effect on the risk of crop failure and livestock loss. Many studies indicate that the high frequency and intensity of some disasters such as floods affect agricultural production as well as food (FAO 2007; Ngoh et al. 2011; Syaukat 2011).

Despite increasing global agricultural production (Ibok et al. 2014), large numbers of people remain malnourished and hungry (Ambali et al. 2015). In 2010, 925 million people were estimated to be undernourished, and about 900 million of these people reside in developing countries (FAO, 2010). According to Bashir et al. (2012), more than 70% of undernourished people depend on agriculture directly or indirectly and live in rural areas. Similarly, IFAD (2009) estimated that about 1.2 billion people are unable to meet the most basic needs for sufficient food daily, and the largest segment of these people are the 800 million poor children, women and men who belong to indigenous populations living in rural areas, especially in developing countries, trying to make a living from subsistence agriculture. These indigenous populations rely heavily on rain-fed agriculture, which makes them vulnerable and adversely affected in times of floods.

In 2019, about 750 million globally were exposed to severe levels of food insecurity. The world has been noted not to be on track to achieving zero hunger by 2030, and over 840 million people have been projected to be affected by hunger by 2030 if this trend continues (FAO et al. 2020). It is expected that flooding will further expose these already vulnerable people to higher levels of food insecurity by reducing their nutritional status due to its negative impacts on food security and agriculture.

Food security, according to FAO (1996, 2008a), exists when there is economic and physical access to safe, sufficient and nutritious food by people at all times, thereby meeting their food preferences and dietary needs for an active and healthy life. Food security in this context consists of four pillars (food affordability, availability, utilization and stability), and these pillars must be fulfilled to realize food security objectives. Attaining a state of food security means sufficient food is available, everyone has access (economic and physical) to the food they need, access and availability are maintained over time (stability) and the food is effectively utilized. This implies that an adequate level of food consumption and good nutrition must be maintained at low risk over time (FAO 2008a); that is, it must be sustainable to achieve a state of food security. However, it has been noted that not all healthy diets are sustainable and not all sustainable diets are always healthy (FAO et al. 2020), and this is usually determined by food affordability.

Floods, like other disasters, have been noted to be linked to all four pillars of food security (Israel & Briones 2012; Ramakrishna et al. 2014; Ajaero 2017; Pacetti et al. 2017; Akukwe et al. 2018; Oskorouchi & Sousa-Poza 2021) in developing countries, as floods disrupt and damage agriculture, which is people's major source of food. According to Pacetti et al. (2017, p. 1), the inner correlations between floods and food security in developing countries are extremely relevant because they depend primarily on agriculture, which can be highly affected by extreme events, damaging their primary access to food. Oladokun and Proverbs (2016) have said flooding is one of the major culprits responsible for the high poverty level of African populations.

Flooding has been noted to negatively impact agriculture and food security by causing loss of livestock, destruction of crops, reduction in quality and quantity of food eaten, increase in price of food, destruction of roads, destruction of farm storage facilities and stream pollution (Israel & Briones 2012; Pacetti et al. 2017; Akukwe et al. 2018, 2020; Week & Wizor 2020). These negative impacts are felt mostly by households in rural areas with a paucity of social amenities such as good roads and pipe-borne water and whose primary source of livelihood is agriculture, as revealed by available literature on impacts of flooding on agriculture and food security in developing countries. This is an indication that rural dwellers residing in the hinterlands experience these impacts first-hand more than urban dwellers, whose primary occupation is not usually farming. Thus, this chapter focuses on bringing to light the areas in which floods have affected agriculture and food security and the relationship between flooding and food security with insights from southeastern Nigeria and on suggesting policy measures to minimize the impacts of flooding on agriculture and food security in developing countries.

Literature review and conceptual framework

Literature review

Over the years, studies on the impacts of natural disasters and climate have been attracting increasing attention among researchers, developers, climatologists, environmentalists and policy makers. This has, however, generated an agreement that there is indeed a relationship between floods, food security and agriculture. A number of studies have shown the many adverse events of floods resulting from climate change in addition to addressing flooding's potential effects on agriculture and food security (Parry et al. 2004; Fischer et al. 2005; Parry et al. 2005; Akukwe et al. 2020). A relationship has been observed between temperature and changing rainfall patterns that influences flooding. Several studies have indicated that raised temperature will not only change rainfall patterns but will increase the severity and frequency of harsh weather (Haines & Patz 2004; Ahern et al. 2006; Ramin & McMichael 2009; Syaukat 2011). The agricultural sector in Africa is already overwhelmed, and an FAO (2016) report stated that the increase in climate-related events is of significant concern to the particularly vulnerable agricultural sectors. The report concluded that the agricultural sector absorbed 25% of the total impact of climate-related disasters in developing countries between 2003 and 2013. The 2017 report, on the other hand, commented that since agriculture is dependent on weather and climate, water and land to thrive, it is natural for it to be particularly vulnerable to natural disaster (FAO 2017).

However, the impacts of climate change on agriculture and food security have been extensively studied, but only a handful of literature exists on in-depth studies on the impacts of flooding on agriculture and food security in developing countries. This chapter reviews some related literature on this subject within countries in Asia and Africa.

In Asia, Ian (2009) studied climate change, flooding and food security in South Asia and noted that the most vulnerable groups in terms of food security during flooding were the poor, women and children, and the study concluded that flooding disrupted food production more frequently and severely. He also noted that severe cases of flooding affected arable land, causing low nutritional levels in the poor in Bangladesh and other countries in South Asia. In the same vein, Israel and Briones (2012) studied the impacts of natural disasters on agriculture, food security and natural resources and environment in the Philippines, and their results revealed that floods, typhoons and drought have an insignificant negative impact on rice production at the provincial level, while floods have a positive impact on rice production at the local level but not at the national level.

Similarly, Ramakrishna et al. (2014) examined the impact of flood on food security and livelihoods of internally displaced person (IDP) households using a binary logit model. Their findings show that floods negatively impact food security by reducing purchasing power and wage income, causing food shortages among households in the Khammam region of India. The extreme floods of Bangladesh and Pakistan in 2007 and 2010 have been used as case studies by Pacetti et al. (2017) to describe how floods hamper agriculture and food security in developing countries where people are predominantly reliant on agriculture. A combination of remote sensing data, water footprint databases and agricultural statistics were used to examine the impacts of flood events on food supply (food availability). They assessed flood damage on agriculture by estimating crop losses and later converted it into water footprint and lost calories, and their results showed a flood-induced reduction of food supply between 8% and 5% in Pakistan and Bangladesh, respectively. Moreover, the work by Oskorouchi and Sousa-Poza (2021) concentrated on the long-term effect of floods on food security as measured by calorie and micronutrient consumption in Afghanistan. They found that persons exposed to flooding within a 12-month period had an approximately 60 kcal decreased daily calorie consumption.

In Africa, Zakari et al. (2014) employed a logit model to show the relationship between flooding and food security, and their results show that floods have a negative and significant impact on household food security in Niger Republic, while Oluoko-Odingo (2006) found that a negative correlation existed between flooding and food crop production at the household level in the Nyando district of Kenya.

Nigeria recorded the most devastating flood events between August and October 2012 that submerged hundreds kilometres of rural and urban lands (Ojigi et al. 2013). UN-OCHA (2012) described this 2012 flood as the worst flood recorded in the past 40 years in Nigeria because over 7,705,378 Nigerians were affected, with 2,157,419 IDPs. Farmlands were massively destroyed within the 2012 period, causing food insecurity in parts of Nigeria (FEWS NET 2012, 2013). Similarly, Sidi (2012) found a significant negative impact of the Nigerian 2012 flood on yam, cassava and potato production areas to the tune of 27.9%, 21.6% and 17.2%, respectively. Ajaero (2017) also noted that the 2012 flood affected the food security status of both female- and male-headed households in the country. In addition, Akukwe et al. (2018) studied the effect of flooding on food security in agricultural communities using principal component analysis, and their findings show that floods adversely affect food security in three areas: food supply and distribution, farm labour and facilities and household income and investment.

Furthermore, Jonathan et al. (2020), using a multistage sampling technique, food security index and logistic regression model, did an economic analysis of flood disasters on food security on arable farm households in the southern guinea savanna zone of Nigeria, and their findings revealed that households that experienced flooding were food insecure.

From the foregoing, it is evident that not many substantial studies have been carried out on the impact of floods on agriculture and food security in developing countries, and the few studies done on this concentrated more on the effects on rural dwellers than urban areas. The empirical evidence of this chapter will contribute to the body of literature on impacts of flooding on agriculture and food security because the study area comprised households in both rural and semi-urban areas.

Conceptual framework

The conceptual framework shown in Figure 3.1 illustrates the relationship between flooding and food security formed predominantly from the Food and Agricultural Organization–Food Insecurity and Vulnerability Information Management Systems (FAO-FIVIMS) framework and sustainable livelihood framework as well as reviewed literature. The frameworks aided the understanding of the connections between the four pillars of food security and their influencing factors at different levels (Verduijn 2005; FAO 2008a, 2008b; FAO/NRCB 2008). Figure 3.1 shows that food security has four pillars: food availability, accessibility, utilization and stability. A state of food security implies that an adequate level of food consumption and good nutrition are sustainable at low risk. Stability stresses the importance of minimizing the risk of the negative effects on the other three pillars (FAO 2008a). Conversely, food insecurity exists whenever there is any negative disturbance or shift (e.g. caused by drought or flooding) in any of the pillars of food security, hence the concept of vulnerability.

Vulnerability refers to those factors that induce the risk of households being affected by flooding or becoming food insecure. Vulnerability degrees are determined by the risk factors that households are exposed to and their ability (IPCC 2007, p. 11) to cope with flooding/food insecurity as well as the degree to which they are affected (sensitivity).

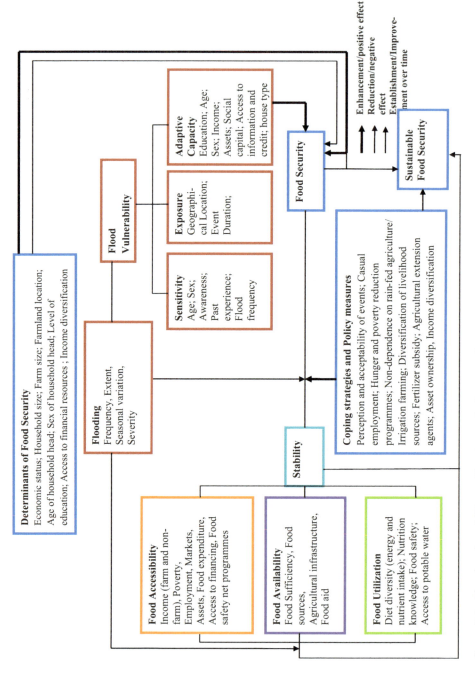

Figure 3.1 Flood, vulnerability and food security framework
Source: Akukwe 2019

On the one hand, flooding is a phenomenon that has a negative impact on food security through reducing labour demand, reducing crop harvest, damaging assets (e.g. houses), reducing farm income derived from crop sales, destroying food/farm storage facilities, destroying road and farmlands and polluting streams that are major water sources. On the other hand, adaptive capacity and coping strategies adopted by households would cushion the effects of flood-induced food insecurity and help improve household food security. Figure 3.1 also shows that food security determinants such as location of farmland and income diversification could either affect or improve food security. In addition, sustainable food security is achieved when the four pillars of food security and coping strategies are maintained over time.

Food security is conceptualized as the dependent variable, and its predictors were drawn from food availability, accessibility, utilization and stability variables measured through the use of indicators (via questionnaire). The categorical variables with an ordinal scale of measurement, such as level of education, classified as no formal education, primary school leaving certificate, secondary school certificate and university degree, were transposed into "no formal education and educated", showing a binary relationship (Akukwe 2019, p. 28).

Research materials and methods

Sample size and sampling method

The study area, southeastern Nigeria, comprises the five states: Anambra, Abia, Enugu, Ebonyi and Imo, located between latitudes 4° 20′ to 7° 10′ north of the equator and longitudes 6° 35′ to 8° 25′ east of the Greenwich Meridian, with a land size of about 28,983 km^2 (Figure 3.2). Southeastern Nigeria experiences the tropical wet-and-dry climate or Aw climate with an average of eight months (March to October) of rainy season and four months of dry season (November to February). Imo and Anambra states were selected for this chapter, as shown in Figure 3.3. Two local government areas (LGAs) that suffer perennial flooding were selected in each state. For equal representation, the Ogbaru and Anambra East LGAs of Anambra state and Ohaji/Egbema and Oguta LGAs of Imo state were purposively selected. According to the National Population Commission (2010), Anambra East has a population of 152,149, Ogbaru has 223,317 persons, Oguta has 142,340 persons and Ohaji/Egbema has 182,891 persons. The sample size was calculated using Yamane's (1967) equation, and according to the equation, any population of 100,000 and more persons would have a sample size of at least 400:

$$n = N \div [1 + N(e^2)] \tag{1}$$

where
n – is the sample size
N – is the population of the four LGAs
e – is the level of precision/sampling error, 0.05
n – 400 households

Two communities (including LGA headquarters, which in most cases are semi-urban areas in Nigeria) were selected in each LGA, giving a total of eight communities (four communities for each state) using a multistage purposive sampling technique. Stratified sampling was employed to determine the number of sampled households at the community and LGA levels, while 400 copies of the structured questionnaire were administered through random sampling.

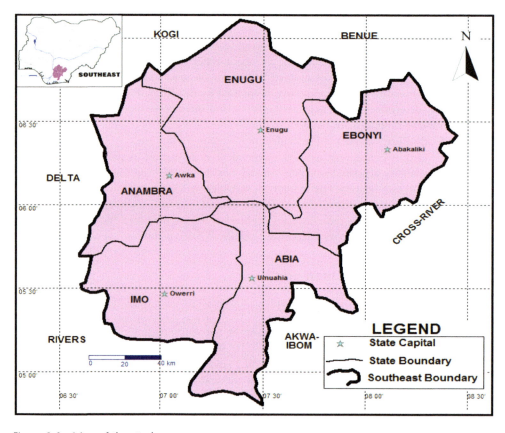

Figure 3.2 Map of the study area
Source: GIS Lab, Department of Geography, University of Nigeria, Nsukka, 2016

Data collection and data analysis

The study was questionnaire based and involved mainly household heads with flood experience between 2012 and 2017. Household food security was calculated using the Household Food Security Survey Module (HFSSM), developed by the USDA. The HFSSM is a set of 18 questions regarding households' food needs (Figure 3.4), and the responses were coded and analysed into food security indices using Rasch analysis. The 18 questions captured the four pillars of food security. The questions had three response categories: often true, sometimes true and never true, in which often true and sometimes true were coded as 1 because they were affirmative responses, and never true was coded as 0 because it depicted a condition of non-occurrence. The scale measures different food security conditions and food insecurity severity ranging from 0 to 10 (Figure 3.5), where 0 represents food secure-households with none or few of the measured conditions, while 10 signifies the most severe condition of food insecurity (Bickel et al. 2000). Household food security status scores were calculated using Rasch analysis, and households were further categorized into four types: food secure, food insecure without hunger, moderately food insecure with hunger and severely food insecure with hunger (Figure 3.5). On the one hand, food-secure households are known to have experienced no or very limited

Impacts of flooding on agriculture and food security

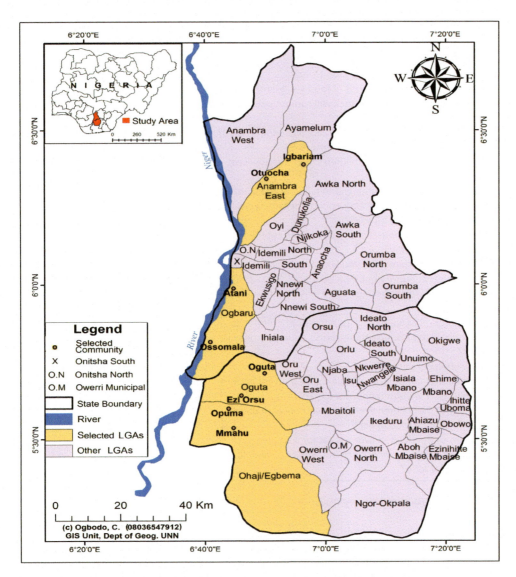

Figure 3.3 Map of the study area showing the sampled LGAs/communities
Source: GIS Lab, Department of Geography, University of Nigeria, Nsukka, 2016

hunger or food insecurity, whereas food-insecure without hunger households either have no or little reduction in food intake without their members experiencing hunger. On the other hand, moderately food-insecure households with hunger had their adults' (with exception of children) food intake reduced, and they experienced hunger repeatedly, whereas severely food-insecure households with hunger repeatedly reduced food intake for both children and adults, and all members experienced hunger (Bickel et al. 2000; Ibok et al. 2014).

Data were collected through a structured questionnaire where the respondents filled out two sets of 18 questions that captured their food security situations before and after flooding, and inferences were made from their responses, majorly extracted from the work of Akukwe (2019).

Question	Responses		
	Often true	Sometimes true	Never true

- Do you always have enough food to eat?
- Do you always have the kinds of food you want?
- Do you worry if your food stock will run out before you get another to eat?
- Do you have enough resources to acquire enough food?
- Could you afford to eat balanced meals?
- Do you supplement your children's feed with low cost foods?
- Can you afford to feed your children balanced meals?
- Were your children not eating enough because you couldn't afford enough food?
- Do adults in your household skip meals or cut the size of their usual meals?
- Do you eat less than what you feel you should?
- Were you ever hungry but didn't eat?
- Did you lose weight because there wasn't enough food to eat?
- Did you or other adults in your household ever not eat for a whole day because there wasn't enough money for food?
- How often did this happen?
- Did you ever cut the size of your children's meal because there wasn't enough money for food?
- Did any of the children ever skip meals because there wasn't enough food to eat?
- Did any of the children ever not eat for a whole day?
- Were the children ever hungry but you just couldn't afford more food?

Figure 3.4 Structured survey questions on household food security

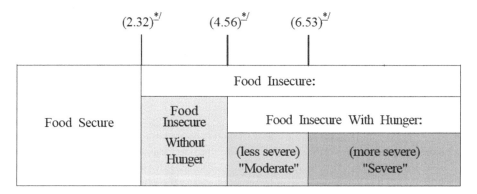

Figure 3.5 Household food security status – categorical measure
Source: Bickel et al. 2000, p. 31
*/ located at midpoint between the two adjacent household scale values.

The extent to which flood negatively affected agriculture and food security was measured on a 5-point Likert scale, with 5 – to a very great extent, 4 – to a great extent, 3 – to a moderate extent, 2 – to little extent and 1 – to no extent.

In order to establish the nature of the relationship that exists between floods and food security status, the determinants of household food security were assessed using multiple binary logistic regression, as seen in the equation

$$[Y/(1-Y)] = b_0 + b_1 x_1 + b_2 x_2 + \ldots b_{25} x_{25} \tag{2}$$

where $[Y/(1-Y)]$ – likelihood that household is food secure/insecure
b_0 – constant that scales the equation
b_1 to b_{25} – coefficients that determine the direction and extent to which the (x) factors affect food security status (Y)

x_1 to x_{25} – factors that influence food security: age, sex of household head, marital status, level of education, literacy rate, diversified income, flood experience, monthly income, off-farm income, dependency ratio, private land ownership, group membership, sufficient own food production, poultry farm ownership, livestock ownership, distance to farm, storage facility availability, distance to market, fertilizer use, food/aid receipt, irrigation, nutrition knowledge, financial support, farm size and access to credit.

Multiple binary logistic regression was selected because of the dichotomous nature of the dependent variable (food secure/food insecure). To show the relationship between food security and its determinants using a binary model, the classified HFSSM food security levels of four were transposed into two: food secure and food insecure (consisting of severely food insecure with hunger, moderately food insecure with hunger and food insecure without hunger) households.

Research results and discussion

Household food security status before and after flooding

The food security status of households before and after flood events is shown in Table 3.1. It was observed that preceding flood disasters, 33.3% of households were food secure, 40.2% were food insecure without hunger, 13% were moderately food insecure with hunger and 13.5% were severely food insecure with hunger in the study area. This indicates that a third of the total population had no food insecurity problem; 40.2% had no or little reduction in food intake; 13% had their adult members' food intake reduced, and they repeatedly experienced hunger; and 13.5% had both their children and adults' food intake reduced, and they repeatedly experienced hunger.

Table 3.1 also reveals that after flooding, there was a negative shift in household food security levels, as the number of food-secure households decreased from 33.3% to 7.2%, causing a 26.1% reduction in food-secure households. Reduced food intake for both adults and children resulting from inadequate resources to acquire food and reduced household food supply caused the extremely higher number of food-insecure households recorded after flooding.

However after flooding, food-insecure without hunger households reportedly reduced to 39.3% from the 40.2% that was recorded before flooding. The table also shows that flooding induced a greater percentage of households into being severely food insecure, with a 2.7% increase in the number of moderately food-insecure with hunger households and a 24.3% increase in severely food-insecure with hunger households. Our findings reveal that flooding affected food

Table 3.1 Before- and after-flood differential household food security status/levels in southeastern Nigeria

Community	Before-flood category of food security status/level (%)				Total
	Food secure	Food insecure without hunger	Moderately food insecure with hunger	Severely food insecure with hunger	
Atani	37.5	37.5	9.4	15.6	64
Ossomala	25.0	45.3	20.3	9.4	64
Otuocha	43.2	40.9	13.6	2.3	44
Igbariam	25.6	34.9	20.9	18.6	43
Oguta	26.8	43.9	9.8	19.5	41
Ezi-Orsu	30.0	35.0	20.0	10.0	40
Mmahu	44.2	34.6	3.9	17.3	52
Opuoma	32.8	48.1	3.9	15.4	52
Total (mean)	33.3	40.2	13.0	13.5	400

Community	After flood category of food security status/level (in percent %)				Total
	Food secure	Food insecure without hunger	Moderately food insecure with hunger	Severely food insecure with hunger	
Atani	6.2	39.1	14.1	40.6	64
Ossomala	1.6	45.3	12.5	40.6	64
Otuocha	9.1	59.1	20.4	11.4	44
Igbariam	0.0	20.9	7.0	72.1	43
Oguta	22.0	19.5	43.9	14.6	41
Ezi-Orsu	7.5	40.0	12.5	40.0	40
Mmahu	7.7	40.4	13.4	38.5	52
Opuoma	7.7	44.2	7.7	40.4	52
Total (mean)	7.2	39.3	15.7	37.8	400

Source: Akukwe (2019)

security by generally causing an increase in the number of food-insecure households to 92.8%, a 26.5% increase over households that were food insecure before flooding in southeastern Nigeria.

The Igbariam community was the most affected after flooding, as 100% of the households became food insecure, where 72.1% became severely food insecure with hunger. The Ossomala community was the second most affected, as after the flooding events, only 1.6% of households remained food secure. After flooding, the Atani, Mmahu, Ezi-Orsu and Opuoma communities also were affected, as fewer than 8% of their households were food secure, revealing that more than 90% of their households experienced food insecurity after flooding. Flooding affected the Oguta community least, because 22% of its households remained food secure after flooding, showing a 4.8% reduction in the number recorded before flooding. Thus, flooding caused a shift that increased the number of *food insecurity hotspots* because most households (except in the Oguta community) moved from food insecure without hunger to either moderately or severely food insecure with hunger after flooding. This implies that a greater number of households will need food assistance to improve their coping ability with food insecurity induced by flooding since a majority of the households repeatedly reduced children's and/or adults' food intake after a flood experience. Atani, Otuocha, Mmahu and Oguta are LGA headquarters and are considered semi-urban areas. Our findings show that floods affect household food security of agrarian communities in both rural and urban areas equally (Akukwe 2019, pp. 155–156).

The negative impacts of flooding on agriculture and food security

Table 3.2 shows the areas and extent to which flooding negatively affected food security and agriculture. Our findings reveal that the impacts of flooding on food security and agriculture were to a great extent as measured on the Likert scale in Table 3.2. The largest percentage of households reported that flooding caused a reduction in farm income derived from crop sales (55.5%), reduction in crop harvest (55.7%), destruction of roads (56%) and increase in food prices (63.5%). Floods were reportedly found to reduce labour demand, affect the quality of food eaten, pollute streams, destroy food/farm storage facilities and reduce the number of times food was consumed and the quantity of food eaten to a great extent, as indicated by 25.3%, 37%, 40.3%, 43.5%, 46.8%, and 47.5% of households respectively. However, 52.5% reported flooding to have caused stream pollution to a very great extent, which affected communities greatly since streams are their major source of water. Our findings are in agreement with those of Devereux (2007), Odufuwa et al. (2012), Ajaero and Mozie (2014), Otomofa et al. (2015), Ikani (2016) and Oskorouchi and Sousa-Poza (2021). Ajaero and Mozie (2014) revealed that flooding caused stream pollution in Ogbaru and Anambra East LGAs, and Ikani (2016) noted loss of farm crops due to flooding in the Gwagwalada area in Nigeria, while Oskorouchi and Sousa-Poza (2021) found flood experiences to be associated with lower diet quantity and quality in Afghanistan.

Flooding was revealed to cause changes in food consumption patterns arising from destruction of farmlands, rise in food prices, disruption of sources of livelihoods, pollution of streams, reduction in crop harvest, decrease in quality and quantity of food eaten, destruction of roads and reduction in meal frequency, which adversely affected food accessibility, availability and utilization over time.

This chapter has revealed that flooding negatively affects agriculture and food security, and this could hamper the achievement of Sustainable Development Goal 2, with an emphasis on achieving food security and promoting sustainable agriculture; thus adequate flood management is necessary in southeastern Nigeria.

Table 3.2 Negative effects of flooding on agriculture and food security

Negative effect of flooding on food security	Extent of effect				
	1	2	3	4	5
Reduces crop harvest	0.0	1.3	2.3	55.7	40.7
Decreases farm income derived from crop sales	0.0	2.5	4.0	55.5	38.0
Destroys road	0.0	10.3	18.0	56.0	15.7
Destroy food/farm storage facilities	0.0	5.0	29.3	43.5	22.2
Reduces labour demand	3.3	24.0	25.7	25.3	21.7
Pollutes streams	0.0	2.5	4.7	40.3	52.5
Reduces the number of times food is consumed	0.0	16.7	27.5	46.8	9.0
Affects the quality of food eaten	6.0	18.5	26.5	37.0	12.0
Affects the quantity of food eaten	0.0	16.0	31.0	47.5	5.5
Increases food item prices	0.0	0.5	19.7	63.5	16.3

Source: Field work, 2017
A 5-point scale with 5 = to a very great extent, 4 = to a great extent, 3 = to a moderate extent, 2 = to little extent, 1 = to no extent.

Relationship between flooding and food security

Table 3.3 shows the 25 variables that predicted food security status in southeastern Nigeria. These variables range from demographic and socio-economic factors to environmental factors, but because this chapter is centred on impacts of flooding on agriculture and food security, only flood experience as a predictor is discussed.

The coefficient of the relationship between flood experience and food security was negative (−1.110) and statistically significant at 5%. This implies that the more a household experiences flooding, the more food insecure that household becomes. This also means that for every unit increase in flood experience, the odds ratio of the household being food secure reduces by a factor of 0.33 (Table 3.3). Our findings agree with those of Ramakrishna et al. (2014) and Zakari et al. (2014), who found floods to have a negative impact on household food security in Khammam (India) and Niger, respectively. The negative effects of floods have been discussed in the preceding pages, and they include destruction of roads leading to farmlands; destruction of farmland and storage facilities, which leads to food shortage and crop failure; reduction in crop

Table 3.3 Logit model output of determinants of food security status

Factors influencing food security	B	S.E.	Wald	df	Sig.	Exp(B)
Sex of household head	.945	.433	4.763	1	.029*	2.572
Age of household head	−.128	.163	.615	1	.433	.880
Marital status	.381	.142	7.239	1	.007*	1.464
Literacy rate	−.616	.581	1.124	1	.289	.540
Level of education	.502	.260	3.715	1	.054*	1.652
Diversified income	.206	.854	.058	1	.810	1.228
Off-farm income	1.432	.679	4.454	1	.035*	4.189
Monthly income	.680	.192	12.533	1	<.0001*	1.973
Dependency ratio	.879	.389	5.115	1	.024*	2.410
Group membership	−.197	.428	.211	1	.646	.822
Private land ownership	−.455	.454	1.004	1	.316	.635
Sufficiency in own food production	1.068	.430	6.175	1	.013*	2.908
Livestock ownership	.893	.410	4.743	1	.029*	2.443
Village/agric. poultry ownership	.839	.386	4.711	1	.030*	2.314
Distance to farm	−.530	.434	1.489	1	.222	.589
Distance to market	.373	.382	.953	1	.329	1.451
Storage facility availability	−.019	.383	.003	1	.959	.981
Fertilizer use	.564	.373	2.293	1	.130	1.758
Irrigation practice	1.997	.659	9.193	1	.002*	7.367
Food/aid receipt	.333	.397	.705	1	.401	1.395
Nutritional knowledge	.807	.536	2.265	1	.132	2.241
Farm size	.036	.300	.014	1	.906	1.036
Financial support	.328	.391	.704	1	.401	1.388
Credit access	−.540	.585	.850	1	.356	.583
Flood experience	−1.110	.337	10.868	1	.001*	.330
Constant	−7.305	1.555	22.084	1	.000	.001
−2 Log likelihood	286.408					
Cox and Snell R^2	.426					
Nagelkerke R^2	.592					

* Significant at 0.05 level of significance ($P \leq 0.05$).

harvest; destruction of roads; and reduction in meal frequency, food quality and food quantity, thereby increasing food insecurity.

As shown in the conceptual framework, the negative impacts of flooding on agriculture and food security could be minimized by putting in place some coping strategies and policy measures such as mechanized and irrigation farming to encourage all-year-round planting and harvest and food safety net and poverty alleviation programmes to help the poor, who are the most vulnerable to the impacts of flooding, as supported by Pacetti et al. (2017).

Conclusion

This chapter analysed the impacts of flooding on agriculture and food security in developing countries. Flooding is an intractable environmental problem that has negative impacts on agriculture and food security in developing countries because their major sources of livelihood are affected. Flooding caused a negative shift in household food security levels by causing an increase in the number of food-insecure households to 92.8%, showing a 26.5% increase over households that were food insecure before flooding in southeastern Nigeria. Reduced food intake for both adults and children resulted from inadequate resources to acquire food and reduced household food supply, causing the high percentage of severely food-insecure households recorded after flooding. Our findings also revealed that floods affect household food security of agrarian communities in both rural and urban areas equally.

Flooding was revealed to cause changes in food consumption patterns arising from destruction of farmlands, rise in food prices, disruption of sources of livelihoods, pollution of streams, reduction in crop harvest, decrease in quality and quantity of food eaten, destruction of roads and reduction in meal frequency, which significantly affected food accessibility, availability and utilization over time. The coefficient of the relationship between flood experience and food security was negative (−1.110) and statistically significant at 5%, with an odds ratio of 0.33. This implies that the more a household experiences flooding, the more food insecure that household becomes. This also means that for every unit increase in flood experience, the odds ratio of the household being food secure reduces by a factor of 0.33.

Consequently, since our findings show that flooding has negative impacts on agriculture and food security, this would hamper the attainment of Sustainable Development Goal 2, which emphasizes achieving food security and promoting sustainable agriculture; thus adequate flood management and policy options geared towards enhancing food security are pertinent in southeastern Nigeria.

Recommendations

1. Irrigation farming to encourage all-year-round and off-season planting, which will boost crop productivity and alter the harvest season, especially for crops like yams that are usually harvested during flooding periods (July–October), thereby reducing food wastage.
2. Food safety net and poverty alleviation programmes to help the poor who are the most vulnerable to the impacts of flooding through making food available and accessible.

Acknowledgements

The authors would like to state that this chapter is part of PhD research submitted at the University of Nairobi by the lead author.

References

Ahern, M, Kovats, RS, Wilkinson, P, Few, R & Malthies, F 2006, 'Global health impacts of floods epidemiologic evidence', *Epidemics Logic Reviews*, vol. 27, no. 1, pp. 36–46, https://doi.org/10.1093/epirev/mxi004.

Ajaero, CK 2017, 'A gender perspective on the impact of flood on the food security of households in rural communities of Anambra state, Nigeria', *Food Security*, vol. 9, no. 4, pp. 685–695.

Ajaero, CK & Mozie, AT 2014, 'Socio-demographic differentials in vulnerability to flood disasters in rural southeastern Nigeria', paper presented at the International Seminar on Demographic Differential Vulnerability to Natural Disasters in the Context of Climate Change Adaptation, Kao Lak, Thailand.

Akukwe, TI 2019, 'Spatial analysis of the effects of flooding on food security in agrarian communities of south eastern Nigeria', PhD thesis, University of Nairobi, Kenya.

Akukwe, TI, Krhoda, GO & Oluoko-Odingo, AA 2018, 'Principal component analysis of the effects of flooding on food security in agrarian communities of south eastern Nigeria', *International Journal of Hydrology*, vol. 2, no. 2, pp. 205–212, http://dx.doi.org/10.15406/jh.2018.02.00070.

Akukwe, TI, Oluoko-Odingo, AA & Krhoda, GO 2020, 'Do floods affect food security? A before-and-after comparative study of flood-affected households' food security status in south-eastern Nigeria', *Bulletin of Geography, Socio-Economic Series*, vol. 47, no. 47, pp. 115–131.

Ambali, OI, Adewuyi, SA, Babayanju, SO & Ibrahim, SB 2015, 'Expansion of rice for job initiative programme: Implications for household food security in Lagos state Nigeria', *Advances in Economics and Business*, vol. 3, no. 3, pp. 99–106, http://dx.doi.org/10.13189/aeb.2015.030303.

Bashir, MK, Schilizzi, S & Pandit, R 2012, 'The determinants of rural household food security for landless households of the Punjab, Pakistan', Working Paper 1208, School of Agricultural and Resource Economics, University of Western Australia, Crawley, Australia.

Bickel, G, Nord, M, Price, C, Hamilton, W & Cook, J 2000, *Guide to Measuring Household Food Security*, revised ed., USDA, Alexandria.

Devereux, S 2007, 'The impact of droughts and floods on food security and policy options to alleviate negative effects', *Agricultural Economics*, vol. 37, no. 1, pp. 47–58, retrieved from www.unicef.org/socialpolicy/files/The_Impact_of_Droughts_and_Floods_on_Food_Security.pdf.

D'Odorico, P & Rulli, MC 2013, 'The fourth food revolution', *Nature Geoscience*, vol. 6, pp. 417–418, http://dx.doi.org/10.1038/ngeo1842.

Famine Early Warning Systems Network 2012, *Third Quarter 2012*, retrieved from www.fews.net/ml/en/info/Pages/fnwkfactors.aspx?l=en&gb=ng&fnwk=factor.

Famine Early Warning Systems Network 2013, *Food Insecurity Increases in Regions Affected by Flooding and Conflict, Nigeria Food Security Update*, retrieved from http://reliefweb.int/sites/reliefweb.int/files/resources/Nigeria%20Food%20Security%20Updated%20March%202013.pdf.

FAO 1996, *Rome Declaration on World Food Security and World Food Summit Plan of Action*, FAO Document Repository, Rome.

FAO 2007, *Climate Change and Food Security: A Framework Document – Summary*, The Food and Agriculture Organization of the United Nations, Rome.

FAO 2008a, *Food Security Concepts and Frameworks: Food Security Information for Action, E-Learning Manual*, viewed 25 June 2015, retrieved from www.fao.org/elearning/Course/FC/en/Course Viewer.asp?language=en.

FAO 2008b, *Climate Change and Food Security: A Framework Document*, The Food and Agriculture Organization of the United Nations, Rome.

FAO 2010, *The State of Food Insecurity in the World: Addressing Food Insecurity in Protracted Crises*, The Food and Agricultural Organization of the United Nations, Rome.

FAO 2011, *The State of Food and Agriculture, 2010–2011: Women in Agriculture – Closing the Gender Gap for Development*, The Food and Agricultural Organization of the United Nations, Rome.

FAO 2016, *Damage and Losses from Climate-Related Disasters in Agricultural Sectors*, The Food and Agriculture Organization of the United Nations, Rome, www.fao.org/3/a-15128e.pdf.

FAO 2017, *The Impact of Disasters and Crises on Agriculture and Food Security*, The Food and Agriculture Organization of the United Nations, Rome, www.fao.org/publications.

FAO, IFAD, UNICEF, WFP & WHO 2020, *The State of Food Security and Nutrition in the World 2020: Transforming Food Systems for Affordable Healthy Diets*, FAO, Rome, https://doi.org/10.4060/ca9692en.

FAO/NRCB 2008, 'Food security and environmental change', Poster presented at the International Conference, Oxford, April.

Fischer, G, Shah, M, Tubiello, FN & Van-Velhuizen, H 2005, 'Socio-economic and climate change impacts on agriculture: An integrated assessment, 1990–2080', *Philosophical Transactions of the Royal Society of London. Series B: Biological Sciences*, vol. 360, no. 1463, pp. 2067–2083, https//doi.org/10.1098/0516.200.1744.

Haines, A, & Patz, JA 2004, 'Health effects of climate change,' *JAMA*, vol. 29, no. 1, pp. 99–103, https://doi.org/10.1001Ijama.291.199.

Ian, D 2009, 'Climate change, flooding and food security in South Asia', *Food Security*, vol. 1, pp. 127–136, https://doi.org/10.1007/5/257-009-0015-1.

Ibok, OW, Idiong, IC, Brown, IN, Okon, IE & Okon, UE 2014, 'Analysis of food insecurity status of urban food crop farming households in Cross River state, Nigeria: A USDA approach', *Journal of Agricultural Science*, vol. 6, no. 2, pp. 132–141.

Ikani, DI 2016, 'An impact assessment of flooding on food security among rural farmers in Dagiri community of Gwagwalada Area Council, Abuja. Nigeria', *Agricultural Development*, vol. 1, no. 1, pp. 6–13.

Intergovernmental Panel on Climate Change 2007, 'Appendix 1: Glossary', in ML Parry, OF Canziani, JP Palutikof, PJ Van der Linden & CE Hanson (eds.), *Climate Change 2007: Impacts, Adaptation and Vulnerability: Contribution of Working Group II to the Fourth Assessment Report of the International Panel on Climate Change*, Cambridge University Press, Cambridge.

International Fund for Agricultural Development 2009, *Dimensions of Rural Poverty*, Rural Poverty Portal, viewed 10 November 2009, retrieved from www.ruralpovertyportal.org/topic.

Israel, DC & Briones, RM 2012, 'Impacts of natural disasters on agriculture, food security and national resources and environment in the Philippines', in Y Sawada & S Oum (eds.), *Economic and Welfare Impacts of Disasters in Easter Asia and Policy Response*, ERIA Research Project Report 2011–8, ERIA, Jakarta, 553–599.

Jonathan, A, Owolabi, MT, Olahinji, IB, Antoye, BT & Henshaw, EE 2020, 'Economic analysis of the effect of flood disaster on food security of arable farming households in southern guinea savanna zone Nigeria', *Journal of Agriculture and Food Sciences*, vol. 18, no. 1, pp. 59–69, https://doi.org/10.4319jafs.v18i1.6.

McCusker, B & Carr, ER 2006, 'The co-production of livelihoods and land use change: Case studies from South Africa and Ghana', *GeoForum*, vol. 37, pp. 790–804.

National Population Commission 2010, *2006 Population and Housing Census: Population Distribution by Sex, State, LGA and Senatorial District (Priority Table Volume III)*, National Population Commission, Abuja.

Ngoh, MB, Teke, MG & Atanga, NS 2011, 'Agricultural innovations and adaptations to climate change effects and food security in Central Africa: Case of Cameroon, Equatorial Guinea and Central African Republic', African Technology Policy Studies Network, Research Paper No. 13, Nairobi.

Odufuwa, BO, Adedeji, OH, Oladesu, JO & Bongwa, A 2012, 'Floods of fury in Nigerian cities', *Journal of Sustainable Development*, vol. 5, no. 7, pp. 69–79.

Ojigi, ML, Abdulkadir, FI & Aderoju, MO 2013, 'Geospatial mapping and analysis of the 2012 flood disaster in central parts of Nigeria', Paper presented at the 8th National GIS Symposium, Dammam, Saudi Arabia.

Oladokun VO & Proverbs D 2016, 'Flood risk management in Nigeria: A review of the challenges and opportunities', *International Journal of Safety and Security Engineering*, vol. 6, no. 3, pp. 485–497.

Oluoko-Odingo, AA 2006, 'Food security and poverty among small-scale farmers in Nyando District, Kenya', PhD thesis, University of Nairobi, Kenya.

Oskorouchi, HR & Sousa-Poza, A 2021, 'Floods, food security and coping strategies: Evidences from Afghanistan', *Agricultural Economics*, vol. 52, no. 1, pp. 123–140.

Otomofa, JO, Okafor, BN & Obienusi, EA 2015, 'Evaluation of the impacts of flooding on socio-economic activities in Oleh, Isoko South local government area, Delta state', *Journal of Environment and Earth Science*, vol. 5, no. 18, pp. 155–171.

Pacetti, T, Caporali, E & Rulli, MC 2017, 'Floods and food security: A method to estimate the effect of inundation on crops availability', *Advances in Water Resources*, vol. 110, pp. 494–504, https://doi.org/10.1016/j.advwatres.2017.06.019.

Parry, ML, Rosenzweig, C, Iglesias, A, Livermore, M & Fischer, G 2004, 'Effects of climate change on global food production under SRES emissions and socio economic scenarios', *Global Environmental Change*, vol. 14, no. 1, pp. 53–67, https://doi.org/10.1010/.groenrcha.2003.10.008.

Parry, ML, Rosenzweig, C & Livermore, M 2005, 'Climate change, and risk global food supply of hunger', *Philosophical Transactions of the Royal Society B-Biological Science*, vol. 360, no. 1463, pp. 2125–2138, https://doi.org/10.1098/rstb.2005.1751.

Ramakrishna, G, Gaddam, SR & Daisy, I 2014, 'Impact of floods on food security and livelihoods of IDP tribal households: The case of Khammam region of India', *International Journal of Development and Economics Sustainability*, vol. 2, no. 1, pp. 11–24.

Ramin, BM & McMichael, AJ 2009, 'Climate change and health in sub-Saharan Africa: A case-based perspective', *Ecohealth*, vol. 6, no. 1, pp. 52–57, https://doi.org/10.1007/510393-009-0222-4.

Sidi, MS 2012, *The Impact of the 2012 Floods on Agriculture and Food Security in Nigeria Using GIS*, United Nations International Conference on Space-based Technologies for Disaster Management – Risk Assessment in the Context of Global Climate Change, Beijing, China.

Syaukat, Y 2011, 'The impact of climate change on food production and security and its adaptation programs in Indonesia', *Journal of ISSAAS*, vol. 17, no. 1, pp. 40–51.

United Nations Office for the Coordination of Humanitarian Affairs 2012, *Nigeria: Floods Situation Report No. 1*, viewed 6 November 2012, retrieved from http://reliefweb.int/report/nigeria/floods-situation-report-no-1-06-november-2012.

Verduijn, R 2005, *Strengthening Food Insecurity and Vulnerability Information Management in Lesotho: FIVIMS Assessment Report*, FAO's Emergency Operations Service (TCEO), Maseru.

Week, DA & Wizor, CH 2020, 'Effects of flood on food security, livelihood and socio-economic characteristics in the flood-prone areas of the Core Niger Delta, Nigeria', *Asian Journal of Geographical Research*, vol. 3, no. 1, pp. 1–17, https://doi.org/10.9734/ajgr/2020/v3i130096.

World Bank 2007, *World Development Report 2008: Agriculture for Development*, World Bank Publications, Washington, DC.

Yamane, T 1967, *Statistics: An Introductory Analysis*, 2nd ed., Harper and Row, New York.

Yaro, JA 2004, 'Theorizing food insecurity: Building a livelihood framework for researching food insecurity', *Norsk Geografisk Tidsskrift*, vol. 58, pp. 23–37.

Zakari, S, Ying, L & Song, B 2014, 'Factors influencing household food security in West Africa: The case of southern Nigeria', *Sustainability*, vol. 6, pp. 1191–1202, https://doi.org/10.3390/su6031191.

4

Adoption pathway for flood-resilient construction and adaptation in Ghana

Eric Kwame Simpeh, Henry Mensah and Divine Kwaku Ahadzie

Introduction

Globally, flooding is the phenomenon with the most impact on the human population (UNDRR, 2015). The frequency of documented flood occurrences has been steadily growing, notably in recent decades. The number of persons killed or badly injured by floods has increased dramatically across the world (United Nations-Water, 2011). According to an Intergovernmental Panel on Climate Change (IPCC) assessment report, the frequency and intensity of flooding are predicted to rise globally (IPCC, 2014). As a result, flood damage is increasing in many parts of the world, and the number of people at danger of flooding is anticipated to rise in future decades. With the increase in flood occurrences, flood-prone communities are at high risk, and as a result, there is an urgent need for collaborative preparatory action to assist in reducing the worst consequences. As such, it is worth noting that proper flood risk management could serve as a reduction mechanism for this phenomenon and also reduce the damages it causes. Flood risk management (FRM) in developing countries has predominantly consisted of putting up structural measures and formulating and implementing flood-related policies, which in most parts of the developing world has been the responsibility of the government (Egbinola, Olaniran & Amanambu, 2017). In addition, various types of urban flood risk may be addressed by not only considering structural measures but a combination of structural and non-structural measures. Structural measures are related to the provision of grey (heavy infrastructure) and green (nature-based, multipurpose interventions) infrastructural measures. On the contrary, non-structural solutions include the capacity to reach a wide population at a cheap cost through education campaigns, the distribution of hazard maps, awareness training, and the creation of early flood warning and evacuation systems (World Bank, 2019). However, with the latest science and philosophies on worsening future scenarios of flood effects, an integrated approach to urban flood risk management should be pursued and incorporated within flood modelling (Nkwunonwo, Whitworth & Baily, 2020).

Similar to many other developing countries, Ghana is faced with an increasing flood occurrence annually. The recent flooding cases which have almost become an everyday event in Ghana (Ahadzie & Proverbs, 2011; Mensah & Ahadzie, 2020) have many impacts on the country. For instance, Golz, Schinke and Naumann (2015) revealed that there is an increasing negative impact on housing as well as the ongoing development of settlements in flood-prone areas in Ghana due to the frequent occurrences of flooding. Also, Asumadu-Sarkodie, Owusu and Jayaweera (2015) found that, within the last three decades, about four million people were affected by flooding with many lives lost and approximately US$780,500,000 of economic damages in Ghana. Asumadu-Sarkodie, Owusu and Rufangura (2015) added that, in Ghana, endemic cases of flooding have been attributed to factors relating to institutional management, lack of prescribed building regulations for flood risk-prone areas, and a lack of consensus in terms of the approach for designing and constructing flood-resilient buildings. For example, Tasantab (2019) noted that institutional incapacities are the major challenges to disaster risk management, especially flood risk management in Ghana. In addition, Essuman (2015) revealed that unauthorized buildings were responsible for the flood which destroyed a lot of properties in Accra in 2015. Hutton and Marsh (2019) also revealed that many of flood occurrences and damage are associated with poor construction of buildings, such as low-level foundations and improper drainage directions.

From the foregoing context, it is clear that the failures and successes of flood risk management in Ghana can be attributed to institutional and regulatory frameworks. In this respect, this chapter aims to examine the institutional and regulatory frameworks for flood-resilient construction and adaptation in Ghana. The chapter is part of an ongoing research agenda to evaluate flood risk management by delving into the area of flood-resilient construction in developing countries such as Ghana. Against this backdrop, the following three research questions are pursued to achieve the aim and objectives of the chapter:

1. What aspects of the building codes/regulations relating to FRM are lacking and require improvement to regulate construction activities in flood-prone areas?
2. What measures have been taken by the National Disaster Management Organisation (NADMO) to improve flood resilience in the construction industry?
3. What are the effective approaches that can be adopted for the design and construction of flood-resilient buildings?

This chapter provides an invaluable contribution to the discourse regarding the quest to adopt a flood-resilient construction and adaptation framework to address the pressing need of managing flood risk holistically to achieve sustainable cities in developing countries. The outcome of the review will not only reveal research implications with the potential to expand the knowledge area but also provide information regarding a pathway for the adoption and promotion of flood-resilient construction and adaptation. This will aid in terms of complementing national building regulations and codes for the design and construction of flood-resilient buildings. In the subsequent sections of the chapter, the role of flood management institutions is described, followed by an overview of flood-resilient construction and adaptation and the influence of building codes and regulations on flood resilience. Afterwards, the methods employed to achieve the aim of the chapter are described. Subsequently, the results in terms of the framework for assessing flood resilience in a building are discussed and followed by an overview of the findings. Finally, the adoption pathways for flood-resilient construction and adaptation and the conclusions are outlined.

The role of flood management institutions in the management of floods in Ghana

Flood management institutions in Ghana are instated with responsibilities for addressing flood-related problems and ensuring that the development of flood resilience in the country is pursued. Almoradie et al. (2020) stated that the influential and vital institutions to flood risk management in Ghana include, but are not limited to, the National Disaster Management Organisation (NADMO), Hydrological Service Department, Water Resources Commission (WRC), Ghana Meteorological Agency (GMet), Environmental Protection Agency (EPA), Water Research Institute (WRI), Land Use and Spatial Planning Authority (LUSPA), Ministry of Sanitation and Water Resources, Red Cross, National Ambulance, National Fire Service, and the Armed Forces. For instance, Poku-Boansi et al. (2020) revealed that the EPA operates at the national level but with offices at the regional level and is responsible for the protection of the natural and built environment, environmental conservation and preservation, environmental sustainability, and climate change. Ahadzie and Proverbs (2010) noted that the Meteorological Services Department is the institution tasked with keeping climatic and rainfall data in Ghana, and NADMO is responsible for making sure that the country is well prepared for the prevention and management of natural disasters. The formation of NADMO was in response to United Nations Declaration GAD 44/236 of 1989 declaring 1990 to 1999 the International Decade for Natural Disaster Reduction (IDNDR). Recognising this, Ghana established NADMO in 1996, backed by a Parliamentary Act (Act 517), to manage catastrophes and emergencies. This constitutional clause, which created NADMO, gives it the legal authority to carry out disaster response and recovery assistance/operations throughout the country. The mandate of NADMO includes developing plans to prevent disasters or mitigate their effects on Ghanaians, as well as coordinating activities before and during emergencies such as victim registration; relief efforts; operations; and ensuring post-disaster rehabilitation, reconstruction, and resettlement, including resource mobilization and provision (NADMO, 2005, 2015).

This means that NADMO is supposed to educate the public on awareness and resilience to natural disasters such as flooding. Working in harmony with other relevant institutions, NADMO has over the years seen to the management of floods in Ghana. However, for some years now, the management of flood vulnerability and menaces in Ghana has been on a narrow path of demolition of affected buildings, evictions of flood victims, and the distribution of relief items (Poku-Boansi et al., 2020). As such, Almoradie et al. (2020) concluded that flood risk management in Ghana is heavily skewed towards reactive approaches rather than preventive approaches to flooding. In the view of Tasantab (2019), institutional incapacities are the major challenges to flood risk management in Ghana. These incapacities have made land users contravene building and land use regulations, thereby putting up structures in swamps, waterways, and other flood-prone areas, creating flood risk. In addition, flood risk management in Ghana is challenging because authorities have no means to plan climate-sensitive approaches, infrastructure designs are not adapted to climate change, there is a lack of collaboration among public and private stakeholders, and there are poor policy processes. Arguably, all these undermine efforts to implement flood management approaches (The International Development Research Centre, 2020).

Flood-resilient construction and adaptation in Ghana

The concept of flood-resilient construction has become a catch-phrase for an integrated approach to flood risk management agreed upon almost everywhere in the world through the concepts of providing space for free movement of water in buildings with no negative impacts (Proverbs & Lamond, 2017). In the view of Liao (2012), flood resilience involves the capacity to tolerate flooding and be able to organise, should physical and economic disruptions occur in the presence of a flood, to prevent injuries and deaths and to maintain the existing socio-economic status. To Djordjević et al. (2011), resilience in the context of flood management is resisting, recovering, reflecting, and responding to flooding and its impacts.

With the weak state interventions for flooding in Ghana, recent flood-prone communities are developing enduring responses and adaptation practices to flooding through residents' networks, political alliances, and sense of place (Amoako, 2018). These efforts made by the communities include restructuring of housing units, construction of communal drains, and creation of local evacuation teams and safe havens. Though many efforts have been made by these communities, Gasparatos et al. (2020) noted that there is a general low community resilience to floods in Ghana. According to Asumadu-Sarkodie, Owusu and Rufangura (2015), it is impossible to eradicate flooding completely in Africa; however, its effects can be minimized by adopting an integrated flood management approach which encourages flood-resilient design, construction and adaptation, and related resources. The adoption of an integrated framework would also help to make the most of the resultant economic and social welfare in an equitable manner without compromising the sustainability of vital ecosystems. In a similar vein, Amoako and Frimpong Boamah (2015) proposed an integrated approach to flood risk management by considering the relationship among urbanization and slum development, rainfall intensity and poor management of surface water, and perceived impacts of climate variability and change that are deemed the major causes of flooding in Ghana.

Building regulations and codes and their influence on flood resilience in Ghana

Every nation's development and progress is dependent on construction. In Ghana, the building sector has severe consequences for the environment, spatial planning, economic planning, land use policy, safety, and risk problems. The construction industry in Ghana contributed an average of 2735.47 million Ghana cedis from 2016 until 2018 to the economy (Owusu-Boateng, 2019). However, improper construction of buildings could be detrimental to the growth of a nation's economy. For instance, after the 3 June floods at Accra in 2015, which destroyed a lot of properties in Ghana, it was heavily discussed that unauthorized buildings were responsible for the problems (Essuman, 2015). This is why the development of regulations and codes that guide the construction sector and its activities in flood-prone areas is of high importance. In this regard, Ghana uses article 41(k) of the 1992 constitution; the Local Governance Act, Act 936; the National Building Regulations 1996, LI 1630; and the Land Use and Spatial Planning Act, Act 925 to regulate the construction sector (Owusu- Boateng, 2019). Nevertheless, Ghana operated without a comprehensive building code until 2018 when the government of Ghana launched its first comprehensive building code, Building Code 2018, GS 1207, expected to transform the entire construction sector (Africa for Africa, 2018). The building code includes the standards, suggestions, planning, management, and practices that will ensure the safe operation and construction of residential and non-residential structures in Ghana. However, with respect to the building codes in Ghana, there is no regulation set to ensure the construction of flood-resilient buildings; that is, the codes do not specify construction designs and materials to be used to ensure the construction of buildings that would be resilient to flooding.

Methodology

This section covers how the systematic review of the literature was conducted. The review was validated with the activities of the National Disaster Management Organization in Ghana. The section further describes the search strategy, data search, method of extraction, and inclusion criteria.

Search strategy

This chapter is based on a systematic review of the literature by exploring the gaps in institutional and regulatory frameworks for flood-resilient construction and adaptation in emerging economies. The chapter reviews published articles and reports from institutions that are germane to the topic under investigation.

This chapter follows a deductive approach. According to Goodwin (2002), a deductive approach takes the form of top-down reasoning from more general (by developing theory) to more specific. The rationale for adopting the systematic review is threefold. First, to establish the relationship between NADMO and the construction industry in terms of flooding. Second, to summarize existing research pertaining to the role of flood management institutions in terms of improving flood resilience in the Ghanaian construction industry, flood-resilient construction and adaptation, and building regulations and codes and their influence on flood resilience. Third, this approach will assist in identifying the missing conceptual construct that should be incorporated into the flood risk management framework to contribute to theory development.

Data search, extraction, and inclusion criteria for the methodology

A systematic analysis approach was adopted from searching for relevant data through to grouping them into areas relevant to the chapter. To delve deep into the thematic areas, the Preferred Reporting Items for Systematic Reviews and Meta Analysis (PRISMA) framework postulated by Moher *et al.* (2009) was adopted as the road map for the literature review. Sadick and Kamardeen (2020) stated that the PRISMA framework consists of four steps: identification, screening, eligibility, and inclusion. These four steps were replicated in this chapter to guide the process of reviewing and selecting journal manuscripts, conference proceedings, and other publications pertinent to the topic under investigation. The selection of publications for the literature review was restricted to those published in the last two decades, that is, from the year 2000 to 2020. The chapter relied on publications and reports that focus on institutional and regulatory frameworks for flood-resilient construction and adaptation in order to develop case studies. The Google Scholar search engine was used to locate the articles and open them via academic–scientific search engines such as the Science Direct and Scopus Distinct databases. Search engines including Google Scholar, PDF Search, and Mendeley were also used to search for data. In the Ghanaian context, the key input phrases into the search engines were: building regulations for flood resilience, flood resilience in the construction industry, flood-resilient buildings, and flood risk management institutions. With respect to the global perspective, the key phrases that were used in the search for germane literature included flood resilience, concepts of flood resilience in construction, design and construction of flood-resilient buildings, indicators for assessing flood resilience, and case studies on adopting flood-resilient construction and adaptation. It is worth mentioning that the key phrases were entered in different combinations in the search engines, as suggested by Mensah *et al.* (2021).

Figure 4.1 presents the process that was taken to obtain the data for the chapter. Consequently, a total of 326 publications were identified and downloaded. The abstracts and titles of these publications were quickly screened, and a total of 216 duplicates were removed. After the screening exercise, content reviews were done on the articles to assess their relevance with regard to the topic. Fifty-four articles were excluded during the content review due to irrelevant content, and a total of 56 germane publications were retained. The germane documents

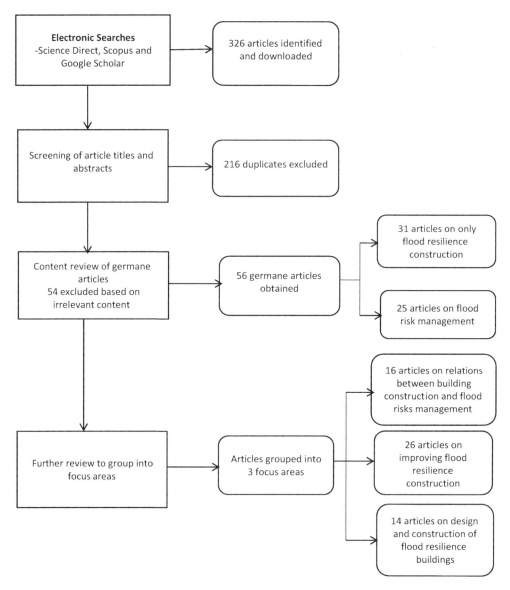

Figure 4.1 PRISMA flow diagram adopted in selecting germane articles
Source: The authors

were then grouped into two categories: articles on only flood-resilient construction and articles on flood risk management. A further review of these articles was finally done to group the articles into three focus areas consisting of 16 articles on relations between building construction and flood risk management, 26 articles on improving flood resilience in construction, and 14 articles on design and construction of flood-resilient buildings. In all, a total of 56 publications were used for the chapter.

Framework for assessing flood resilience in buildings

What aspects of the building codes/regulations relating to FRM are lacking and require improvement in order to regulate construction activities in flood-prone areas?

Building regulations and codes are important and should be imposed on every construction undertaking in every country of the world. Building regulations are aimed at protecting people's safety, health, and welfare. Restrictions are also intended to encourage fuel and electricity conservation, protect and improve the environment, and promote sustainable development (Proverbs & Lamond, 2017). Also, the inclusion of building regulations and codes in construction undertakings could prevent disasters such as flooding that could occur due to poor construction and structural development in flood-prone areas. Building regulations cover many of the major concerns of a society relating to structural setups, including safety and public health and environmental preservation. Building regulations establish quality, flood adaptability capacity, and safety and energy performance of a building in years to come, as original design and building decisions set the operating and maintenance costs of the building for life. This means that ineffective or no comprehensive building regulations could lead to flooding.

Unfortunately, the growing rural–urban migration in many developing countries takes place with corresponding construction developments in urban areas without effective building regulations, resulting in the construction of unsafe, vulnerable settlements (World Bank & GFDRR, 2015). For instance, in Ghana, endemic cases of flooding have been attributed to newly unregulated constructed buildings and factors relating to institutional management, lack of prescribed building regulations for flood risk–prone, areas and a lack of consensus in terms of the approach for designing and constructing flood-resilient buildings (Bowker, Escarameia & Tagg, 2007). Thus, the availability of effective building codes and regulations will serve as a flood risk management tool for putting up structures in flood-prone areas. As a result, flood control systems that apply to all forms of floods usually attempt to guarantee that construction in flood-prone regions can resist specified levels of flood risk by enforcing suitable building regulations (World Bank, 2019). Despite the contribution of existing building codes and regulations in Ghana to safe environmental existence, aspects of these codes and regulations that relate to flood risk management are limited. This means that there is still a need for improvement in the building codes and regulations with regard to flood risk management through the regulation of construction activities in flood-prone areas.

Inadequate funding and support for the building codes and regulations at the local levels in Ghana is one of the aspects of the flaws that impede efforts to regulate construction activities in flood-prone areas. This is in line with the World Bank report that the major issue in middle- and low-income countries with regard to construction management is lack of funding and support (World Bank & GFDRR, 2015). This situation has aggravated the problem and allowed people to build unauthorized structures in flood-prone areas, thereby increasing the risk of flooding in such areas. In this regard, the World Bank suggest that tackling the major factors limiting the effectiveness and efficiency of regulatory frameworks will help in ensuring flood resilience through construction activities (World Bank & GFDRR, 2015).

Again, the building regulations (Regulation 6 of the National Building Regulations 1996, LI 1630) indicate that the design of buildings should only be done by architects and engineers such as civil engineers and structural engineers. This, however, fails to give a specific design that these construction experts should adopt when designing for building in flood-prone areas or buildings that can withstand floods. This indicates that the regulation side lined the issue of flood risk management in the design of buildings and hence the need for improvement.

What measures have been taken by NADMO to improve flood resilience in the construction industry?

The loss of lives and property caused by natural catastrophes, especially flooding, undermines the efforts to achieve sustainable development and the Millennium Development Goals (United Nations Centre for Regional Development ([UNCRD], 2009). Hence, it is critical to ensure that there is enhancement in the construction of flood-resilient buildings (Adebimpe et al., 2018). In response to flood disasters and the rising cost of damages associated with them (Ahadzie, Mensah & Simpeh, 2021), organizations in many countries, such as NADMO of Ghana and FEMA of the United States, have been working in partnership with local and state governments for decades to reduce natural disaster losses by developing risk-based hazard maps and supporting and improving building hazard mitigation technologies in the country (Ansah et al., 2020). In Ghana, NADMO is tasked with carrying out all activities, from preparedness to response and recovery; preventing disasters; raising awareness in disaster-prone communities and institutions concerning all hazard/disaster types; and training and motivating communities, particularly volunteers, to take action to prevent and respond to disasters. However, while NADMO is mandated by law to be responsible for a wide range of activities in disaster management, resource constraints have limited their actions to largely post-disaster management activities, typically in providing relief items to victims of disasters, including flooding.

Performing its duties in the past decades, NADMO has been improving flood resilience in the construction industry. For example, in 2009, NADMO urged construction sector stakeholders to play their parts in ensuring that building projects and drainage systems are constructed in a way that will withstand floods (Ghana News Agency [GNA], 2009). Thus, the construction industry through various means has been charged by NADMO to incorporate strategies in the construction process that could help buildings withstand floods. This would aid in limiting disasters like flooding, which frequently result in the loss of property and people. However, not much effort in flood risk management has concentrated on the construction of resilient buildings in Ghana. Almoradie et al. (2020) argued that flood risk management in Ghana still depends heavily on reactive measures rather than preventive measures. Thus, flood risk management institutions do not put in enough effort to fight the issue except after flooding occurs.

What are the effective approaches that can be adopted for the design and construction of flood-resilient buildings in Ghana?

Structures and the built environment in general play a vital role in flood risk management; therefore, when physical structures are built on or near flood plains, it is evident that they must be protected (Proverbs & Lamond, 2017). In this regard, the need to develop effective approaches that can be adopted to ensure flood-resilient techniques are incorporated in buildings, especially those constructed at flood zones, has become a paramount concern in the construction industry (Adebimpe et al., 2018). This is to serve as a contributor to the global fight against the rampant cases of flooding worldwide. Resilient structures are designed and built in such a way that they avoid, minimize, or lessen the damage caused by flooding. The World

Bank provided some suggestions to ensure that the construction of flood-resilient buildings is enhanced (World Bank & GFDRR, 2015), including:

- Effective land use and land governance systems;
- Enhancement of legislative foundations for effective building code regulatory regimes at the national level;
- Focusing on creating building standards appropriate to the poor and vulnerable;
- Clear specifications of building code design and building standards of construction; and
- Ensuring effective and efficient building code administration and institutional capacity.

Also, as recommended by DELWP (2019), to ensure that buildings are flood resilient, the following approaches should be adopted by the construction industry in carrying out construction activities in flood-prone areas.

- Using elevated footings in preference to slabs on the ground for buildings;
- Locating or realigning buildings to allow for a free flow path parallel to the direction of the flow of running water; and
- Restricting the size of a building's footprint in flood-prone areas to appropriate heights.

In addition, Proverbs and Lamond (2017) and Golz, Schinke and Naumann (2015) opine that approaches such as the building of long-lasting structures using rain screens, waterproof interior coatings, and impact-resistant windows; carrying out critical system analysis of disasters on proposed areas before structures are put in those areas; and determining the capacity of buildings to withstand disasters after construction is done is essential for constructing flood-resilient buildings. The design and construction of flood-resilient buildings also in some ways rely on the flood risk management strategies adopted by the country. Effective and proper flood risk management would include the construction of flood-resilient buildings. Based on this, Almoradie et al. (2020) provided some ways forward and opportunities for improvement of flood risk management which will have an equally important influence on the construction of flood-resilient buildings in the country. These suggestions include enhanced socio-technical capacity and resilience where community individuals are allowed to be included in flood risk management, effective stakeholder involvement and cooperation in flood risk management, and improvement in government policies with regard to flood management. If all these approaches are effectively incorporated into the building of structures in Ghana, enhanced flood resilience will prevail.

Research findings

Based on the preliminary literature review, it was evident that:

- Flooding is prevalent in Ghana, and though there have been many suggested solutions to mitigating this menace, little improvement is seen since the phenomenon has now become an annual tragedy in Ghana. The situation is ascribed to the fact that flood management institutions such as NADMO play a limited role in improving resilience in the construction industry;
- The implementation of flood-resilient construction and adaption has been slow in the Ghanaian construction industry, the reason being that flood risk management in Ghana is heavily skewed towards reactive approaches instead of preventive approaches to flooding.

This situation has allowed land users to contravene building and land use regulations, thereby putting up buildings in flood-prone areas, causing floods. There is also a lack of planning regarding climate-sensitive approaches and infrastructural designs that are resilient to the impact of climate change;

- The lack of collaboration among public and private stakeholders contributes to the prevalence of floods in Ghana. The study revealed that collaboration between and among flood management organizations in Ghana is paper based and merely happens on the ground (Gyireh, 2015). For instance, the Meteorological Services Department of Ghana is supposed to feed NADMO climatic and rainfall data for preparedness for heavy rainfalls and flooding events; however, they do not have robust data that will help NADMO and other flood management bodies to better plan ahead of flooding (Ahadzie & Proverbs, 2010); and
- Issues related to flooding in Ghana can be tracked to institutional challenges and regulatory flaws with regard to the management of the built environment. For this reason, the construction industry has been inadequately monitored, leading to the construction of buildings which are not flood resilient (Ahadzie et al., 2016; Tasantab, 2019). For instance, there is no regulation set to ensure that the construction of flood-resilient buildings is promoted.

Designing the adoption pathway

It is evident that institutions, actors, and regulatory systems exist that prescribe and implement strategies for addressing flood challenges in Ghana. However, it requires a collaborative effort from institutions and actors to drive the process. Arguably, the wider construction and built environment have a significant role to play in shaping the drive towards adopting flood-resilient construction and adaptation. According to Proverbs and Lamond (2017), the concept of property-level flood resilience is renewed global attention towards prioritizing the assets of the built environment and the construction industry to effectively combat flooding. Proverbs and Lamond (2017) categorized flood-resilient technology under three strategies: avoidance technology, water exclusion technology, and water entry technology. In a similar vein, Gabalda et al. (2012) identified a broad variety of flood-resilient technology and assigned them to three predefined categories: perimeter technologies, building aperture technologies, and building technologies, grouped into flood-resilient building materials and flood-resilient building constructions. According to Proverbs and Lamond (2017), future developments in terms of flood-resilient technology are likely to concentrate on encouraging the application of flood-resilient materials and technologies both in the design and construction of new and the modification of existing buildings. Moreover, Proverbs and Lamond (2017) and Golz, Schinke, and Naumann (2015) contend that a greater understanding of performance standards for resilient materials and technologies will aid in providing confidence in such measures and support uptake.

Drawing from the works of Proverbs and Lamond (2017) and Golz, Schinke, and Naumann (2015), a pathway for promoting and adopting flood-resilient construction and adaptation is presented in Figure 4.2. The objective of this strategic pathway is to create an enabling environment to promote flood-resilient construction and adaptation. This roadmap indicates the sequential steps that must be followed by all role-players to facilitate the adoption and application of flood-resilient materials and technologies in the construction industry. As shown in Figure 4.2, the starting point to promoting and implementing a flood-resilient construction and adaptation framework is the identification of flood-resilient materials and technologies. Second, performance standards for resilient materials and technologies need to be established through laboratory testing and certification to determine the resilience properties of building materials/products and technologies. It is important to highlight that standard and certification

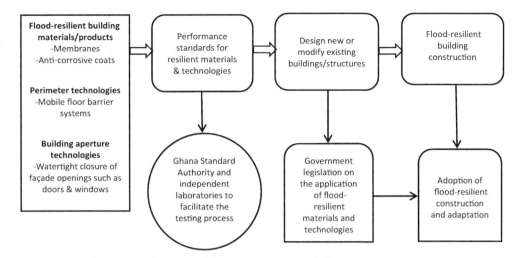

Figure 4.2 A pathway for adoption and promotion of flood-resilient construction and adaptation

authorities and independent laboratories need to come to the fore to support this initiative by facilitating the testing process. As the third stage, certified building materials and products as well as technologies may be adopted in terms of the design of new and modification of existing buildings. This strategy should be supported by government-enacted legislation that would complement national building codes and regulations with respect to design and construction of buildings in flood-prone areas. This strategy is linked to the fourth step and would encourage built environment professionals and other stakeholders to interrogate the applicability of flood-resilient materials and technologies in the design of new and in the retrofit and adaptation of existing buildings.

Conclusion

This chapter aimed to examine the institutional and regulatory frameworks for flood-resilient construction and adaptation in a developing country context, focusing on Ghana. The chapter focused on the role that flood management institutions play in improving flood resilience in the construction industry, flood-resilient construction and adaptation, and the importance of building regulations and codes in promoting property-level flood resilience.

The chapter revealed that flood management institutions in Ghana play a limited role in improving resilience in the construction industry. More so, the construction industry is lagging in terms of the adoption of flood-resilient construction and adaptation. This is because the construction industry is extremely conservative, and solutions to addressing flood-related issues are based on reactive approaches as opposed to preventive approaches. There is also a lack of regulation promoting the implementation of flood-resilient construction and adaptation.

The chapter contributes to the literature by developing a flood-resilient construction and adaptation framework (FRCAF) which encourages flood-resilient design, construction and adaptation, and related resources. The chapter also delves into the grey area of preventive approaches to flooding in the developing country context by contributing to the body of knowledge regarding the discourse in flood management approaches. Therefore, the chapter suggests that the effective management of the presented approaches will require national coordination, incorporating governmental development planning efforts to identify and enhance institutions, regulatory systems, and structures to strengthen adaptation efforts to the annual floods in Ghana.

References

Adebimpe, O. A., Oladokun, Y. O., Odedairo, B. O., & Oladokun, V. O. 2018, 'Developing flood resilient buildings in Nigeria: A guide'. *Journal of Environment and Earth Science*, vol. 8, no. 3, pp. 143–150.

Africa for Africa. 2018, 'Ghana launches its first building code regulation'. Construction Review Online website. Accessed March 22, 2021. Available from: https://constructionreviewonline.com/news/ghana/ghana-launches-its-first-building-code-regulation/#:~:text=Thebuildingcodesetsregulations,andelectricalsystemsamongother.

Ahadzie, D. K., Mensah, H., & Simpeh, E. 2021, 'Impact of floods, recovery, and repairs of residential structures in Ghana: Insights from homeowners'. *GeoJournal*, pp. 1–16.

Ahadzie, D. K., Dinye, I., Dinye, R. D., & Proverbs, D. G. 2016, 'Flood risk perception, coping and management in two vulnerable communities in Kumasi, Ghana'. *International Journal of Safety and Security Engineering*, vol. 6, no. 3, pp. 538–549. https://doi.org/10.2495/SAFE-V6-N3-538-549.

Ahadzie, D. K., & Proverbs, D. G. 2010, 'Flooding and post flooding response strategies in Ghana'. *WIT Transactions on Ecology and the Environment*, pp. 281–291. https://doi.org/10.2495/FRIAR100241.

Ahadzie, D. K., & Proverbs, D. G. 2011, 'Emerging issues in the management of floods in Ghana'. *International Journal of Safety and Security Engineering*, vol. 1, no. 2, pp. 182–192.

Almoradie, A., Brito, M. M., Evers, M., Bossa, A., Lumor, M., Norman, C., & Hounkpe, J. 2020, 'Current flood risk management practices in Ghana: Gaps and opportunities for improving resilience'. *Journal of Flood Risk Management*, vol. 13, no. 4, p. e12664. https://doi.org/10.1111/jfr3.12664.

Amoako, C. 2018, 'Emerging grassroots resilience and flood responses in informal settlements in Accra, Ghana'. *GeoJournal*, vol. 83, no. 5, pp. 949–965.

Amoako, C., & Frimpong Boamah, E. 2015, 'The three-dimensional causes of flooding in Accra, Ghana'. *International Journal of Urban Sustainable Development*, vol. 7, no. 1, pp. 109–129.

Ansah, S. O., Ahiataku, M. A., Yorke, C. K., Otu-Larbi, F., Yahaya, B., Lamptey, P. N. L., & Tanu, M. 2020, 'Meteorological analysis of floods in Ghana'. *Advances in Meteorology*. https://doi.org/10.1155/2020/4230627.

Asumadu-Sarkodie, S., Owusu P. A., & Jayaweera H. M. P. C. 2015, 'Flood risk management in Ghana: A case study in Accra'. *Advances in Applied Science Research*, vol. 6, no. 4, pp. 196–201.

Asumadu-Sarkodie, S., Owusu P. A., & Rufangura, P. 2015, 'Impact analysis of flood in Accra, Ghana'. *Advances in Applied Science Research*, vol. 6, no. 9, pp. 53–78.

Bowker, P., Escarameia, M., & Tagg, A. 2007. 'Improving the flood performance of new buildings'. *Construction Industry Research and Information Association*. Accessed March 22, 2021. Available from: www.planningportal.gov.uk/uploads/br/flood_performance.pdf.

Department of Environment Land Water and Planning (DELWP). 2019, *Guidelines for Development in Flood Affected Areas*. Melbourne: Government of Victoria, pp. 1–60. Accessed March 22, 2021. Available from: www.water.vic.gov.au/__data/assets/pdf_file/0025/409570/Guidelines-for-Development-in-Flood_finalAA.pdf.

Djordjević, S., Butler, D., Gourbesville, P., Mark, O., & Pasche, E. 2011, 'New policies to deal with climate change and other drivers impacting on resilience to flooding in urban areas: The CORFU approach'. *Environmental Science & Policy*, vol. 14, no. 7, pp. 864–873.

Egbinola, C. N., Olaniran, H. D., & Amanambu, A. C. 2017, 'Flood management in cities of developing countries: The example of Ibadan, Nigeria'. *Journal of Flood Risk Management*, vol. 10, no. 4, pp. 546–554.

Essuman, K. A. 2015, 'Complying with the building laws of Ghana'. Modern Ghana website. Accessed March 22, 2021. Available from: www.modernghana.com/news/633189/complying-with-the-building-laws-of-ghana.html.

Gabalda, V., Koppe, B., Kelly, D., Hunter, K., Florence, C., Golz, S., Diez, J., Monnot, J. V., & Marquez, P. 2012, 'Tests of flood resilient products (report to the EU Commission)'. *SMARTeST Project*. Available from: https://www.schutz-vor-naturgefahren.ch/files/Downloads/Downloads_DE/Ueberschwemmung/SMARTeST-WP2-DeliverableD2.2-Dec12.pdf.

Gasparatos, A., Ahmed, A., Naidoo, M., Karanja, A., & Fukushi, K. 2020, 'Sustainability challenges in sub-Saharan Africa in the context of the sustainable development goals (SDGs)'. In A. Gasparatos, A. Ahmed, M. Naidoo, A. Karanja, K. Fukushi, O. Saito, & K. Takeuchi (Eds.), *In Sustainability Challenges in Sub-Saharan Africa I*. Singapore: Springer, pp. 3–50.

Ghana News Agency (GNA). 2009, 'NADMO calls for proper construction of drainage systems'. Accessed March 22, 2021. Available from: www.myjoyonline.com/nadmo-calls-for-proper-construction-of-drainage-systems/.

Golz, S., Schinke, R., & Naumann, T. 2015, 'Assessing the effects of flood resilience technologies on building scale'. *Urban Water Journal*, vol. 12, no. 1, pp. 30–43.

Goodwin, C. J. 2002, *Research in Psychology – Methods and Design*. New York: John Wiley & Sons.

Gyireh, P. F. V. 2015, *Sustainable Management of Flood Disasters in the Upper East Region*. Ghana: Kwame Nkrumah University of Science and Technology, vol. 4. Available from: http://ir.knust.edu.gh/bitstream/123456789/2186/1/PASCHALFANG-VIELGYIREH%2CTHESIS 2011.pdf.

Hutton, T., & Marsh, C. 2019, 'Flood damage in historic buildings'. Cathedral Communications Limited website. Accessed March 19, 2021. Available from: www.buildingconservation.com/articles/flood/flood_damage.htm.

The International Development Research Centre. 2020, 'Integrated climate smart flood management for Accra, Ghana'. International Development Research Centre website. Accessed March 22, 2021. Available from: www.idrc.ca/en/project/integrated-climate-smart-flood-management-accra-ghana.

IPCC. 2014, 'Climate change 2014: Impacts, adaptation, and vulnerability. Part A: Global and sectoral aspects'. In C. B. Field et al. (Eds.), *Contribution of Working Group II to the Fifth 5AR of the IPCC*. Cambridge: Cambridge University Press, p. 1132.

Liao, K. H. 2012, 'A theory on urban resilience to floods – A basis for alternative planning practices'. *Ecology and Society*, vol. 17, no. 4. https://doi.org/10.5751/ES-05231-170448.

Mensah, H., & Ahadzie, D. K. 2020, 'Causes, impacts and coping strategies of floods in Ghana: A systematic review'. *SN Applied Sciences*, vol. 2, no. 5, pp. 1–13.

Mensah, H., Amponsah, O., Opoku, P., Ahadzie, D. K., & Takyi, S. A. 2021, 'Resilience to climate change in Ghanaian cities and its implications for urban policy and planning'. *SN Social Sciences*, vol. 1, no. 5, pp. 1–25.

Moher, D., Liberati, A., Tetzlaff, J., Altman, D. G., & Prisma Group. 2009, 'Preferred reporting items for systematic reviews and meta-analyses: The PRISMA statement'. *PLoS Medicine*, vol. 6, no. 7, p. e1000097.

National Disaster Management Organization (NADMO). 2005, *Framework for Disaster Management in Ghana*. Accra: Ghana Book Press.

National Disaster Management Organization. 2015, *National Disaster Management Organization Amended Act, 2015*. Accra: Ghana Book Press.

Nkwunonwo, U. C., Whitworth, M., & Baily, B. 2020, 'A review of the current status of flood modelling for urban flood risk management in the developing countries'. *Scientific African*, vol. 7, p. e00269. https://doi.org/10.1016/j.sciaf.2020.e00269.

Owusu-Boateng, D. 2019, 'K-ARCHY & COMPANY website'. Accessed March 23, 2021. Available from: www.karchyconsults.com/uncategorized/legal-framework-for-construction-built-environment-in-ghana/.

Poku-Boansi, M., Amoako, C., Owusu-Ansah, J. K., & Cobbinah, P. B. 2020, 'What the state does but fails: Exploring smart options for urban flood risk management in informal Accra, Ghana'. *City and Environment Interactions*, vol. 5, p. 100038. https://doi.org/10.1016/j.cacint.2020.100038.

Proverbs, D., & Lamond, J. 2017, 'Flood resilient construction and adaptation of buildings'. In *Oxford Research Encyclopedia of Natural Hazard Science*. Oxford: Oxford University Press, pp. 1–32. https://doi.org/10.1093/acrefore/9780199389407.013.111.

Sadick, A. M., & Kamardeen, I. 2020, 'Enhancing employees' performance and well-being with nature exposure embedded office workplace design'. *Journal of Building Engineering*, p. 101789.

Tasantab, J. C. 2019, 'Beyond the plan: How land use control practices influence flood risk in Sekondi-Takoradi'. *Jàmbá Journal of Disaster Risk Studies*, vol. 11, no. 1, pp. 1–9. https://doi.org/10.4102/jamba.v11i1.638.

UNDRR 2015, *Sendai Framework for Disaster Risk Reduction 2015–2030 United Nations Office for Disaster Risk Reduction*. Geneva, Switzerland: UNDRR.

United Nations Centre for Regional Development (UNCRD). 2009, 'Disaster-resilient regional development strategies in the world – from Hyogo to the world'. International Disaster Management Symposium 2009. Available from: https://www.uncrd.or.jp/index.php?page=view&type=400&nr=171&menu=229.

United Nations-Water. 2011, 'Cities coping with water uncertainties'. Media Brief, UN-Water Decade Programme on Advocacy and Communication. Available from: https://www.un.org/waterforlifedecade/pdf/unwdpac_biennial_report_2010_2011.pdf.

World Bank. 2019, *Learning from Japan's Experience in Integrated Urban Flood Risk Management: A Series of Knowledge Notes*. Washington, DC: World Bank.

World Bank and Global Facility for Disaster Reduction and Recovery (GFDRR). 2015, 'Investing in urban resilience: Protecting and promoting development in a changing world'. Accessed March 23, 2021. Available from: www.gfdrr.org/sites/default/files/publication/Investing%20in%20Urban%20Resilience%20Final.pdf.

5

Resettlement as a flood-preventive measure in Sri Lanka

Investigation into the socio-economic impacts

Senuri Siriwardhana, Udayangani Kulatunga and Bingunath Ingirige

Introduction

In the past decade, urban flood events have been on the rise, mainly due to global warming and unplanned urbanization (Kulp & Strauss, 2019). Nofal and van de Lindt (2020) reported in their recent study that flooding has been accepted as the costliest natural disaster that can cause a number of fatalities as well as disruption of a considerable portion of communities' infrastructure. In 2019, for instance, natural disasters had a severe impact on 25 million people around the world, causing US$17 billion worth of damages to the world's economy, and floods accounted for more than 60% of that figure (Global Humanitarian Assisstance [GHA], 2019). Kraemer (2016) stated that the predictions of worst-case scenarios in future flood events have brought the world to realize the importance of flood-preventive measures now more than ever.

Generally, flood-preventive measures can be divided into structural (technological approaches) and non-structural measures (planning approaches) (Meyer et al., 2012). Structural measures mainly focus on standards and the installation of disaster prevention facilities, while non-structural measures are related to urban and architectural planning (Hansson et al., 2008). Flood-preventive resettlements can be identified as one of the best non-structural preventive measures since there is a growing importance of this approach from the socio-economic and institutional perspectives compared to structural measures (Kang et al., 2018; Moore & Acker, 2018). Per the findings by Correa (2011) and Thaler and Fuchs (2020), the resettlement of communities located in high-risk areas can be considered the best option when it is difficult and unfeasible to mitigate the risk factors involved with severe flooding that cannot be controlled. Acknowledging this, Badri et al. (2006) and Schindelegger (2018) asserted that a well-planned resettlement process can provide positive development outcomes such as a more favourable socio-economic environment consisting of new job opportunities and better access to education and health facilities. By appreciating the value of the aforementioned features, most wealthy countries successfully implement resettlement as a flood-preventive measure while addressing community needs and other socio-economic factors (Claudianos, 2014). However, there are many failures in developing countries in establishing resettlement as a flood-preventive measure due to a lack of resources and a lack of understanding of the socio-economic needs of the relocating community (Arnall, 2014; Khalid et al., 2017; Kang et al., 2018).

As an island in the Indian Ocean, Sri Lanka is prone to frequent floods during monsoon rainy seasons (Sri Lanka Disaster Knowledge Network, 2012; Weerasinghe et al., 2018). Although mortality is low, floods annually displace thousands of people across the country, contributing to significant damage to their houses and infrastructure (Farley et al., 2018). For example, from the latest records, due to the floods that struck in May 2017, 2,093 houses were completely damaged, and 11,056 houses were partially damaged, making the establishment of resettlements as a flood-preventive measure a priority (Wanninayake & Rajapakshe, 2018).

In their study, Siriwardhana et al. (2021) identified planned resettlement as a key component of vulnerability reduction in areas at risk of flooding in Sri Lanka. Furthermore, community resettlement has become the ultimate choice for communities devastated by floods if there are no other options available. Nevertheless, little is known about the long-term effects of such resettlement programmes on social, cultural, economic and political aspects. There is a dearth of studies that discuss the success of resettlement as a flood-preventive measure that resolves the problems posed by flooding in developing countries like Sri Lanka. Accordingly, this chapter addresses this gap by presenting a case study on how, and why, the social, cultural and economic activities of the resettled community have changed as a result of resettlement by examining the resettled communities' views on such changes. The case study investigated the recent flood resettlement project implemented following the 2017 floods in the Kalutara District of Sri Lanka when populations living in the Kalu river valleys were moved to areas that are better protected from potential future flooding. As part of this case study, the resettlement policies and procedures in Sri Lanka as a whole were investigated in detail to ascertain how they have impacted the socio-economic factors of the resettled community. The outcomes of the study contribute to the knowledge and practice of resettlement as a flood-preventive measure in developing countries.

Post-flood resettlement policies in Sri Lanka

The government of Sri Lanka has introduced several resettlement policies focusing on both flood and landslide disasters with the aim of relocating vulnerable communities and thereby rebuilding their lives in secure locations (Fernando, 2018). The policy framework with reference to flood and landslide resettlement in Sri Lanka is discussed in this section.

With regard to resettlement policies, a macro-level policy setting needs to be accompanied by micro-level political engagement since land and property relations are highly contextual and often highly politicized (Weerasinghe, 2014). The policy of land acquisition for resettlement was launched by the government of Sri Lanka to provide three options for the community, as follows.

1) Provision of 1.6 million LKR to buy a house with land
2) Provision of 0.4 million LKR to buy land and 1.2 million LKR for housing construction
3) Provision of land and 1.2 million LKR for housing construction

The policies which are designed for resettlements in new locations include several prerequisites for site selection which should be applicable for both government and identified benefiting sites. According to the National Building Research Organization (NBRO, 2018), the disaster risk factor of the new site for possible hazards, access to social and physical infrastructure, and environmental compatibility of the site are the major criteria that are assessed by NBRO, while the final decision is taken based upon the overall suitability of the site in terms of human settlement. In order to obtain the financial allocations by the government, the relevant information that is required to evaluate the aforementioned criteria should be presented for the lands chosen by the beneficiaries.

Beneficiary selection is a community-driven process that is closely monitored by the local authorities. Communities displaced by floods who are living in temporary shelters are given priority. Per the report of NBRO (2018), beneficiaries who lived in Restricted Zones or Warning Zones of the flood plains will be considered for reconstruction with resilient features, while the beneficiaries located in Prohibited Zones of the flood plains are considered for safer locations through resettlement.

The reconstruction policies for these flood resettlements are mainly focused on Build Back Better (BBB) principles. "BBB is an important concept which incorporates adopting a holistic approach to improve a community's physical, social, environmental and economic conditions during post-disaster reconstruction and recovery activities to create a resilient community" (Mannakkara & Wilkinson, 2014, p. 338). Per the BBB concept, in the owner-driven approach (where the owners are allowed to make the decisions over their housing constructions), the beneficiaries are provided with several options for accessing sustainable financial protocols in the long term. Technical support is provided by technical staff from divisional secretariats to ensure the application of disaster risk reduction standards developed under the BBB concept. Most importantly, these disaster risk reduction standards not only consider floods but also all potential disasters, such as landslides, tsunami, storms and so on (Ministry of Disaster Management, 2017).

In addition, according to NBRO (2018), funds are released to the house owners in four instalments. Accordingly, a fixed amount of money is released at each stage of construction. The fund releasing procedure for housing construction is illustrated in Table 5.1.

The funds in Table 5.2 were determined considering the standard housing designs (per the design guidelines) through the evaluation of actual market rates of building materials excluding the labour component. Per the policy, the labour component of the construction projects should be covered by the community. Since the market prices of the building materials and labour fluctuate with economic changes in the country, the completion of projects with the allocated funds was somewhat challenging. When it comes to the design, several government authorities jointly designed feasible housing layouts to be chosen by the beneficiaries (National Council for Disaster Management, 2010; NBRO, 2018). Another option provided is for the beneficiaries to design their own houses following the guidelines and structural standards imposed by the government authorities. Per the guidelines, the housing designs should consist of a minimum floor area of 650 sq. ft., a disaster-resilient foundation and superstructure (with approval from

Table 5.1 Fund releasing procedure

Instalment	Stage	Amount in LKR
1	Land preparation, excavation and completion of foundation	150,000
2	Completion of walls up to roof level and supply of timber	300,000
3	Completion of roof	450,000
4	Completion of core house – doors and windows, plastering, finishing and painting	300,000
TOTAL		1,200,000

Table 5.2 Respondent profile

Code	Gender	Occupation
R1	Male	Local leader of the resettled village
R2	Male	Retired government officer
R3	Female	Housewife
R4	Female	Housewife
R5	Male	Government servant
R6	Male	Labourer
R7	Male	Carpenter
R8	Female	Student
R9	Male	Labourer
R10	Female	Office worker
R11	Female	Housewife
R12	Female	Labourer (garment factory)
R13	Male	Village elder
R14	Male	Retired army officer
R15	Male	Government officer

NBRO), two bedrooms with 100 sq. ft. or one bedroom with floor area of more than 120 sq. ft., with a kitchen, permanent roof, watertight toilet and septic tank (NBRO, 2018).

According to the Housing Policy, communities are not allowed to build or occupy houses in flood-prone areas (National Housing Authority, 2010). Therefore, in order to avoid the migration of a community back to previous flood-prone areas, water and electricity facilities are disconnected by the government and beneficiaries are asked to tear down their previous houses in order to receive the financial allocations.

Kalu Riverbank resettlement: empirical investigation

An empirical investigation was conducted using a case study based upon the resettlement project in which the community that originally lived in the Kalu river valley in Kalutara District, Sri Lanka was moved to an area that is better protected from future flooding (Case of Kalu River Bank Resettlement). The data collection was undertaken by conducting semi-structured interviews with 15 beneficiaries of this specific resettlement project. The respondent sample consisted of representatives from several social groups including local leaders, retired officers, housewives and daily workers. The respondent profile is presented in Table 5.2. The purpose was to gather meaningful insights on post-flood resettlement programmes and their success from the end-users' perspective and thereby evaluate any factors that hindered the progress with regard to the social, cultural, economic and political aspects. To this end, the interview guideline covered aspects relating to the respondents' pre- and post- resettlement living conditions and their family composition, livelihoods, land ownerships and social interactions. Content analysis was undertaken for the study since the researchers wanted to become more familiar with the data set and the data gathered from the interviews were manageable to analyse.

Per the respondent profile, representatives of several community groups were selected for the data collection process in order to gather distinct experiences according to their social, cultural and economic perceptions.

Background to the case of the Kalu Riverbank resettlement

According to the NBRO guidelines, the Kalu Riverbank resettlement project belongs to the third option of land acquisition, whereby residents construct their own houses with allocated finances on land provided by the government. Therefore, the infrastructure facilities were supplied by the government, while LKR 1.2 million was provided for each housing construction per the estimations of NBRO based upon the market prices of building materials at the time of policy development.

The beneficiaries who lived in the Kalu Riverbank were given priority for this particular project. Since there were some dissimilarities between the size of the land slots, a lottery system was conducted to select the lands for the beneficiaries.

With regard to the house designs, some beneficiaries selected the design options provided by NBRO, while some designed their own housing layout with the aid of a consultation provided by the technical officers in the divisional secretariat. However, those who selected the second option were required to obtain the necessary approvals for their designs by adhering to the design standards of the NBRO. These NBRO guidelines were issued to ensure the minimum requirements for a disaster-resilient house.

The construction process was conducted by the beneficiaries under the close supervision of technical officers. The fixed amount was released in four instalments (Table 5.1) with the purpose of ensuring productive utilization of the finances for the construction process. Beneficiaries were given the freedom to select building materials and construction technologies as long as they adhered to the quality requirements introduced by the authorities. Furthermore, a few training programmes were conducted focusing on the carpenters, masons and labourers among the beneficiaries to educate them about the choice of cost-effective materials and construction methods and the incorporation of disaster-resilient features.

After the community was resettled, several workshops were conducted under the Grama Shakthi programme by the divisional secretariat to encourage the residents to find self-employment opportunities in the resettled area. Accordingly, home gardening competitions were arranged to promote gardening as one such earning opportunity for housewives while staying at home.

In order to enhance social interaction, a community hall was constructed on the site which can be occupied by the beneficiaries for their social events.

Socio-economic impacts of the Kalu Riverbank resettlement

The lived experience of the community on the original Kalu Riverbank area

Flooding has become a frequent phenomenon for those who live near the Kalu Riverbank. The lived experience of the respondents with regard to their original location is presented in this section in order to evaluate the necessity for resettlement.

The majority of the participants have seen flooding as an annual event rather than a rare event, and thus they have had plenty of flood experiences connected with their original location. One of the government servants (R2) expressed that generally, they receive flood warnings. The respondent further expressed that "Aside from the warnings, everyone is aware that when the monsoons begin in May, with constant rain, the water level in the river rises, potentially causing flooding". R6 recounted his experience, stating that during the monsoon season, they move their transportable properties to safer places, while non-movable items such as furniture are tied down with ropes to avoid them being washed away by water. R5 and R6 explained how they acted when they became aware of impending floods. According to R5, "we were always awake at midnight, and from time to time we checked the water levels. We were ready to be evacuated from our homes at any time". R6 claimed that after each flood, they had to clear up debris and rubbish, clean the furniture and return to their usual living conditions, which took around three months and cost them a significant amount of money. Each year, as a result of this recurring phenomenon, the community faced long-term difficulties such as property destruction, loss of livelihood owing to the disruption of routine lives and, most significantly, psychological anxiety. In addition, floods have become more severe in recent years, with some homes being completely destroyed, leaving people homeless. R9, who is such a flood victim, expressed his experience as "the floods of 2017 took everything away from us. We never expected this because the regular floods aren't quite as severe as this".

Per the opinions of the respondents, the community has clearly suffered a great deal as a result of flooding in the original location. They are also able to understand the force of an extreme event as opposed to a normal event because of the recent 2017 flooding. However, when the respondents were questioned regarding whether they needed to be relocated due to severe floods in the original location, their responses were twofold. One part of their responses depicts that, despite the severity of the flood damage, leaving the original location was not preferred for a variety of reasons. A number of the respondents (R1, R2, R5, R14, R15, R10, R11) expressed this opinion, offering different justifications. R1 and R2 stated that living conditions were good in the original location, with easy access to public services such as hospitals, schools, banks and government offices. Additionally, the respondents were quite concerned about their social status. The majority of the respondents stated that the government's financial provisions are insufficient to build a house that conforms to their social status. Therefore, they felt that it was better to live in the original house and undertake regular repairs rather than living in a small house that does not fulfil their requirements. Furthermore, some of the respondents were reluctant to change their comfortable environment and friendly neighbourhood. In addition, R14, who was a government servant, elaborated that the community has become accustomed to the annual floods and was confident in handling the situation.

On the other hand, some of the respondents were of the opinion that resettlement is the only option to overcome the recurring flood hazard. R7, whose house was completely damaged by the floods, stated, "we were hesitant to build the new residences in the original area because we didn't want to go through the same difficulties again". Furthermore, R9 considered that receiving decent dwellings in a safer area was an opportunity, as they used to live in slum dwellings.

Community perspective on the Kalu Riverbank resettlement

The beneficiaries had different viewpoints about the policies and procedures adopted in the resettlement programme in terms of site selection, beneficiary selection and design and construction of the resettlement project.

Resettlement

In terms of the new location, the community placed a high priority on primary facilities in order to maintain their daily routines. The site was located in a hilly area, which caused many lifestyle difficulties. For example, R5 stated that "Traveling for our day-to-day activities is quite tough because the community do not have their own vehicles. The topography of the resettlement in a hilly area has further worsened it". Since the primary facilities such as hospitals, pharmacies, schools and supermarkets were in the town that were far from the site, the community faced many challenges. R6, who was a student, explained that his school was near his original location, while now it is exhausting to travel daily from the new site. Due to this reason, his parents are trying to change his school. The problem with transportation facilities became more serious as the community struggled to go to their usual jobs. R10, who is a labourer, explained that "If we travel for our old job which was nearby the previous site, two-thirds of our daily income would be spent on travel causing plenty of financial problems". As a result, the majority of the daily income earners gave up their previous employment and engaged in alternative jobs with less income than the previous jobs. Additionally, R12, who is a female worker in a textile factory, shared her experience as "Previously there was a staff transport service running by our original location. But now since there is no such system, I am forced to get up early and walk to work alone in the dark". Due to this difficulty, one of her friends (R3) who worked at the same factory had to quit her job and became a housewife. R3 commented that "Although there are many economic difficulties, we are helpless with these limited facilities".

With reference to the environment of the new site, the majority admired the natural environment, which is calm and quiet compared to the original location. In terms of social environment, as the community was all from the same original location, there was good social interaction from the start per the view of R10. However, there were some troublesome scenarios such as having the same neighbours, who were rivals, from the previous location. In addition, one of the retired officers stated that the host community expressed their opposition to this resettlement, initially believing that immoral acts may increase through the resettled community. Additionally, freedom is felt to be limited in the new location due to its high density of housing (with houses very close to each other). This was expressed by R10 as "Previously, we had enough room to call our own. However, we are now all living extremely near to one another which has a negative impact on our freedom and privacy".

Beneficiary selection and ownership

All the respondents accepted that the government had successfully engaged in the beneficiary selection process. Per their opinion, everyone relocated under the programme deserved a home since they were truly suffering from floods as a result of living in a flood-prone region. R1, who is a local leader, expressed this as "The grama niladharis (village officers) with the support of divisional secretariat conducted a thorough screening process for selecting beneficiaries while no political involvement was observed". The lottery system which was performed for selecting the land slots for the beneficiaries was also praised by the respondents, stating it "minimises the conflicts and enhances the social interaction and the equality". Furthermore, the ownership policy was successfully implemented whereby land deeds were provided to the beneficiaries after they were resettled in the new house.

Preventing the community from returning to the original location

Another extensively debated policy among the beneficiaries was the one adopted to prevent the community from returning to their original location. The majority criticized the disconnection of water and electricity facilities of the previous houses, while R10 argued, "it is so unfair to implement such a policy because new settlements do not offer the required services. Now we can't go back to our previous house, and we can't live here with these poor facilities either". However, from the authorities' perspective, they implement this as a policy to prevent the community from moving back to their original houses, which, in turn, would increase their vulnerability to future floods; avoiding this is the major objective of resettlement.

"House for a House" policy

People who lived within extended families at the original location were severely impacted by the introduction of the "House for a House" policy. R4, who lived with her daughter's family, claimed that the offered residence does not have sufficient space for two families to live in comfortably. Supporting this, R12 stated, "it would be preferable if each family was provided with a separate house, or if there was enough space for multiple families to live in a single house as before".

Design and construction policies

With regard to the design and construction policies, the majority appreciated the fact that the community was allowed to create their own homes with financial assistance. Furthermore, they indicated that giving multiple design alternatives or allowing users to design their own layout was quite beneficial. Accordingly, five of the respondents designed their own layout, while others opted for a choice from a list of 12 options. R5 and R10, who designed their own layout, explained that the design options were not compatible with their family background and social status. R10 further commented, "this allowed us to design our own home as per our expectations taking architectural and astrological considerations into account". Other respondents, on the other hand, who had chosen one of the government design options, accepted that, although there were some financial challenges, having multiple options from which to choose and not having to worry about receiving approval was a fantastic opportunity.

The majority appreciated the enhanced community participation in the construction policies. Nonetheless, since the government provided a standard guideline to be followed for beneficiaries in the design and construction phases, R2 stated "it is hard to balance the required conditions with the limited financial allocation". Adding to this, R3 and R4 criticized the inclusion of disaster-resilient features under the BBB concept: "It is good to have those features, but they were cost consuming and government allocations were not enough". Furthermore, the majority thought that building columns and beams with reinforcement, which increases their strength against disasters, was a useless additional cost. However, R7 and R12 appreciated being allowed to use materials such as doors and window frames from previously damaged houses, if these were of the necessary standard, which reduced their overall cost. Furthermore, they appreciated the technical guidance provided by the authorities throughout the process. In the main, the overall opinion about the reconstructions was satisfactory because beneficiaries were able to design and construct their houses to meet their own expectations.

Community recovery processes

The gardening programmes that were conducted by the government aimed at economic recovery were appreciated by the community. Nonetheless, R3 and R4 emphasized that "we don't have good practice over gardening and it is expected to have more space in our lands to successfully conduct gardening as an additional source of income". Furthermore, R3, who quit her job due to the resettlement, suggested that it would be very helpful if sewing machines could be provided to housewives with the necessary training in order to start self-employment on their

Table 5.3 Summary of the community's opinions regarding the Kalu Riverbank resettlement

Policy	Opinion	Occupation													
		Government officers/office worker					Local leader	Labourers/carpenter					Housewives			
		R1	R2	R5	R14	R15	R1	R6	R7	R9	R12	R3	R4	R11
Resettlement	Difficulty of accessing primary services	✓		✓			✓	✓	✓	✓	✓	✓	✓	✓
	Travelling difficulties for daily livelihood	✓	✓	✓	✓			✓	✓	✓	✓	✓		✓
	Travel by personal vehicles	✓		✓	✓				✓					
	Calm and quiet natural environment	✓		✓	✓		✓	✓				✓	✓	
	Host community objections						✓							
	Less conflicts among the neighbours		✓									✓	✓	✓
Beneficiary selection and ownership	Fair screening process by the authority		✓	✓	✓		✓	✓	✓		✓			
	No political involvement						✓							
	Land deeds were provided properly	✓	✓	✓	✓	✓	✓	✓		✓	✓	✓	✓	✓
Prevention of returning	Unfair due to the unavailability of primary facilities		✓	✓	✓			✓		✓		✓		
	Resettlement objective can be achieved						✓						✓	
House for a House policy	Extended families suffer from space limitations										✓			
	Even for own designs, land area is insufficient for expansions				✓									
Design policies	Own design	✓	✓	✓	✓	✓								
	Architectural and astrological considerations	✓	✓	✓	✓									
	Compatible designs with family background and social status	✓	✓	✓	✓	✓								
	Financial insufficiency		✓			✓								
	Design options by government						✓	✓	✓	✓	✓	✓	✓	✓
	Having multiple options to choose						✓	✓	✓	✓	✓	✓	✓	✓
	No need of approvals							✓	✓	✓				✓
	Financial insufficiency							✓	✓	✓	✓	✓	✓	✓
Construction policies	Regular consultation by technical officers			✓		✓		✓	✓	✓	✓	✓	✓	✓
	Preferred materials and construction methods			✓	✓			✓	✓	✓	✓	✓	✓	✓
	Difficulty balancing financial allocations			✓	✓	✓		✓	✓	✓	✓			
	Allowed to use materials from previous house							✓	✓	✓	✓			
	Additional cost for resilient features							✓				✓	✓	✓
Community recovery processes	The necessity of a community hall did not emerge		✓				✓							
	Need more space and training for conducting gardening											✓		✓
	Require sewing machines to initiate self-employment											✓		✓

own. Moreover, the community hall which was constructed to enhance social interaction was not frequently used by the beneficiaries.

A summary of the opinions of the community regarding resettlement is presented in Table 5.3.

Per the findings, Figure 5.1 illustrates the overall summary of the opinions of the community regarding resettlement.

The cognitive map shown in Figure 5.1 has been divided into three levels (refer to the legend). Level 1 denotes the main resettlement policies that are implemented in the flood resettlement project. In the second level, the sub-themes are presented with regard to each policy. The opinions of beneficiaries under each sub-theme are presented in the third level of the map. Furthermore, the impact of each resettlement policy on the social, cultural, political and economic perspectives is demonstrated through the external level by connecting the respective opinions to each aspect.

Per the opinions of beneficiaries in Figure 5.1, it is evident that resettlement policies have affected the social, cultural, economic and political aspects of the community which determine the success level of the overall resettlement.

Social aspects

A consideration of the social aspects is very important for the success of post-disaster resettlement, especially for settlements that have been permanently relocated. Accordingly, the enhancement of social interaction among the community, minimizing social conflicts and the social status of each community groups were considered satisfactory by the community, although some viewpoints expressed that some of these actions have been unsuccessful. As an example, a community hall was constructed as an initiative to enhance social interaction, but it is currently useless to this particular community, as no initiatives have been taken to encourage the community to organize social gatherings. Social conflicts were successfully minimized by giving equal treatment to everyone in the beneficiary selection and the land slot selections. This resulted in creating trust among the community with regard to the authorities, which is typically lacking in most of the policy implementation scenarios in Sri Lanka.

The maintenance of social status is a desired human characteristic, and it affects the overall social standing of a community. Through the findings, the community can be categorized into a middle-income community and low-income community who depend on daily wages. It can be observed that there is a clear difference in social status between these two groups. It appears that the policies were successful in balancing the requirements of these social groups in many instances. For example, allowing beneficiaries to design and construct their houses as one option was mostly appreciated by the middle-income community, as they had the opportunity to have designs compatible with their social status by adding preferred architectural and aesthetic features. On the other hand, the provision of common design options was appreciated by the low-income community since the designs were compatible with their own social level. However, resettlement policies with regard to site selection were not successful in terms of social wellbeing because the site location created many difficulties in terms of accessing primary services such as supermarkets, hospitals and schools, and this negatively affected social living conditions.

Economic aspects

It is evident from Figure 5.1 that the economic aspects of the community were severely affected by the resettlement, where most of the policies were unsuccessful in fulfilling economic

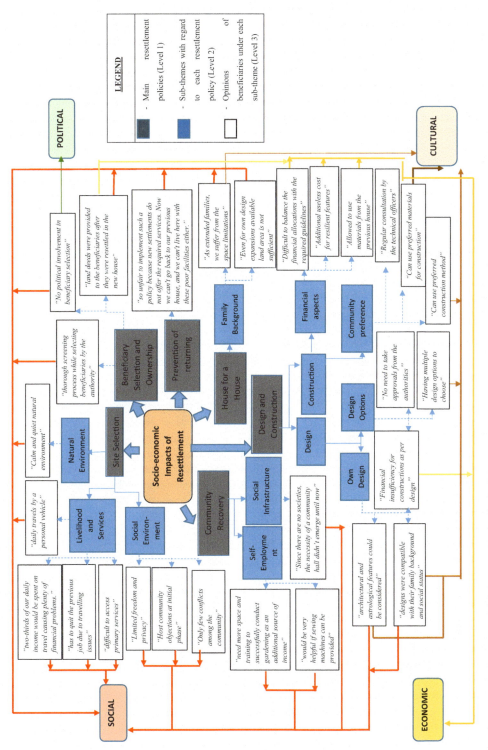

Figure 5.1   Socio-economic impacts of resettlement

65

wellbeing. The major reason for this was that the resettled site was far away from the community's livelihoods, and daily access to work was challenging due to the insufficient transportation system. This issue, to a certain degree, caused a downward trend in the day-to-day economy of low-income families who were unable to spend a considerable amount on travelling compared with their daily income. However, the middle-income community successfully engaged in their livelihoods, as they had private vehicles. This indicates that the resettlement placed the poor community in more severe conditions economically, as no successful livelihood opportunities were provided by the policies within the resettled area. In addition, it is worth mentioning the introduction of gardening as an additional income generation opportunity for the community; this was unsuccessful due to the practical issue of space limitations. However, the community believes that rather than gardening, there are plenty of other self-employment opportunities that can be implemented within the area if the necessary resources are provided, such as sewing machines.

Another major challenge was the insufficiency of financial allocations for constructions. The construction policies include several criteria that are required to be followed, which were expensive. Therefore, the community on certain occasions had to bear the additional cost themselves in order to complete the construction. This financial insufficiency caused the community to have misconceptions about the disaster-resilient features (that were compulsory to be included) and to view them as worthless. However, the ownership policy was of economic benefit for the beneficiaries. As there were no issues concerning the land deeds, which were delivered timely, beneficiaries were able to keep them as collateral to acquire financial loans from banks. For many, this provided relief, as they were then able to have further economic assistance.

Cultural aspects

Culture covers such aspects as the way of life of a group, the specific way of behaving, and a set of strategies adopted for existence linked to the ecological setting (Boen & Jigyasu, 2005). Cultural considerations were taken into account to a satisfactory extent within the design and construction policies. Since the policies mostly promoted an owner-driven approach, beneficiaries were allowed to design and construct their houses per their own cultural requirements. For example, beneficiaries were able to include astrological features per their beliefs and utilize conventional building materials with user-friendly construction methods per their preference. Since the beneficiaries were able to include their own cultural requirements, the policies were highly appreciated from the cultural perspective.

However, resettlement policies had both successful and unsuccessful outcomes in cultural aspects. First, the resettlement of the community (who had similar cultural preferences) from the same original location together minimized many conflicts with satisfactory outcomes. On the other hand, the community was accustomed to the culture that was prevalent within the surroundings of the river banks at their original site, with close proximity to primary services and livelihoods. Some members of the community engaged in land cultivation at the original site to earn an extra income, which added further value to the aspect of proximity to livelihoods. Furthermore, there was an independent lifestyle with increased privacy. However, the resettlement caused a complete change from the original cultural background to one where primary services and livelihoods were difficult to achieve, alongside limited space for cultivation and limited freedom and privacy. Therefore, the community suffered long-term consequences as a result of this cultural difference, which they had to adjust to over time.

With regard to the extended family culture, the "House for a House" policy was a serious challenge to extended families due to significant space limitations. This suggests that when designing policies, it is necessary to develop alternatives for these exceptional scenarios in order to successfully implement flood resettlement policies.

Political aspects

It is a common belief that political involvement in any type of resettlement project cannot be avoided. Even though there might be political involvement when developing policies and changing them from time to time, in this case, the authorities were able to avoid political involvement at the implementation stage, and this is a factor which should be appreciated. Per the findings, the authorities stuck to the policies which kept illegal activities at a minimum level, resulting a corruption-free resettlement programme.

Therefore, it can be deduced that these flood-resettlement policies impacted the community, resulting in both negative and positive outcomes in terms of social, economic, cultural and political aspects.

Conclusion and recommendations

Resettlement of communities away from flood zones can be identified as one of the most successful non-structural flood-preventive measures that can keep a community away from vulnerable areas while building up their living standards in a safer location. However, in order to successfully implement resettlement as a flood-preventive measure, it is necessary to fulfil the social, economic and cultural aspects of the community through the development of a strategic resettlement policy framework devoid of any political favouritism.

In the Sri Lankan context, as flooding has become more intense and frequent, flood resettlement has been given satisfactory consideration with the development of flood resettlement policy frameworks. Resettlement policies have been introduced for site selection, land acquisition, ownership, housing design and construction and community recovery. It is noteworthy that these policies have both positive and negative impacts upon a community in terms of social, economic, cultural and political perspectives. However, the success of these policy implementations is not at an adequate level in terms of the community's perspective. Improper policy implementation without proper strategic plans has led to a number of long-term issues for beneficiaries that may result in resettlement failures. Therefore, future resettlement policies should pay more attention to practical implementation, taking on board the community's expectations.

Multiple recommendations can be suggested, which have been deduced through this study, to overcome the existing challenges. In order to minimize social issues within the resettled zone, it is highly recommended to develop infrastructure facilities (such as roads, water, electricity and other amenity schemes), even in rural areas, followed by proper land development. Furthermore, an alternative option that can be suggested to resolve the issue of land scarcity is construction of multi-storied housing complexes and letting the community choose such complexes according to their preference. Moreover, it is highly recommended to enhance the bottom-up approach in policy development, whereby community proposals can be put forward to create a more sensible policy framework. In addition, the development of policy guidelines per funding availability, undertaking fundraising programmes locally and internationally focusing on these types of resettlements and supporting the uneducated within the community on financial management in the owner-driven approach are some of the recommendations to enhance the economic aspects of policies. Furthermore, it is essential to provide training and awareness programmes which target different community groups to encourage and educate them on the importance and means of community empowerment and recovery.

Acknowledgements

The authors would like to acknowledge the financial assistance provided by the Global Challenges Research Fund (GCRF) and the Economic and Social Research Council (ESRC) under the Grant ES/T003219/1 entitled "Technology Enhanced Stakeholder Collaboration for Supporting Risk-Sensitive Sustainable Urban Development".

References

Arnall, A. (2014) 'A climate of control: Flooding, displacement and planned resettlement in the lower Zambezi River valley, Mozambique', *The Geographical Journal*, 180(2). Available at: www.jstor.org/stable/43868599.

Badri, S. A. et al. (2006) 'Post-disaster resettlement, development and change: A case study of the 1990 Manjil earthquake in Iran', *Disasters*, 30(4), pp. 451–468. doi: 10.1111/j.0361-3666.2006.00332.x.

Boen, T., & Jigyasu, R. (2005). 'Cultural considerations for post disaster reconstruction post-tsunami challenges', *UNDP Conferences*, 2005, 1–10.

Claudianos, P. (2014) 'Out of harm's way; preventive resettlement of at risk informal settlers in highly disaster prone areas', *Procedia Economics and Finance*, 18, pp. 312–319.

Correa, E. (2011) *Preventive resettlement of populations at risk of disasters*, The World Bank, Washington.

Farley, J. M. et al. (2018) 'Evaluation of flood preparedness in government healthcare facilities in Eastern Province, Sri Lanka', *Global Health Action*, 10(1). doi: 10.1080/16549716.2017.1331539.

Fernando, N. (2018) 'Voluntary or involuntary relocation of underserved settlers in the city of Colombo as a flood risk reduction strategy: A case study of three relocation projects', *Procedia Engineering*, 212(2017), pp. 1026–1033. doi: 10.1016/j.proeng.2018.01.132.

Global Humanitarian Assistance (GHA) (2019) *Global humanitarian assistance report*. Available at: www.globalhumanitarianassistance.org/report/gha-report-2015.

Hansson, K., Danielson, M. and Ekenberg, L. (2008) 'A framework for evaluation of flood management strategies', *Journal of Environmental Management*, 86(3), pp. 465–480. doi: 10.1016/j.jenvman.2006.12.037.

Kang, S. J., Lee, S. J. and Lee, K. H. (2018) 'A study on the implementation of non-structural measures to reduce urban flood damage-focused on the survey results of the experts', *Journal of Asian Architecture and Building Engineering*, 8(2), pp. 385–392. doi: 10.3130/jaabe.8.385.

Khalid, M. S. et al. (2017) 'Resettlement housing for flood victims: Motivation for the contractors', *International conference on Energy, Environment and Economics*, pp. 11–13.

Kraemer, A. (2016) *Resettlement from flood-prone areas*, Ecologic. Available at: www.ecologic.eu.

Kulp, S. A. and Strauss, B. H. (2019) 'New elevation data triple estimates of global vulnerability to sea-level rise and coastal flooding', *Nature Communications*, 10(1), p. 12. doi: 10.1038/s41467-019-12808-z.

Mannakkara, S. and Wilkinson, S. (2014) 'Re-conceptualising "building back better" to improve post-disaster recovery', *International Journal of Managing Projects in Business*, 7(3), pp. 328–341. doi: 10.1108/IJMPB-10-2013-0054.

Meyer, V., Priest, S. and Kuhlicke, C. (2012) 'Economic evaluation of structural and non-structural flood risk management measures: Examples from the Mulde River', *Natural Hazards*, 62(2), pp. 301–324. doi: 10.1007/s11069-011-9997-z.

Ministry of Disaster Management (2017) 'Floods and landslides', in *Reduction and predictability of natural disasters*, London: Routledge, p. 284. doi: 10.4324/9780429492549-6.

Moore, J. and Acker, L. (2018) 'Recurrent flooding, sea level rise, and the relocation of at-risk communities: Case studies from the commonwealth of Virginia', *William & Mary Law School Scholarship Repository*. Available at: https://scholarship.law.wm.edu/vcpclinic/26%0ACopyright.

National Building Research Organization (NBRO) (2018) *Hazard resilient housing construction manual*, Sri Lanka: NBRO.

National Council for Disaster Management (2010) *National policy on disaster management*. Available at: www.disastermin.gov.lk/web/images/pdf/sri lanka disaster management policy english.pdf.

National Housing Authority (2010) *Mainstreaming disaster risk reduction into housing sector in Sri Lanka*. Available at: https://www.preventionweb.net/publication/mainstreaming-disaster-risk-reduction-housing-sector-sri-lanka.

Nofal, O. M. and van de Lindt, J. W. (2020) 'Understanding flood risk in the context of community resilience modeling for the built environment: Research needs and trends', *Sustainable and Resilient Infrastructure*, pp. 1–17. doi: 10.1080/23789689.2020.1722546.

Schindelegger, A. (2018) 'Relocation for flood retention in Austria', in *Opportunities and constraints of land management in local and regional development*, p. 274. Available at: https://books.google.lk/books?hl=en&lr=&id=42FuDwAAQBAJ&oi=fnd&pg=PA111&dq=Importance+of+planned+resettlements+for+floods&ots=4IO3HTNlzY&sig=QyZYtQ74PMurB61aWXlS_GZnlF0&redir_esc=y#v=onepage&q=relocation for flood&f=false.

Siriwardhana, S. D. et al. (2021) 'Cultural issues of community resettlement in post-disaster reconstruction projects in Sri Lanka', *International Journal of Disaster Risk Reduction*, 53(May 2020), p. 102017. doi: 10.1016/j.ijdrr.2020.102017.

Sri Lanka Disaster Knowledge Network (2012) *South Asian disaster knowledge network*, New Delhi: Sri Lanka Disaster Knowledge Network.

Thaler, T. and Fuchs, S. (2020) 'Financial recovery schemes in Austria: How planned relocation is used as an answer to future flood events', *Environmental Hazards*, 19(3), pp. 268–284. doi: 10.1080/17477891.2019.1665982.

Wanninayake, S. B. and Rajapakshe, P. S. (2018) 'Increasing trend of flood and landslide disasters in Sri Lanka: Socio-economic perspective', *6th International Conference of Sri Lanka Forum of University Economists*, 2017, pp. 107–110.

Weerasinghe, K. M. et al. (2018) 'Qualitative flood risk assessment for the western province of Sri Lanka', *Procedia Engineering*, 212(2017), pp. 503–510. doi: 10.1016/j.proeng.2018.01.065.

Weerasinghe, S. (2014) *Planned relocation, disasters and climate change: Consolidating good practices and preparing for the future*, The UN Refugee Agency. Available at: www.unhcr.org/54082cc69.html.

6

A survey-based approach to estimating the downtime of buildings damaged by flood in developing countries

Rayehe Khaghanpour-Shahrezaee, Melissa De Iuliis and Gian Paolo Cimellaro

Introduction

Natural and human-made disasters have a severe impact on communities. Over the years, managing and minimizing the risk of natural disasters has been considered a goal of resilient communities (De Iuliis et al., 2019b; Cimellaro et al., 2010, 2009). Buildings and communities, especially in developing countries, are not resilient enough to extreme natural disasters, such as earthquakes, floods, and so on. They need to be prepared and less vulnerable to achieve a high resilience (De Iuliis et al., 2019a; Cimellaro, 2016). Significant properties of residential buildings in developing countries are non-engineered, and most of them do not conform to the prevailing building codes. In some developing countries, the proportion of non-engineered buildings is believed to exceed 85% (Shrestha et al., 2005). Population growth, especially in developing countries, has increased vulnerability and risk as well as subsequently decreasing the resilience against natural hazards (Atrachali et al., 2019; Hallegatte et al., 2016). The lack of resilience in infrastructures; high vulnerability of buildings; and non-integrated prevention, preparedness programs, and policies are the reasons for extensive damage and loss observed in past events (Benson et al., 2007). Considering these problems in developing countries, quantification of risk and resilience can play an essential role in providing a resilient community.

Determining resilience has many challenges, especially in developing countries. Different aspects, such as predicting a hazard occurrence, vulnerability assessment, coupling effects, and interdependencies, make it difficult to calculate the resilience (Cimellaro and Solari, 2014; Cimellaro et al., 2014, 2018).

The concept of resilience has several definitions, but in general, resilience is the ability of systems to rebound after severe disasters (Renschler et al., 2010; Bruneau et al., 2003; Cimellaro et al., 2010; Chang and Shinozuka, 2004). Different frameworks have been developed to quantify resilience. For instance, Cimellaro et al. (2010) provided a framework for the quantitative definition of resilience using an analytical function that may fit both technical and organizational issues. A quantitative method to evaluate resilience at the state level was presented by

Estimating the downtime of buildings damaged by flood

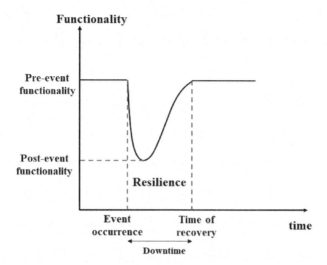

Figure 6.1 Resilience function and downtime

Kammouh et al. (2018b). Graphically, resilience is the normalized shaded area underneath the functionality function, as shown in Figure 6.1 (Cimellaro et al., 2010).

The main parameter to estimate resilience is the downtime, that is, the time required to restore the functionality of a given system. Comerio (2006) defines downtime as 'the time necessary to plan, finance, and complete repair facilities damaged by earthquakes or other disasters', and it consists of rational and irrational components. The rational components are the required time and cost to repair damaged facilities that are easily predictable and quantifiable. The irrational components consider the time required to start repairs, such as inspection, engineering mobilization, finance process and planning, and so on. Determining the uncertain parameters is still tricky, as they could vary depending on the condition of the affected area.

Several studies have been performed on developing loss and downtime estimation techniques. Researchers have struggled to develop a practical downtime estimation method, despite the overwhelming number of variables and uncertainties (Kircher et al., 2006; Almufti and Willford, 2013; Hamburger et al., 2012).

In this work, a fuzzy inference system is exploited to deal with the uncertainties that may affect the recovery process. Existing methodologies for quantifying downtime implement probabilistic approaches and need data for their implementation that are usually unavailable and uncertain, especially in developing countries. Therefore, it is paramount to have a simple method for predicting the downtime and resilience of buildings in developing countries. This chapter aims at quantifying the downtime of residential buildings by deploying a novel method that benefits from the previous work introduced by De Iuliis et al. (2019b) and adapting it to developing countries. In the methodology proposed by De Iuliis et al. (2019b), the downtime of residential buildings is calculated based on three main parameters: repairs, delays, and utility disruptions. Information about the number of workers and time needed for repairing each component is used to calculate the downtime due to repair. This information is extracted from the Performance Assessment Calculation Tool (PACT) (Hamburger et al., 2012) and the Resilience-based Earthquake Design Initiative (REDi™) methodology (Almufti and Willford, 2013). Inspection, engineering mobilization, financial process, and so on are the source of delays considered in the methodology. The main challenge of assessing downtime in developing countries is collecting information about the available workforce, financial resources, and required time for repairing structural and non-structural components. This information is collected by surveys, interviews, and data available from previous events.

The proposed approach can help governments, owners, and decision-makers to minimize the impacts of natural hazards, allow damaged buildings to recover promptly, and improve resilience. The remainder of this chapter is structured as follows: The second section is dedicated to describing the downtime model designed for residential buildings. In the third section, an illustrative example is presented to clarify the approach of assessing downtime. Conclusions, advantages and limitations of the methodology, and recommendations for future work are presented in the final section.

Downtime model for residential buildings

Developing a downtime model for residential buildings in developing countries starts by selecting the indicators that affect downtime. The indicators refer to the building structure, the repair processes and financial resources, and the lifelines that may be damaged after a flood occurrence. The indicators can be aggregated into three main factors: downtime due to repairs (DT repairs), downtime due to delays (DT delays), and downtime due to utility disruptions (DT utilities) (see Figure 6.2). The total downtime (DT) will be calculated as the combination of these three factors:

$$DT = \max((DT\ repairs + DT\ delays); DT\ utilities) \quad (1)$$

The methodology proposed in this work can be divided into the following:

1. Gathering information about the damage level of buildings affected by flood;
2. Evaluating the required repairs based on the damage level;
3. Estimating the impeding factors (downtime due to delays);
4. Evaluating the required time for restoring lifelines (downtime due to utility disruption);
5. Calculating the total downtime.

The process of collecting data and performing each step will be expounded upon in the following sections. The procedure of adapting the previous methodology introduced by De Iuliis et al. (2019b) for developing countries is explained in each step.

Step 1: The first step of the methodology is determining the damage level of buildings. Information about the damage level is necessary to identify the design and the components of the analyzed building, and it is considered the main parameter to quantify the downtime. For developing the methodology, it is assumed that such information about the damage level is available. Generally, a rapid visual screening survey is used to have a first estimation of the building's damageability, as it does not require any structural analysis and provides an initial assessment of the damage level.

Step 2: Required time for repairs is calculated in the second step. Downtime due to repairs depends on the number of damaged components, the damage state of each component, and the number of workers assigned for the repair. In De Iuliis et al., (2019b), component repair times (in terms of worker-days) are obtained by PACT, a calculation tool produced by FEMA (2012). Given that PACT is provided for developed countries, it is necessary

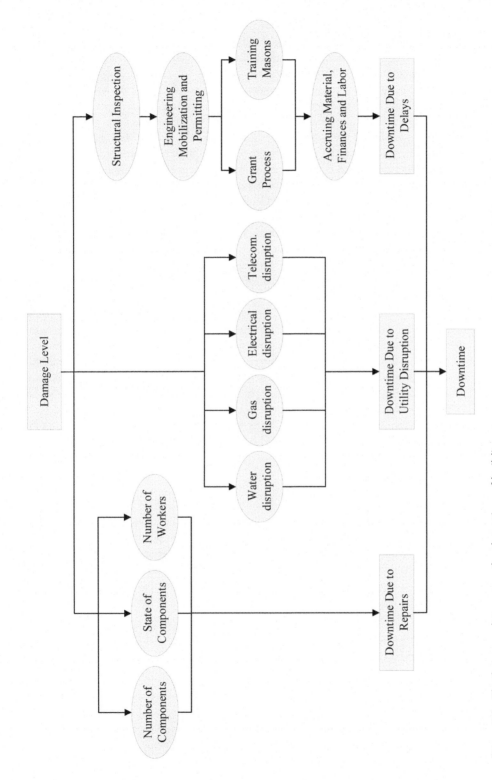

Figure 6.2 The framework to assess the downtime of buildings

to calculate a penalty factor based on developing countries' data to verify repair time calculated by PACT for developing countries. The initial data about the number of workers and the time needed for repairing the damaged components are collected through a survey. By using this penalty factor, the number of worker-days for each component in developing countries can be found.

Step 3: The third step is determining the delays. Downtime due to delays is derived from irrational components and depends on the reconstruction process after the flood occurrence. These delays can increase the time required to achieve a recovery state. The irrational components in the previous methodology presented by De Iuliis et al. (2019b) are considered based on the REDi methodology, considering knowledge and information regarding developed countries. Also, the time needed for completing each process is modeled based on the facilities, technology, and workforce in developed countries. According to the fact that the available facilities and material and the way of managing the available workforce and other financial sources in developing countries are different, it is clear that sources of delays are different and need to be considered in the present methodology. Inspection, engineering mobilization, finances and grant process, training masons and accruing laborers, material, and finances are recommended irrational components in developing countries. Nevertheless, they can be changed based on the study region's situation and the government.

Step 4: In this step, the time needed for repairing disrupted infrastructures is calculated. Predicting the downtime due to utility disruption has many challenges, as utility systems are widely distributed in a region, so they endure a wide range of local site effects and different damage levels. Downtime due to utility disruptions is estimated using restoration functions derived based on the best distribution to fit the data. The data are collected based on the information available from previous events such as earthquakes, floods, and so on in developing countries.

Step 5: In the final step, the total downtime is computed as the maximum of the sum of the time needed for repairs and delays and the time of utility disruption (equation 1).

Building damageability

Building damage is the main parameter to quantify downtime. Information about the level of damage and state of each building component after an event is effective in calculating the time required for repairing the damaged components and delays. Components with a lower level of damage will be repaired in less time than components with a higher level of damage. On the other hand, delays such as inspection, engineering mobilization, and so on can be carried out in a shorter time compared to a higher level of damage. For developing the methodology, it is assumed that such information is available. It can be estimated through a rapid visual screening survey that does not require any structural analysis.

Downtime due to repairs

As mentioned before, downtime due to repairs depends on the damage state of each building component as well as the recovery time and the number of workers needed for repairing each structural and non-structural component. So the first step in calculating downtime due to repairs is gathering information about the number of workers and required time for repairing each component. In the previous methodology, De Iuliis et al. (2019b) have used the number of workers suggested by Almufti and Willford (2013) for each repair sequence and also the repair time of each component extracted by PACT. In this methodology, a brief survey is provided to get information about the number of workers and the time required for repairing each structural and non-structural building component in developing countries.

The repair sequences, which define the order of repairs, are considered based on REDI methodology (Almufti and Willford, 2013) to quantify the repair time. These repair sequences consist of structural and non-structural repairs. Non-structural repairs are classified into six sequences: sequence A (pipes, HVAC distribution, partitions, ceiling tiles), sequence B (exterior partitions, cladding/glazing), sequence C (mechanical equipment), sequence D (electrical system), sequence E (elevators), and sequence F (stairs). Structural repairs are carried out before non-structural repairs (see Figure 6.3). The structure is repaired only one floor at a time, starting from the lower floor, since the lower floor's structural integrity must first be guaranteed before the upper-floor repairs are started. After finishing structural repairs, non-structural repairs can be started. It is assumed that non-structural repair sequences (A to D) can be carried out simultaneously.

In the survey, questions are prepared in the form of each repair sequence so that the answers can be compared with the corresponding numbers in REDI methodology. Participants should indicate the required time and the number of labor allocations for completing each repair sequence in developing countries, based on their knowledge and experiences from previous events such as earthquakes, floods, and so on. The average number of workers for each repair sequence calculated from the survey results is used to estimate the downtime due to repair.

As mentioned before, a penalty factor is used in this methodology to adapt each component's repair time from PACT to developing countries. The results obtained from the survey are used for calculating the penalty factor. Since PACT provides the repair time in terms of worker-days, the average repair time and number of workers for each repair sequence obtained from the survey are multiplied to calculate the average repair time in terms of worker-days. For example, based on survey results, five workers are required to repair each damaged unit in repair sequence C in 31 days. Thus, 155 worker-days are needed to repair each unit. Comparing the resulting number of worker-days for each repair sequence with the corresponding number in the example in REDI methodology, the penalty factor can be calculated. For instance, based on

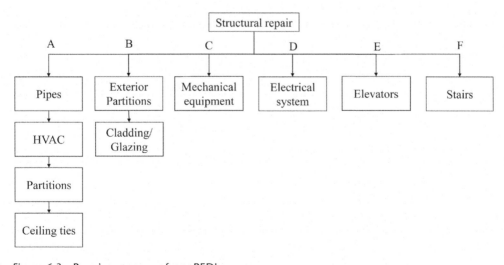

Figure 6.3 Repair sequences from REDI

the example in REDI methodology, 73 worker-days are required to repair each unit of repair sequence C. By dividing 155 by 73, the penalty factor for repair sequence C is calculated as about 2.13. By multiplying the penalty factor by the repair time of each component extracted from PACT, the required repair time for each damaged component in developing countries can be obtained.

Finally, the time needed to complete each repair sequence is obtained by dividing the total number of worker-days of each sequence by the corresponding number of workers (obtained from survey results). Since it is assumed that the non-structural repairs are carried out in parallel, the maximum required time calculated for these repair sequences is determinative in total repair time. The downtime due to repairs is eventually calculated by summing the repair time for structural and non-structural repairs.

Downtime due to delays

Downtime for a particular building is not limited to the time necessary to complete repairs. Several impeding factors cause delays that can affect the required time for achieving a recovery state. As mentioned previously, downtime due to delays is derived from irrational components. Data needed to calculate downtime due to delays, such as types of irrational components and required time for each process, are collected based on previous events, surveys, and interviews with engineers and governments. Inspection after flood occurrence, engineering mobilization, finances process, training masons and accruing laborers, material, and finances are common irrational components in developing countries.

The types of irrational components and the amount of delay they cause in the reconstruction process have many sources of uncertainties. Uncertainties in calculating the required time for completing each process are considered by modeling each process as a random variable with a statistical distribution (Almufti and Willford, 2013; Longman and Miles, 2019). Each irrational component and method of estimating the delay is described in the following sections.

Inspection process

After an event, inspectors are needed to inspect potentially damaged buildings. Delays due to inspection depend on the availability of inspectors and the importance of the building. Inspectors are expected to arrive earlier if the building is an essential facility. Also, if the number of available inspectors is less than required, it takes more time to inspect all of the buildings affected by the flood. This factor commonly affects the time needed for inspecting buildings. For instance, in Nepal, approximately one inspector is available for every 200 damaged buildings after an earthquake, which can increase the time of inspection (Dhital, 2016). Based on the available data from Nepal, Longman and Miles (2019) modelled the duration of the inspection process as a uniform distribution between 10 and 30 days (Figure 6.4). It can be concluded that with a probability of 50%, the inspection process is completed after 20 days. The required time for inspection corresponding to 90% of the non-exceedance probability of inspection is 28 days.

Engineering mobilization process

After the building is inspected and the damage level is determined, the engineering mobilization process starts to design the repair and retrofit process. The time required for finding engineers and the time needed to carry out engineering review and re-design are delays that affect the downtime. In developing countries, this impeding factor is affected by the number of

Estimating the downtime of buildings damaged by flood

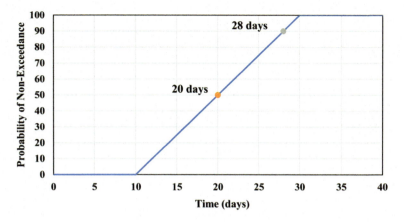

Figure 6.4 Example of the impeding curve for inspection in developing countries

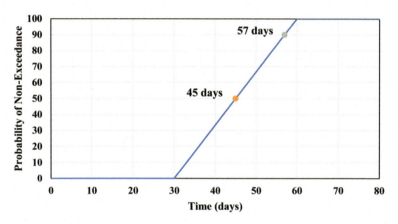

Figure 6.5 Example of the impeding curve for engineering mobilization in developing countries

available engineers and the process of engineering calculations. For instance, the engineering mobilization and permit process's duration can be presented as a uniform distribution between 30 and 60 days, as shown in Figure 6.5 (Longman and Miles, 2019).

Financing process

Significant delays can occur due to the time required to obtain the proper financing needed for the repair process. The delays depend on the damage level and the amount of needed financing, as well as the availability of finances. They are predicated based on the method and condition of financing in developing countries. Financing may be procured through loans or insurance payments or government grants. The process of gaining finances and grants takes time and has many uncertainties. This information can be obtained from data available from previous events and surveys. The grant may be paid in several installments, based on the repair's progress (Longman and Miles, 2019); this process can be modeled as a uniform distribution between 20 and 40 days (Figure 6.6).

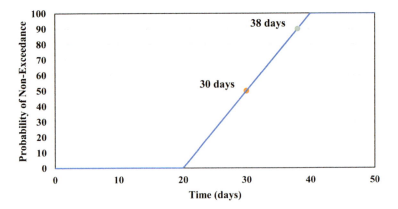

Figure 6.6 Example of the impeding curve for grant process in developing countries

Training masons

Education is one of the main aspects of making developing countries more resilient, and the governments in developing countries should invest in large-scale training programs. It is anticipated that local masons play a significant role in the building construction process. Any enhancement in their skills in resistant construction can significantly help improve the resilience of buildings and communities. One of the most significant delays in the reconstruction process in developing countries is the time needed for training masons and laborers. Based on experiences and information from past events in developing countries, there is not enough skilled workforce for the repair and reconstruction process.

Training skilled masons that can handle the repair process takes significant time that affects the downtime after events such as floods, earthquakes, and so on. A seven-day course of training masons is considered in Nepal, in which about 350 masons can be trained at the same time in seven days (Cluster, 2019). Accruing material, finances, and labor required for each building's repair process also takes time that can cause additional delays.

Downtime due to utility disruption

Downtime due to utility disruption is determined through restoration curves of the lifelines such as power systems, water systems, gas systems, and telecommunication systems. Data used to construct the restoration curves can be collected from information about past events. In the previous research proposed by Kammouh et al. (2018a) and De Iuliis et al. (2021), information about the disruption time of infrastructure services published in the literature has been used for calculating the disruption time of infrastructures. The information included data related to the earthquakes that occurred after the 1960s and has been classified based on the earthquake magnitude. In the present methodology, the published information with numerical data reporting the actual time needed to restore the infrastructure service in developing countries has been considered in the analysis. In addition to the data related to developing countries presented in Kammouh et al. (2018a), new information about other earthquakes in developing countries has been collected. The list of the databases used to create the restoration curves of lifelines in developing countries is presented in Table 6.1. As can be recognized from the table, each earthquake may cause damage to more than one infrastructure system at the same time. Based

Estimating the downtime of buildings damaged by flood

Table 6.1 Database of earthquakes, number of affected infrastructures, and the corresponding downtime for each lifeline

| Earthquakes | Lifelines affected ||||||||| References |
| | Power systems || Water systems || Gas systems || Telecom. systems || |
	No.	DT (days)	No.	DT (days)	No.	DT (days)	No.	DT (days)	
Gorkha (2015)	16	(7), (1), (0), (180), (9h), (9h), (12h), (9h), (9h), (90), (12), (12), (9h), (1), (67), (4)	1	1–10	–	–	1	14	(Aihara et al., 2018; Didier et al., 2017; Zhu et al., 2017)
Pisco (2007)	4	(14), (28), (7), (42)	1	42	–	–	–	–	(Tang and Johnsson, 2010)
Kashmir (2005)	1	6	1	10	–	–	–	–	(Hussain et al., 2006)
Manjil-Rudbar (1990)	–	–	–	–	–	–	1	100	(Eshghi, 1992)
Maule (2010)	6	(14), (1), (3) (10), (14)	4	(42), (4), (16), (6)	2	(10), (90)	4	(17), (7), (3), (17)	(Evans and McGhie, 2011)
Michoacan (1985)	4	(4), (10), (3), (7)	4	(30), (14), (40), (45)	–	–	1	160	(O'Rourke, 1996)
Illapel (2015)	1	3	1	3	–	–	–	–	(The government of Chile, 2015)
Luzon (1990)	3	(7), (20), (3)	3	(14), (14), (10)	–	–	3	(5), (10), (0.4)	(Sharpe, 1994)
El Asnam (1980)	–	–	1	14	–	–	–	–	(Nakamura et al., 1983)
Valdivia (1960)	1	5	1	50	–	–	–	–	(Edwards et al., 2003)
Bam (2003)	1	4	3	(14), (10)	–	–	1	1	(Ahmadizadeh and Shakib, 2004)
Samara (2012)	1	1	1	2	–	–	1	1	(Sismo, 2012)
Arequipa (2001)	1	1	3	(32), (34)	–	–	–	–	(Edwards et al., 2003)
Izmir (1999)	1	1	2	(50), (29)	1	1	1	10	(Gillies et al., 2001)
Chi-chi (1999)	3	(40), (14), (19)	1	9	1	14	1	10	(Soong et al., 2000)

79

on the damage level, the damaged systems need different times to recover even when the infrastructures are of similar types. There were some cases where the damage information was not available or no damage was recorded. Such cases are marked with a dash (-) in the table.

A probabilistic distribution with the best fit is assigned to the empirical data to design the restoration curves for infrastructure systems. The empirical disruption times can be fitted with statistical distributions. Based on previous research (De Iuliis et al., 2021; Kammouh and Cimellaro, 2017; Kammouh et al., 2018a), gamma, exponential, and lognormal cumulative distributions are recommended distributions that can fit the data.

Goodness-of-fit tests were used to identify the appropriate distribution for the empirical data. There are several tests to verify the goodness of fit in the literature. The Kolomogorov-Smirnov (or K-S) and the chi-square test were performed to choose the best fit distribution. The Kolmogorov-Smirnov test compares the experimental cumulative frequency with the cumulative distribution function (CDF) of an assumed theoretical distribution. The chi-square goodness-of-fit test compares the observed frequencies of the variate with the corresponding theoretical frequencies calculated from the assumed theoretical distribution model (Kammouh et al., 2018a).

Figure 6.7 shows the cumulative frequencies with three theoretical CDF fitting distributions for power and telecommunication systems. Figure 6.8 shows the frequency histogram and the

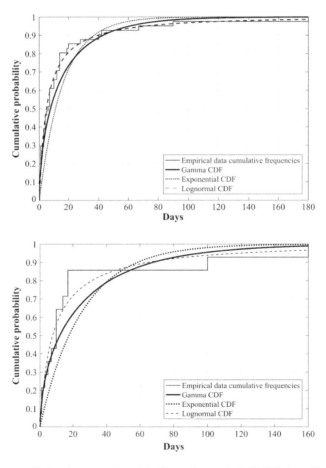

Figure 6.7 The cumulative frequencies with three theoretical CDF fitting distributions for a) power systems and b) telecommunication systems

Estimating the downtime of buildings damaged by flood

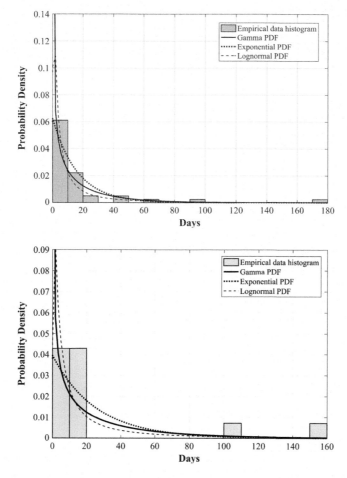

Figure 6.8 Histogram and theoretical PDF fitting distributions for a) power systems and b) telecommunication systems

Table 6.2 Kolmogorov-Smirnov goodness-of-fit test for power and telecommunication systems

Theoretical distribution	Power system		Telecom. system	
	D_n	D_n^{α} ($\alpha = 0.05$, $n = 41$)	D_n	D_n^{α} ($\alpha = 0.05$, $n = 4$)
		0.212		0.349
Gamma	0.161		0.239	
Exponential	0.256		0.317	
Lognormal	0.063		0.140	

probability density function (PDF) of the theoretical distributions for power and telecommunication systems. The same approach has been performed for other infrastructures. Results of the goodness-of-fit tests are presented in Table 6.2 and Table 6.3. The results show that the lognormal distribution fits the data. Finally, the restoration functions are performed in terms of probability of recovery versus time (Figure 6.9). The downtime corresponding to 95% of exceedance probability of recovery can be used as a deterministic downtime for each infrastructure.

Calculating total downtime

In the final step, the downtime due to delays, repairs, and utility disruption are combined to calculate the total downtime. Downtime due to delays and utility disruption starts simultaneously after the flood occurrence, and downtime due to repairs follows the downtime due to delays.

Table 6.3 Chi-square goodness-of-fit test for the power and telecommunication systems

Theoretical distribution	Power system			Telecom. system		
	Chi-square	$f = k - 1$	$C_{1-\alpha, f}$ ($\alpha = 0.05$)	Chi-square	$f = k - 1$	$C_{1-\alpha, f}$ ($\alpha = 0.05$)
Gamma	9.60	15	25.00	6.88	5	11.07
Exponential	21.69	14	23.68	9.87	4	9.488
Lognormal	10.99	15	25.00	5.74	5	11.07

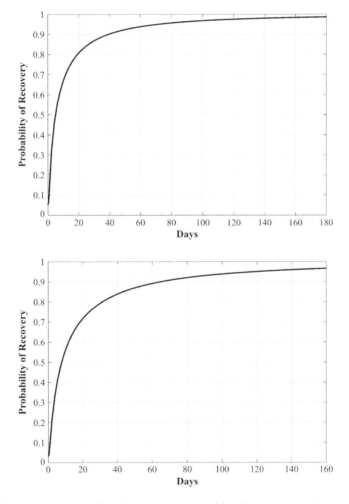

Figure 6.9 Restoration curves for a) power systems, b) telecommunication systems, c) gas systems, and d) water systems

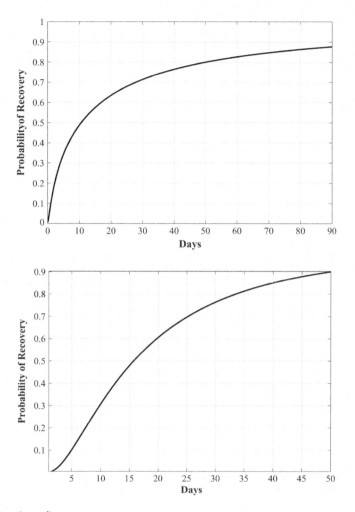

Figure 6.9 (Continued)

The repair process on each floor starts with the structural repair. After finishing the structural repairs, non-structural repairs start simultaneously. The downtime due to repairs is calculated as the sum of structural repairs and the maximum of non-structural repairs (repair sequence A to F). Total downtime is computed as the sum of downtime due to repairs and the maximum of downtime due to delays and downtime due to utility disruption (equation 1).

Illustrative example

In this section, the proposed approach of estimating downtime in developing countries is illustrated through an example. A three-story residential building located in Nepal with floor area $A = 4800$ sq. ft. per floor is assumed. It is a reinforced concrete system. A rapid visual screening survey is conducted after a flood occurrence that shows that the building is at a *medium* damage level. In the following, the steps for estimating downtime are illustrated in detail.

Downtime due to repairs

Repair sequences considered in the example are structural repair, repair sequences A, B, C, D, and F. Table 6.4 presents the corresponding number of workers per floor. It should be noted that the number of workers per 1000 sq. ft. or damaged unit is calculated based on survey results in developing countries.

Repair times for building components are summarized in Table 6.5. The number of worker-days for each unit (EA, each unit) or area (SF, square feet) of components is obtained from the modified values from PACT for developing countries. By multiplying the unitary worker-days value by the corresponding number of units or area of components, the number of worker-days of each component can be computed.

Table 6.4 Number of workers for structural and non-structural repairs

Repair sequence	Number of workers per floor
Structural repair	#workers = (4800 sq.ft) (4 workers/500 sq.ft) = 39 workers
Repair sequence A	#workers = (4800 sq.ft) (10 workers/1000 sq.ft) = 48 workers
Repair sequence B	#workers = (4800 sq.ft) (10 workers/1000 sq.ft) = 48 workers
Repair sequence C	#workers = (1 damaged unit) (5 workers/damaged unit) = 5 workers
Repair sequence D	#workers = (1 damaged unit) (4 workers/damaged unit) = 4 workers
Repair sequence F	#workers = (4 damaged unit) (7 workers/damaged unit) = 28 workers

Table 6.5 Component repair times and worker days

Floor	Repair	Component type	Worker-days per unit or area	EA or SF	Total worker-days
Floor 1	Structural repairs	Concrete beam	64.633	2 units	129.265
		Link beams < 16 in.	49.297	1 unit	49.297
	Repair sequence A	Interior partitions	14.2	215.3 sq. ft	3057.26
		Ceiling	48.28	30 sq. ft	1448.4
	Repair sequence B	Exterior partitions	90.88	20 sq. ft	1817.6
	Repair sequence D	Transformer < 100 kVA	5.163	1 unit	5.163
		Low-voltage switchgear	6.322	1 unit	6.322
	Repair sequence F	Stairs	39.661	4 units	158.642
Floor 2	Structural repairs	Concrete beam	64.633	1 unit	64.633
		Link beams < 16 in.	49.297	1 unit	49.297
	Repair sequence A	Interior partitions	14.2	220 sq. ft	3124
		Ceiling	48.28	10 sq. ft	482.8
	Repair sequence B	Exterior partitions	90.88	5 sq. ft	454.4
	Repair sequence D	Transformer < 100 kVA	5.163	1 unit	5.163
		Low-voltage switchgear	6.322	1 unit	6.322
	Repair sequence F	Stairs	39.661	4 unit	158.642

Estimating the downtime of buildings damaged by flood

Floor	Repair	Component type	Worker-days per unit or area	EA or SF	Total worker-days
Floor 3	Structural repairs	Concrete beam	64.633	3 unit	193.898
	Repair sequence A	Interior beam	14.2	190 sq. ft	2698
		Ceiling	48.28	15 sq. ft	724.2
	Repair sequence D	Transformer < 100 kVA	5.163	1 unit	5.163
		Low-voltage switchgear	6.322	1 unit	6.322
	Repair sequence F	Stairs	39.661	4 unit	158.642
Roof	Repair sequence C	Chiller	31.49	1 unit	31.49

Table 6.6 Repair times for each repair sequence

Floor	Repair	Repair time (days)
Floor 1	Structural repairs	RT = (178.56 worker days)/39 workers = 4.58 days
	Repair sequence A	RT = (4505.66 worker days)/48 workers = 93.87 days
	Repair sequence B	RT = (1817.60 worker days)/48 workers = 37.87 days
	Repair sequence D	RT = (11.48 worker days)/4 workers = 2.87 days
	Repair sequence F	RT = (158.64 worker days)/28 workers = 5.67 days
Floor 2	Structural repairs	RT = (113.93 worker days)/39 workers = 2.92 days
	Repair sequence A	RT = (3606.80 worker days)/48 workers = 75.14 days
	Repair sequence B	RT = (454.40 worker days)/48 workers = 9.47 days
	Repair sequence D	RT = (11.48 worker days)/4 workers = 2.87 days
	Repair sequence F	RT = (158.64 worker days)/28 workers = 5.67 days
Floor 3	Structural repairs	RT = (193.90 worker days)/39 workers = 4.97 days
	Repair sequence A	RT = (3422.20 worker days)/48 workers = 71.30 days
	Repair sequence D	RT = (11.48 worker days)/48 workers = 2.87 days
	Repair sequence F	RT = (158.64 worker days)/28 workers = 5.67 days
Roof	Repair sequence C	RT = (31.49 worker days)/5 workers = 6.30 days

By summing the total number of worker-days corresponding to structural components and non-structural components of each repair sequence (last column in Table 6.5) and dividing by the number of workers defined from Table 6.4, the days required for each repair sequence in each floor are obtained (Table 6.6). Finally, the downtime due to repairs is calculated as:

$$DT_{repairs} = 12.47 + 240.31 = 252.78 \text{ days} \qquad (2)$$

Downtime due to delays

Delays considered in this example are inspection, engineer mobilization, financing process, and training masons. Delay time corresponding to the 50% of non-exceedance probability from Figures 6.4 to 6.6 is used as the required time for inspection, engineering mobilization, and financing. Seven days of training masons are considered based on Cluster (2019). Impeding factor

Table 6.7 Impeding factor delays

Impeding factor	Inspection	Engineering mobilization	Financing	Training masons
Delays (days)	20	45	30	7

Table 6.8 Recovery time of infrastructures

Lifeline	Power systems	Water systems	Gas systems	Telecom. systems
Recovery time (days)	5	16	11	9

delays are presented in Table 6.7. Based on the sequence of delays illustrated in Figure 6.2, the process of financing and training masons can be carried out simultaneously after engineering mobilization. The total delays can be calculated as:

$$DT\, delays = 20 + 45 + \max(30, 7) = 95 \text{ days} \qquad (3)$$

Downtime due to utility disruption

The 50% probability of non-exceedance values for utility disruption from Figure 6.9 are selected for each utility system (Table 6.8). Downtime due to utility disruption starts simultaneously after the flood occurrence. As a result, the maximum time needed for repairing infrastructures is assumed as the downtime due to utility disruption:

$$DT\, utilities = 16 \text{ days} \qquad (4)$$

Finally, the total downtime is computed as the combination of downtime due to delays, repairs, and utility disruption:

$$DT = \max(DT\, utilities, DT\, delays) + DT\, repairs = 95 + 252.78 = 347.78 \text{ days} \qquad (5)$$

Conclusion

This chapter introduces a new survey-based approach for estimating the downtime for flood risk management in developing countries. Downtime is divided into three main components: downtime due to repairs, downtime due to delays, and downtime due to utility disruption. A present approach for assessing downtime in developed countries is used, and this approach is adapted for developing countries. Calculating the downtime requires essential information about facilities' availability and financial and human resources for the reconstruction process in developing countries. Therefore, the main challenge of assessing downtime is collecting this initial information. This information is obtained through a survey prepared by the authors to get information about the recovery process in developing countries. A penalty factor is calculated based on survey results to modify the repair time calculated by PACT for developing countries. Available data from past events are used to calculate downtime due to delays and utility disruption.

Downtime estimation has many uncertainties. It requires a comprehensive analysis of parameters that contribute to different uncertainties, such as the time needed for passing inspection, engineering mobilization, or other processes; the delay time for each infrastructure; and so on. These parameters are modeled as random variables with a probabilistic distribution.

Governments and decision-makers can use the proposed downtime assessment to improve flood risk management in their communities, minimize the impacts of natural hazards, allow damaged buildings to recover promptly, improve resilience, and make decisions for future natural hazards during the planning phase.

The main limitation of the proposed methodology is the availability of a database to test and verify the approach for each developing country. This work can be improved in future works by considering a real database for any specific region. Further research work will also be oriented towards implementing the methodology to a real case study to illustrate its applicability and verify which component shows a high impact on total downtime.

Acknowledgments

The research leading to these results has received funding from the European Research Council under the Grant Agreement n° 637842 of the project IDEAL RESCUE Integrated Design and Control of Sustainable Communities during Emergencies.

References

Ahmadizadeh, M. & Shakib, H. 2004. On the December 26, 2003, Southeastern Iran earthquake in Bam region. *Engineering Structures*, 26, 1055–1070.

Aihara, Y., Shrestha, S., Rajbhandari, S., Bhattarai, A. P., Bista, N., Kazama, F. & Shindo, J. 2018. Resilience in household water systems and quality of life after the earthquake: A mixed-methods study in urban Nepal. *Water Policy*, 20, 1013–1026.

Almufti, I. & Willford, M. 2013. *REDi™ rating system: Resilience-based earthquake design initiative for the next generation of buildings*. Arup Co [Online]. Available: https://static1.squarespace.com/static/61d5 bdb2d77d2d6ccd13b976/t/61e85a429039460930278463/1642617413050/REDi_Final+Version_ October+2013+Arup+Website+%288%29.pdf.

Atrachali, M., Ghafory-Ashtiany, M., Amini-Hosseini, K. & Arian-Moghaddam, S. 2019. Toward quantification of seismic resilience in Iran: Developing an integrated indicator system. *International Journal of Disaster Risk Reduction*, 39, 101231.

Benson, C., Twigg, J. & Rossetto, T. 2007. *Tools for mainstreaming disaster risk reduction: Guidance notes for development organisations*. ProVention Consortium [Online]. Available: https://www.preventionweb. net/files/1066_toolsformainstreamingDRR.pdf.

Bruneau, M., Chang, S. E., Eguchi, R. T., Lee, G. C., O'Rourke, T. D., Reinhorn, A. M., Shinozuka, M., Tierney, K., Wallace, W. A. & Von Winterfeldt, D. 2003. A framework to quantitatively assess and enhance the seismic resilience of communities. *Earthquake Spectra*, 19, 733–752.

Chang, S. E. & Shinozuka, M. 2004. Measuring improvements in the disaster resilience of communities. *Earthquake Spectra*, 20, 739–755.

Cimellaro, G. P. 2016. *Urban resilience for emergency response and recovery*. Cham, Switzerland: Springer. DOI: 10.1007/978-3-319-30656-8.

Cimellaro, G. P., Arcidiacono, V. & Reinhorn, A. 2018. Disaster resilience assessment of building and transportation system. *Journal of Earthquake Engineering*, 1–27.

Cimellaro, G. P., Fumo, C., Reinhorn, A. M. & Bruneau, M. 2009. *Quantification of disaster resilience of health care facilities*. Buffalo, NY: MCEER.

Cimellaro, G. P., Reinhorn, A. M. & Bruneau, M. 2010. Framework for analytical quantification of disaster resilience. *Engineering Structures*, 32, 3639–3649.

Cimellaro, G. P. & Solari, D. 2014. Considerations about the optimal period range to evaluate the weight coefficient of coupled resilience index. *Engineering Structures*, 69, 12–24.

Cimellaro, G. P., Solari, D. & Bruneau, M. 2014. Physical infrastructure interdependency and regional resilience index after the 2011 Tohoku earthquake in Japan. *Earthquake Engineering & Structural Dynamics*, 43, 1763–1784.

Cluster, G. S. 2019. *A.18 – Nepal, 2016–2017, earthquake recovery (Case study). Shelter projects: 2017–2018. Fédération internationale des sociétés de la Croix Rouge* [Online]. Available: https://www.shelterprojects.org/shelterprojects2017-2018/SP17-18_A18-Nepal-2016-2017.pdf.

Comerio, M. C. 2006. Estimating downtime in loss modeling. *Earthquake Spectra*, 22, 349–365.

De Iuliis, M., Kammouh, O., Cimellaro, G. & Tesfamariam, S. 2019a. Resilience of the built environment: A methodology to estimate the downtime of building structures using fuzzy logic. In *Resilient structures and infrastructure*. Singapore: Springer.

De Iuliis, M., Kammouh, O., Cimellaro, G. P. & Tesfamariam, S. 2019b. Downtime estimation of building structures using fuzzy logic. *International Journal of Disaster Risk Reduction*, 34, 196–208.

De Iuliis, M., Kammouh, O., Cimellaro, G. P. & Tesfamariam, S. 2021. Quantifying restoration time of power and telecommunication lifelines after earthquakes using Bayesian belief network model. *Reliability Engineering & System Safety*, 208, 107320.

Dhital, M. 2016. *Reconstruction woes, Kathmandu post* [Online]. Available: http://kathmandupost.ekantipur.com/news/2016-12-22/reconstruction-woes.html [Accessed 27 March 2017].

Didier, M., Grauvogl, B., Steentoft, A., Ghosh, S. & Stojadinovic, B. 2017. *Seismic resilience of the Nepalese power supply system during the 2015 Gorkha earthquake*. 16th World Conference on Earthquake Engineering, 16WCEE, Santiago, Chile.

Edwards, C., Eidinger, J. & Schiff, A. 2003. Lifelines. *Earthquake Spectra*, 19, 73–96.

Eshghi, S. 1992. *Behavior of lifeline systems during Manjil-Iran earthquake of June 20, Î990*. Earthquake Engineering, 10th World Conference, Balkema, Rotterdam.

Evans, N. & Mcghie, C. 2011. *The performance of lifeline utilities following the 27th February 2010 Maule earthquake Chile*. Proceedings of the Ninth Pacific Conference on Earthquake Engineering Building an Earthquake-Resilient Society, Auckland, 14–16 April.

FEMA, P. 2012. *Seismic performance assessment of buildings: Methodology*. Redwood City, CA: Applied Technology Council for the Federal Emergency Management Agency, 1.

Gillies, A. G., Anderson, D. L., Mitchell, D., Tinawi, R., Saatcioglu, M., Gardner, N. J. & Ghoborah, A. 2001. The August 17, 1999, Kocaeli (Turkey) earthquake lifelines and preparedness. *Canadian Journal of Civil Engineering*, 28, 881–890.

The Government of Chile. 2015. *ONEMI. Anlisis multisectorial eventos 2015: Evento hidrometeorolgico Marzo Terremoto/Tsunami Septiembre* [Online]. Available: https://reliefweb.int/report/chile/lisis-multisectorial-eventos-2015-evento-hidrometeorol-gico-marzo-terremototsunami.

Hallegatte, S., Vogt-Schilb, A., Bangalore, M. & Rozenberg, J. 2016. *Unbreakable: Building the resilience of the poor in the face of natural disasters*. Washington, DC: World Bank Publications.

Hamburger, R., Rojahn, C., Heintz, J. & Mahoney, M. 2012. *FEMA P58: Next-generation building seismic performance assessment methodology*. 15th World Conference on Earthquake Engineering [Online]. Available: https://www.iitk.ac.in/nicee/wcee/article/WCEE2012_4156.pdf.

Hussain, S., Nisar, A., Khazai, B. & Dellow, G. 2006. *The Kashmir earthquake of October 8, 2005: Impacts in Pakistan*. Earthquake Engineering Research Institute [Online]. Available: https://reliefweb.int/report/pakistan/kashmir-earthquake-october-8-2005-impacts-pakistan.

Kammouh, O. & Cimellaro, G. P. 2017. *Restoration time of infrastructures following earthquakes*. 12th International Conference on Structural Safety & Reliability (ICOSSAR 2017), Los Angeles, CA.

Kammouh, O., Cimellaro, G. P. & Mahin, S. A. 2018a. Downtime estimation and analysis of lifelines after an earthquake. *Engineering Structures*, 173, 393–403.

Kammouh, O., Dervishaj, G. & Cimellaro, G. P. 2018b. Quantitative framework to assess resilience and risk at the country level. *ASCE-ASME Journal of Risk and Uncertainty in Engineering Systems, Part A: Civil Engineering*, 4, 04017033.

Kircher, C. A., Whitman, R. V. & Holmes, W. T. 2006. HAZUS earthquake loss estimation methods. *Natural Hazards Review*, 7, 45–59.

Longman, M. & Miles, S. B. 2019. Using discrete event simulation to build a housing recovery simulation model for the 2015 Nepal earthquake. *International Journal of Disaster Risk Reduction*, 35, 101075.

Nakamura, S., Aoshima, N. & Kawamura, M. 1983. *A review of earthquake disaster preventive measures for lifelines*. Proceedings of Japan Earthquake Engineering Symposium [Online]. Available: https://www.iitk.ac.in/nicee/wcee/eighth_conf_California/.

O'Rourke, T. 1996. *Lessons learned for lifeline engineering from major urban earthquakes*. Proceedings, Eleventh World Conference on Earthquake Engineering [Online]. Available: https://www.iitk.ac.in/nicee/wcee/article/11_2172.PDF.

Renschler, C. S., Frazier, A. E., Arendt, L. A., Cimellaro, G. P., Reinhorn, A. M. & Bruneau, M. 2010. *A framework for defining and measuring resilience at the community scale: The peoples resilience framework*. Buffalo, NY: MCEER.

Sharpe, R. 1994. *July 16, 1990 Luzon (Philippines) earthquake*. Cupertino, CA [Online]. Available: https://www.phivolcs.dost.gov.ph/index.php/earthquake/destructive-earthquake-of-the-philippines/2-uncategorised/212-1990-july-16-ms7-9-luzon-earthquake.

Shrestha, S. N., Acharya, S. P., Kandel, R. C., Upadhyay, B. & Bothara, J. K. 2005. *Earthquake resistant construction of buildings: Curriculum for mason training*. Asian Disaster Preparedness Center (ADPC) [Online]. Available: http://lib.riskreductionafrica.org/handle/123456789/1296.

Sısmo, C. 2012. *7.6 Mw (Magnitud de Momento) Samara, Region de Guanacaste, sector peninsula de Nicoya*. Comisin Nacional de Prevencin de Riesgos y Atencin de Emergencias Gobierno de Costa Rica [Online]. Available: http://www.pgrweb.go.cr/scij/Busqueda/Normativa/Normas/nrm_texto_completo.aspx?param1=NRTC&nValor1=1&nValor2=74378&nValor3=91835&strTipM=TC.

Soong, T. T., Yao, G. C. & Lin, C. 2000. *Damage to critical facilities following the 921 Chi-Chi, Taiwan earthquake*. Buffalo, NY [Online]. Available: https://www.semanticscholar.org/paper/Damage-to-Critical-Facilities-Following-the-921-%2C-Soong-Yao/7d3062e4558077f4fa142c4c14127240d7faf673.

Tang, A. K. & Johnsson, J. R. 2010. *Pisco, Peru, earthquake of August 15, 2007: Lifeline performance*. Japan Society of Civil Engineers [Online]. Available: http://ascelibrary.org/doi/pdf/10.1061/9780784410615.fm.

Zhu, J., Manandhar, B., Truong, J., Ganapati, N. E., Pradhananga, N., Davidson, R. A. & Mostafavi, A. 2017. Assessment of infrastructure resilience in the 2015 Gorkha, Nepal, earthquake. *Earthquake Spectra*, 33, 147–165.

Section II
Preparedness, prevention, responses and recovery

7

Application of spatial planning and other tools to support preparedness for flooding in developing countries; example from Nigeria

Oluseyi O. Fabiyi

Introduction and problem description

Flood disasters have been one of the major impacts of global climate change and some of the most significant disasters that our world faces today. Flooding in rural areas damages crops and constrains the food security of the populace (Pidgeon & Butler, 2009; Dawson et al., 2011). However, in most developing countries, urban floods have become more devastating and frequent in recent times. The huge economic losses from urban flood disasters are raising considerable concerns among experts as they limit potential for economic growth among developing nations (Butler & Pidgeon, 2011; Butler, 2008). Cities are typically engines of growth and the catalyst for sustainable national development. It is estimated that 50% of the world population will live in urban areas in the year 2030. Further, a majority of the 3 billion people who will dwell in the cities by the year 2030 will be in sub-Saharan Africa.

Most cities in Africa have had terrible experiences of urban flood disasters that are not only dangerous but also reoccurring. For example, in July 2021, more than 40% of Lagos state's land area was submerged after rainfall that fell for more than 5 hours. Many commuters were trapped and houses submerged. These flood disasters are usually followed by loss of lives and properties. It is estimated that the majority of urban population in poor countries have no capacity to adapt to the effects of climate change or even the potential to rebound after a flood disaster before another one occurs. The root cause of most flood disasters in Nigeria, like any developing country, is largely due to poor planning, inadequate flood control mechanisms and uncontrolled solid waste disposal in cities.

Causes of flooding include natural factors such as heavy rainfall, high tides, storm surges, embankment failures and low infiltration capacity of the soil, among others (White, 2010). But principally urban flooding in developing countries is due to poor drainage, disorganized building layout, unplanned development, improper management of solid waste and a host of other anthropogenic activities.

Flood control refers to all methods adopted to reduce or prevent the negative effects of flood waters (Alcoforado, 2018). The role of spatial paining in the control and management of floods in urban areas is very germane both in preparedness and the response to flood disasters. Unfortunately, however, physical planning in many Nigerian cities is very weak, and political actors often do not consider planning a major ingredient of effective city management. Most cities in developing countries grew organically, which means they have a carryover of poor planning from the old cities. Investors and developers often feel above the law when it comes to obeying building code and planning standards in the development of their building and rebuilding activities. The majority of buildings in cities have one infraction or contravention or another in the building. The population increase from urbanization results in more pressure on the demand for accommodation to house new arrivals and those who are moving to higher social status. Consequently, the increased urban hard landscape results in more impervious surfaces and reduced infiltration capacity of the soils.

The two major approaches for managing urban floods are structural and non-structural approaches. The structural approach includes construction of dams, drainage, dikes, ramps and other control measures to retain excess water and allow it to gradually flow out without causing any damage to the human population (Dawson et al., 2011; Butler & Pidgeon, 2011). The other approach to flood control is spatial planning, development control flood forecasting, early warning systems, flood risk and hazard management strategies and urban solid waste control, among other measures.

However, central to both structural and non-structural strategies/approaches is the effective deployment of spatial planning tools and institutional mechanisms.

Spatial planning tools in flood management

Spatial planning has been described as the art and science of ordering the use of land, siting of buildings and managing building character for the purpose of securing economic convenience, beauty or branding for human settlements (Keeble, 1969; Fabiyi, 2016). It is a rational and systematic process of guiding public and private actions and influencing the future by identifying and analyzing alternatives and outcomes. Flooding is a natural event that is within the purview of projection and predictive modelling. When the future possible occurrence of flood disaster is understood, it is essential to integrate this into spatial planning for the future growth of cities.

Spatial planning arranges the physical space in urban areas for the current population and future inhabitants of cities. The future space requirements of cities are projected through spatial analytical techniques. As the city grows, the hard landscape also increases, which will inevitably increase the run-off and the potential for increased flooding. Scientists and climate experts have predicted intensified flooding globally from increased frequency and magnitude of extreme precipitation events (Dankers & Feyen, 2008; Hirabayashi, Kanae, Emori, Oki & Kimoto, 2008). The climatic parameters resulting in extreme flood events have also been well understood, and the potential disasters from flooding have been adequately modelled to the effect that the world is targeting reducing greenhouse gases (GHGs) to pre-industrial level of less than $2°$ per annum.

The combined knowledge of the climatic projection and the future land consumption of cities could greatly help in deploying an effective flood preparedness systems for the city through a combination of urban flood management and urban planning (Price & Vojinovic, 2008; Chang, Tseng & Chen, 2007; Macharis & Crompvoets, 2014; Alexander, 2013).

Spatial planning, when effectively carried out, can contribute to flood mitigation because it can influence, re-order or divert flooding events through river training to locations that have less negative effects and shelter developments from flooding events (Neuvel & Van Den

Brink, 2009; Howe & White, 2004; White & Richards, 2007; Van Heezik, 2008). Spatial planning in flood control can apply to local or regional planning to control river systems. Usually river channels pass through different regions, and therefore spatial planning can prevent or moderate the effects of flooding at the national or regional scale.

Spatial planning has been categorized into different types based on the major ingredients and tools deployed for urban systems administration. These are:

- *Strategic planning*
- *Land use planning*
- *Master planning*
- *Urban revitalization*
- *Infrastructure planning*

The first two spatial planning types are germane to achieving an effective flood preparedness program in cities.

Strategic planning usually focuses on setting high-level goals and a desired future, a picture of a utopian ideal future that urban gatekeepers have for the settlement. It itemizes the core strategies to achieve the mission and vision of the urban actors and gatekeepers. A typical example of a vision could be a city that is free of flooding and minimizing preventable natural disasters. The strategic objectives to achieve the identified vision may include efforts, activities and the institutions concerned to achieve the objectives, vision and mission designed for the urban future.

Land use planning concerns the setting up of regulatory policy, government statutes, building codes, planning standards and the general layout of the zoning. It may include maps of broad classification of the entire area into residential, commercial, industrial and institutional zones. It may allocate broad spaces for these issues and may not require public acquisition of the land area, but the plan when approved by the government shall guide developers in the choice of land use and the limitations for different types of development in those areas. There are standards for setbacks to roads, rivers and other natural features, Building codes would also specify the facilities to be provided in different types and forms of buildings. This planning tool could be used to make provision for the free flow of excess storm water in flood-prone areas of the city during projected heavy downpours.

Development should be arranged in such a way that waterways are not impeded by developers, while land uses that could be resilient to flooding could be located close to floodable water. Such land uses include trails, recreation areas and agroforestry. This seeks to arrange the physical space and guide future activities according to sustainability and other accepted principles.

Specific tools in planning for flood preparedness

There are provisions in urban planning space standards and building codes which are useful for the prevention of flooding in urban areas. The following are major planning tools that are used in urban areas to prevent flooding.

River setbacks

Setbacks are planning provisions that prescribe distances around river banks where building developments are prohibited. There are different prescribed setback distances depending on the categories of river order. The specification of a river setback is the minimum distance to the river bank where development is allowed, which in Nigeria varies between 15 and 45 meters.

The river setback requirements ensure that the flood plane is protected from building activities so that there can be free flow of excess storm water during excessive rainfall or release of excess water from upper stream embankments. It is noted that excess water in rural areas and the forest has fewer negative effects due to low resources at risk compared to excess water in cities. Flooding in cities also results in great discomfort to urban residents, loss of life and material resources. Enforcement of river setback regulations ensures that the potential for a flood disaster in the city is substantially reduced. Town planners are charged with the responsibility and police power to prevent developers from building within the approved setback minimum distance from the river channels.

Zoning

One of the major spatial planning tools often used to prevent flooding in urban areas are zoning ordinances, which often take off from the master plan or land use plan for the city. Zoning ordinances or a land use plan consist of a master plan for land use within the municipality. It divides the municipality into residential, commercial, industrial, administrative and other zones. These zones usually arrange urban land uses in such a way that the flood zones in the city regions are allocated to land uses that are resilient to flooding such as trails, recreation parks, and orchards, among others.

Municipal spatial data infrastructure

Municipal spatial data infrastructure (SDI) refers to the repository of all spatial data useful for effective management of urban systems. The SDI is often backed by a spatial data sharing protocol or arrangement to ensure responsible use of spatial data and to introduce spatially enabled governance in the city. Most urban gatekeepers in Nigeria have data sets in their archives that are produced by them or acquired to assist them in the management of the city systems. Unfortunately, however, most of these data operate in silos and are utilized by their respective ministries, departments and agencies without interactions with other data sets. The individualization of acquisition, storage and use of spatial databases in the city is one of the major reasons urban governance in developing countries is still very weak and incapable of preventing flood disasters. It also hinders timely response to flood disasters in the city. A spatial-enabled city will produce effective flood governance and is an effective way of reducing the incidence of flooding in the city. When data are shared among the relevant agencies, early response to flood events and flood warnings would be more effective.

Development control

Development control relates to the process of implementing and maintaining building and land sub-division regulations, space standards and building code specifications. It involves the activities of the regulating agency to secure orderly and efficient urban systems. Nigerian Urban Development Decree no. 88 of 1992, amended as the Nigerian Urban and Regional Planning Act 1999, defined development as

> carrying out of any building, mining or other operations in, on over or under any land; the making of any material change and the use of any land or building structure, or conversion of land, building structures from its established or approved use and/or including the

placing or display of advertisement and urban furniture on; the making of any environmentally significant change in use of any or demolition of building including felling of trees.
(Nigerian Urban and Regional Planning Act, 1999; Fabiyi, 2006; Ezra, Kyom & Balasom, 2013)

Development control is a tool used by planners to maintain standards and enforce specifications provided by relevant legislation which regulates the development of land and building. It is professional activities carried out by urban and regional planners to ensure that development follows the provisions laid down by the city master or land use plan in the city.

The main objectives of development control are to regulate building and rebuilding operations, to ensure orderly growth of cities, protection and enhancement of the built environment and coordination of both public and private investment in land and property to achieve efficiency (Ogunsesan, 2004; Ezra, Kyom & Balasom, 2013). Development control is the legal authority conferred on the local town planning officer to enforce compliance and achieve efficient functioning of urban systems. Development control serves as a frontline tool by town planners to ensure that building and rebuilding activities are not taking place in flood-prone areas and to ensure that there is free flow of excess storm water in cities

Urban waste management infrastructure plan

The importance of an urban waste management infrastructure plan cannot be overemphasized in the control of flood disasters. When waste collection arrangement and the deposition of waste is properly synchronized in the city, it will enhance efficient control of storm water in the city and reduce possible flood occurrence. A majority of drainage blockage and canal siltation is caused by indiscriminate dumping of household solid waste. Solid waste management is a major problem in Nigerian cities because most urban poor in the cities cannot afford the fee for waste collection; therefore they practice dumping instead. It is usual practice for urban poor to dump their waste into natural and humanmade drainage channels. The resultant effect is blockade of waterways and narrowing of urban drainage networks. One of the effective tools to limit flood disasters is a comprehensive urban solid waste infrastructure master plan. This will be done after a critical study of the waste generation of the city and waste disposal behavioral pattern of the urban residents and industries in the city and periphery.

An urban solid waste infrastructure master plan will specify the appropriate locations of bins, skips, depots, dumps and sanitary landfill sites in the city region. There would be provision for an effective and efficient waste collection mechanism to ensure that every household and waste generation point in the city is well captured in the plan. The cost recovery for the private sector involved in waste collection should also be considered to ensure sustainability.

Flood risks and early warning systems

Flood risk and early warning systems are a major tool that can be used to prevent urban flooding and significantly reduce the impact of potential flood disasters. When residents are warned of impending doom from flooding, the casualties and damages could be minimized. A spatially enabled flood warning system will predict possible flood disasters before they occur and alert the people about the nature and location of impending flooding. The framework for flood risk management must of necessity take into consideration flood prediction modelling, flood risk vulnerability measurement, flood protection measures, flood prevention measures and

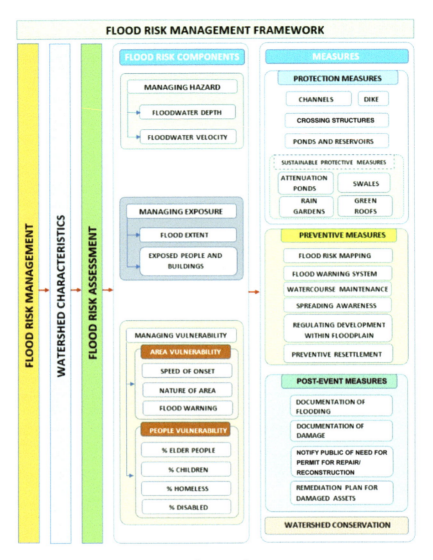

Figure 7.1 Typical flood risk management framework

post-flood disaster measures (see Figure 7.1). Managing flood risk begins with modeling of the potential of flooding and ends with activities to respond to the flood disaster and rehabilitate flood victims.

Flood risk early warning systems are a geospatial information–based solution that systematically predict upcoming flood disasters with spatial reference and alert residents some hours before the flood. It is usually in order to alert urban dwellers between 6 and 12 hours prior to an impending flood disaster and indicate the magnitude of the flooding. Some warning systems will recommend possible actions to take. An effective flood risk and early warning systems will predict spatial differentiation of flooding and the level of severity at different parts of the city. It will provide advice on what the residents should do to minimize the effects or damage and rapidly itemize resources at risk in a geospatial platform.

Flood risk and early warning systems involve meteorological and hydrogeological modeling together with the relevant land use analyses and land use forecasting. The following parameters are critical to efficient flood risk and early warning systems:

- *Rainfall:* Rainfall intensity, frequency of rainfall, onset of rain, cessation of rain and so on. This can be modelled through meteorological satellite data.
- *Hydrology:* River competence; river channel depth; river channel capacity; percentage of silt; land use around the river channels, tributaries and distributaries; river catchment and basin characteristics; and lakes and dams, estuaries and lagoons, creeks and other hydrology land forms that may affect the capacity of the existing river channels to contain excess storm water.
- *Land use*: The major land uses in the city, including urban build-up, rural hinterlands, recreation and urban green spaces and so on. These land use types determine the capacity of the urban area to absorb excess storm water. These parameters are built into the flood risk model, which could be deployed on on-site servers or cloud-based servers to monitor the environmental conditions of the project area with respect to flood disasters. There are three components of a typical flood risk and early warning system, as shown in Figure 7.2: (i) cloud-based modelling, (ii) laboratory analysis and assessments and (iii) output and public flood alert systems.

Operationalization of spatial planning tools for flood preparedness in Nigerian cities

Spatial planning tools currently operational in Nigeria have been those prescribed by the Town and Country Planning Law of Nigeria, which was amended to the Urban and Regional Planning Act in 1999. The law put the management of spatial planning activities in the concurrent schedule of Nigeria's three layers (tiers) of government. The three tiers of government have major roles to play in spatial planning of cities and rural areas. The Land Use Act also provided for ownership and governance of lands in urban and rural areas in Nigeria. The Land Use Act vested the custodianship of land in the governor of the state, who hold all land in trust on behalf of all citizens of the state. However, the control of land use and prescription of uses to which lands are put were both in the purview of both the local government and the state government through the town planning ministries and authorities. The National Planning Department of the Ministry of Works Housing and Urban Development is responsible for planning decisions on all federal lands and capital territories. There are also national building codes that prescribe space standards, planning standards, specifications for buildings and building operations in Nigeria.

The Town and Country Planning Act of 1992, amended in 1999, provided for physical development planning in Nigeria and for the protection of environment from obnoxious or unwanted development. The law provided for the establishment of development control departments and urban and regional planning tribunals. The law also prescribed the preparation and implementation of physical development plans at the federal, state and local levels. The law prescribed the establishment of the National Urban and Regional Planning Commission, state urban and regional planning boards in each of the states and federal capital territories and local planning authorities in each of the 774 local government area councils in Nigeria.

While the National Urban and Regional Planning Commission is expected to formulate national policies for urban and regional planning and prepare and implement a national physical development plan and regional and subject plans, the commission is also expected to establish and maintain urban and regional planning standards.

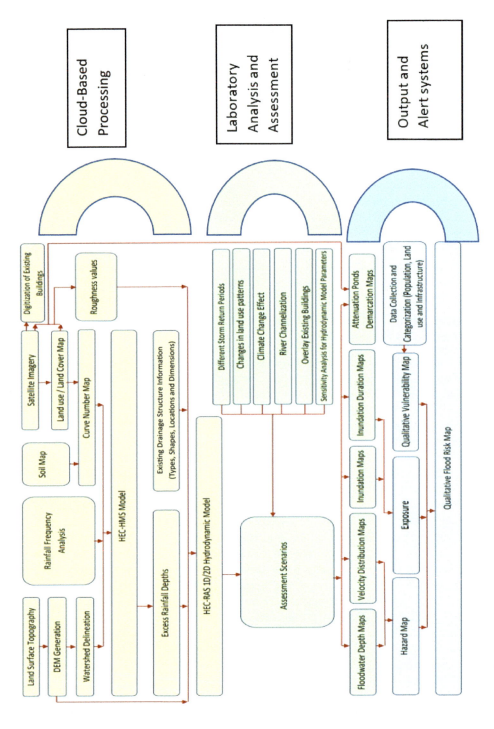

Figure 7.2 Flood risk assessment framework

The development control departments at the state and the local government planning authority are expected to control the use of land and supervise physical development from the informal and corporate sector to ensure compliance with the regulatory provisions of the space standards and building codes, among others.

Part of the duties of development control units of both the state and the local government areas is that every developer should submit, together with development plans/building plans, a detailed environmental impact statement. Developers are expected to submit a site plan of the development which shows the specific setbacks from rivers and other infrastructure such as roads, high-tension power lines and so on.

The officers empowered to operationalize the laws and provisions are given police power to enforce compliance and to demolish any building that contravenes the provision of the law. The police power is backed up by the executive power of the governor and the chairman of the local government area, respectively.

No developer can commence building operation without clearance from the statutory agencies that are empowered by the law to grant permits. Any building that fails to comply with regulations would be demolished and the cost of demolition paid by the contravening developers. The demolition can be partial or total, depending on the nature of the contravention.

When the setback of rivers is encroached upon or a building development blocks the free flow of river channels, the responsible agencies are empowered to demolish the building and evacuate the debris to allow the free flow of storm water.

Spatial planning practice in Nigeria and its implications for flood control

The bulk of spatial planning in Nigeria takes place at the local planning authority. The local town planning authorities in the local government area council are primarily responsible for the control of development at the city level, including land use ordering for development on private and communal lands, approval of building plans, preparation of land use schemes, provision of infrastructure and upgrading of slums in the city together with the rehabilitation of urban infrastructure. Despite these enormous responsibilities, the planning authorities in the local government area council are often poorly staffed and inadequately equipped and lack the financial capacity to effectively carry out their responsibilities. In order to overcome the weak financial status of most local government authorities, some states in Nigeria have established urban development boards or commissions which draw resources and operational expenses directly from the governor's office. These commissions are often charged with the responsibilities of holistic physical development and planning for all cities within the state.

Though the state government's Ministry of Lands and Urban Planning is better equipped with requisite personnel and material resources, its activities are limited to making planning policies and laws for the state and management of state-owned estates and new towns. Some states have a capital development authority to handle planning and development issues in the state capital.

The local government planning authority is responsible for the following, among other responsibilities:

- Process and issue all planning permits based on the extant laws and regulations guiding physical planning
- Monitor and ensure compliance with the provision of existing development plans (if available) guiding development of towns and rural areas

- Prepare district plans
- Development of guide plans
- Town and local plans
- Keep records of planning permit applications granted, rejected or withdrawn and publish the list in the official gazette of the state
- Evaluate physical planning technical reports
- Prevent illegal development
- Engage in stakeholder consultation, enlightenment and publicity

These roles of the local planning authority to ensure that cities are without flood disasters cannot be realized if the staff are not equipped to deploy the various tools highlighted in the section 'Specific Tools in Planning for Flood Preparedness'. Unfortunately, however, many Nigerian cities still continue to experience flood disasters on a regular, annual basis due to some of the reasons highlighted in the next section.

Causes and factors or urban flooding in Nigerian cities

Rapid urbanization and need for space for urban poor housing

Rapid urbanization in developing countries is a major factor in the increasing vulnerability of urban land to flooding. As urban population increases, the concentration of humans and materials in areas previously dedicated to other land uses because of their vulnerability to flooding becomes a contentious issue for urban gatekeepers. The poor population that migrates to the city from rural areas or smaller cities is attracted by the economic opportunities in the city but are ill equipped for the job options in cities. They thus become unemployable or have less capacity to participate in the urban economy. Therefore, due to low financial resources, they cluster in (colonize) available unoccupied territories in the city, usually abandoned mines, unoccupied river setbacks or designated green zones in the city or city fringes. Once they are settled in a river setback, the government usually lacks the political will to remove them from the places of illegal occupation, and human right activists seem to be on their side, insisting on alternatives before relocations can be effected. This result in a buildup of squatter development and encroachment on flood plains.

Sizes of sanitation and storm water infrastructure in the city

The sizes of storm water infrastructure such as culverts, drainages and bridges in the city are usually outdated and old in view of the dynamics and the requirements of the rapidly changing urban systems. The design and implementations of storm water infrastructure in the city do not adequately provide for the growth of cities in the future; therefore, as the city grows, the runoff increases and the storm water goes beyond the capacity of the drainage systems and river channels. Consequently, storm water overtops drainages and canals, while excess storm water fills the road network and houses.

Weak urban governance and development control

The enforcement of planning standards and strict compliance with the space standards in most cities are often weak. Developers are allowed to block humanmade drainage and even natural drainage during building operations. Officials often turn a blind eye as a result of the developers' position in society. The developers involved usually build heavy embankments on river channels to prevent water from coming into their properties, but this results in narrowing the drainage channels and provides velocity to storm water, to cause havoc downstream.

Low finance to provide enduring flood risk infrastructure

Preventing flooding in cities through a structural approach requires heavy and continuous investments from the government and private sectors. City administrations in developing countries usually lack the required capacity to invest in preventing flood disaster for cities. There is also low human capital among urban dwellers to organize themselves into a formidable team to tame the scourge of urban flooding. Most organized private sectors are often overtaxed to spare resources for high-cost social responsibilities.

Poor solid waste management

Poor solid waste management in the city is a major factor in urban flooding in Nigeria. Household waste is often indiscriminately dumped in drains and incidental open spaces. There are many accidental dump sites and unmanned skips in cities where people dump solid waste, which can remain un-evacuated for long periods. During excessive rainfall, storm water pushes the waste into drainage canals and rivers, and it gets trapped at the entrances of bridges and culverts. Once the waste is trapped, it rapidly blocks the free flow of water in river channels, and result is inundation of the upper stream. The canals and drainage systems are blocked during rainfall by waste, and the accumulation can cause flood disasters in cities.

Lack of synergy among relevant agencies

There is duplication of functions among agencies responsible for control of flooding in cities. The Ministry of Land, Ministry of Physical Planning and Urban Development, Ministry of Environment, Ministry of Works and Transport, emergency management agencies, solid waste management board and public health are the affiliated agencies and ministries that are responsible for different aspects of flood control in the city. Unfortunately, however, these agencies do not work in tandem to ensure prevention of flooding due to strict adherence to their respective mandates.

Lack of land use plan or master plan that addresses the issue of flooding

Only a few cities in Nigeria have a comprehensive master plan that specifically addresses flooding. This is required to ensure that the issues of flooding are taken into consideration in the future growth of the city through massive public infrastructure and strict adherence to legal provisions and statutes that guide urban development.

Spatial data infrastructure for planning data sharing among multiple agencies for flood control

According to Kuhn (2005), SDI is a coordinated series of agreements on technology standards, human resources and related activities necessary to acquire, process, distribute, use, maintain and preserve spatial data. The implementation of SDI enables relevant government officials to share data that can be used to manage flooding in cities (Williamson, Rajabifard & Feeney, 2003). A robust SDI provides the location of spatial data sets that could be used to model and predict

Oluseyi O. Fabiyi

Figure 7.3 Ibadan metropolitan area in Nigeria

scenarios for floods in cities. It will also provide a flood preparedness action plan for handling excessive flooding in cities.

There is usually poor coordination between the ministry responsible for planning and environmental management as well as the meteorological agencies, emergency management agency and waste management in the city. When a robust SDI is deployed on a robust SDI platform, distribution and access to data by Ministries Departments and Agencies (MDAs) and the general

Spatial planning to support preparedness for flooding

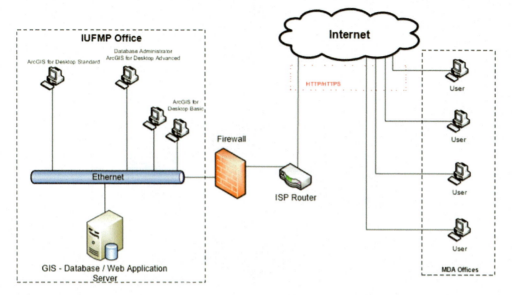

Figure 7.4 Architecture of spatial data infrastructure deployed in Ibadan by IUFMP

public will be enhanced. The SDI will also enable an efficient flood alert system that is based on location specificity.

A typical example of an SDI for flood management is the Ibadan Urban Flood Management Project (IUFMP). The project is a flood management project financed by the World Bank credit facility to address the issues of perennial flooding problems in Ibadan city. Figure 7.4 shows the deployment of the SDI for data sharing among MDAs in Oyo state for flood-related events and data.

Implementation of SDI in Ibadan city for flood control

The implementation of SDIs in Ibadan city was supported with the establishment of three GIS laboratories in different agencies. The laboratories have commentary hardware and software for spatial data management. The facilities in each laboratory include three workstations, one server and five standalone systems. These system are linked with a LAN network and also have broadband internet facilities. Each laboratory has a separate public IP address through which data can be accessed by different departments in the government and the public that have view or download access to any data set. The laboratories were also looped through their hardware servers. A geoportal was developed to publish data that are for public consumption and to specifically inform the public about the flooding situation in the city. Figure 7.5 shows the linkages between different agencies and the GIS data hub provided by the three laboratories. Training was provided for different levels of personnel to manage the SDI and the upload of fresh data to the platform for effective flood control and management.

Challenges for implementing spatial planning tools for urban flood management in developing countries

From the foregoing, there are adequate legal provision and administrative instruments to ensure orderliness and effective control in Nigerian cities; however, there are obvious reasons many of

Figure 7.5 Data sharing among MDAs for flood management in Ibadan city

our cities are still being inundated yearly by excess flood water, resulting in human and material losses.

The next section presents some reasons effective flood preparedness is elusive in many Nigerian cities.

i. There are weak human and material resources at local planning authorities to secure proper urban administration and enforce compliance with town planning regulations. Nigeria does not practice mayoral systems where a single authority would be in control of urban management. Most cities have been fractionated into different local government areas, which mean the city is under many local administrations and may have conflicting goals and policies.
ii. The local planning authorities in the local government areas that make a city are often working at cross purposes or at best competing with each other for revenue generation while the issues of effective city administration, especially in the area of flood control, are left to suffer.
iii. Controlling floods in the city requires multiple agencies in a seamless interdisciplinary arrangement; unfortunately, most government agencies restrict their activities to their mandates and are usually reluctant to share data or responsibilities with other agencies. The budget approved for flooding is usually spent on unrelated activities such as off-shore training and ephemeral contracts. Government agencies are reactionary to flood disasters since it gives visibility to the government as an entity that cares when palliative items are distributed to flood victims.
iv. High rate of staff turnover in MDAs through promotions, transfers and retirements. Flood control is a specialized disciplinary area, and it requires extensive and intensive training of qualified personnel; however, when trained staff leave the office, investments on capacity building are often wasted, and training has to begin anew.

v. The political will to enforce spatial planning provisions and specifications is lacking among political office holders; therefore, developers disobey the extant rules with reckless abandon while the residents downstream are faced with the consequences of the actions of developers upstream.

Conclusion

This chapter presented the major spatial planning tools that could be deployed for flood management in cities, especially in a low-resource economy. Spatial planning tools and geospatial technologies have been integrated in flood management in most developed countries (Jing Ran & Nedovic-Budic, 2016). The relevance of spatial tools and other geospatial techniques to effective flood control was succinctly presented in the chapter. The issues surrounding effective flood control and management of city flood plains were specifically addressed.

Effective deployment of spatial planning tools and other geospatial instruments on the backbone of efficient SDI in flood management in cities will result in the following achievements in cities in developing countries:

- A formidable platform for monitoring city growth and changes in the urban landscape with respect to the capacity to absorb shocks such as excessive storm water and flooding
- More resilient cities will emerge with adequate provisions to guide development of spatial projections including demographic, economic and land consumption and thus plan for containment of excess storm water in the future
- Appropriate deployment of spatial planning tools will prevent new development in flood-prone areas
- It will improve management of upstream catchment areas so that watersheds can be protected against entrapment
- Urban slum and squatter settlements will be rehabilitated and revitalized based on the modern planning approach that seeks to help the poor in the city upgrade their residences through the integration of corporate private developers in urban upgrading and renewal

References

Alcoforado, F. (2018). Flood control and its management. *Journal of Atmospheric & Earth Sciences*, 1, 005. Printed from Oxford Research Encyclopedias, Natural Hazard Science.

Alexander, D. E. (2013). Emergency and disaster planning. In A. López-Carresi, M. Fordham, B. Wisner, I. Kelman, & J. C. Gallard (Eds.), *Disaster management: International lessons in risk reduction, response and recovery* (pp. 125–141). London: Routledge.

Butler, C. (2008). Risk and the future: Floods in a changing climate. *21st Century Society*, 3, 159–171.

Butler, C., & Pidgeon, N. (2011). From 'flood defence' to 'flood risk management': Exploring governance, responsibility, and blame environment and planning. *Environment and Planning C: Government and Policy*, 29(3), 533–547.

Chang, M. S., Tseng, Y. L., & Chen, J. W. (2007). A scenario planning approach for the flood emergency logistics preparation problem under uncertainty. *Transportation Research Part E: Logistics and Transportation Review*, 43(6).

Dankers, R., & Feyen, L. (2008). Climate change impact on flood hazard in Europe: An assessment based on high-resolution climate simulations. *Journal of Geophysical Research: Atmospheres*, 113(D19).

Dawson, R. J., Ball, T., Werritty, J., Werritty, A., Hall, J. W., & Roche, N. (2011). Assessing the effectiveness of non-structural flood management measures in the Thames Estuary under conditions of socio-economic and environmental change. *Global Environmental Change*, 21(2), 628–646.

Ezra, V. L., Kyom, B. C., & Balasom, M. K. (2013). Nature, scope and dimensions of development control, tools and machineries in urban planning in Nigeria. *International Journal of Innovative Environmental Studies Research*, 1(1), 48–54.

Fabiyi, O. O. (2006). Urban land use change analysis of a traditional city from remote sensing data: The case of Ibadan metropolitan area, Nigeria. *Humanity & Social Sciences Journal*, 1(1), 42–64.

Fabiyi, O. O. (2016). Urban space administration in Nigeria: Looking into tomorrow from yesterday. *Urban Forum*, 28(2), 45–68. DOI:10.1007/s12132-016-9299-3

Hirabayashi, Y., Kanae, S., Emori, S., Oki, T., & Kimoto, M. (2008). Global projections of changing risks of floods and droughts in a changing climate. *Hydrological Sciences Journal*, 53(4), 754–772.

Howe, J., & White, I. (2004). Like a fish out of water: The relationship between planning and flood risk management in the UK. *Planning Practice and Research*, 19(4), 415–425.

Jing Ran, J., & Nedovic-Budic, Z. (2016). Integrating spatial planning and flood risk management: A new conceptual framework for the spatially integrated policy infrastructure. *Computers, Environment and Urban Systems*, 57, 68–79.

Keeble, L. (1969). Principles and practice of town and country planning. *Estates Gazette*, 72.

Kuhn, W. (2005). *Introduction to Spatial Data Infrastructures*. Presentation held on March 14 2005. Available from: http://www.docstoc.com/docs/2697206/Introduction-to-Spatial—DataInfrastructures. Accessed July 17, 2022.

Macharis, C., & Crompvoets, J. (2014). A stakeholder-based assessment framework applied to evaluate development scenarios for the spatial data infrastructure for Flanders. *Computers, Environment and Urban Systems*, 46, 45–56.

Neuvel, J. M. M., & Van Den Brink, A. (2009). Flood risk management in Dutch local spatial planning practices. *Journal of Environmental Planning and Management*, 52(7), 865–880.

The Nigerian Urban and Regional Planning Act, 1992, *The Nigerian Urban & Regional Planning Law Number 88*, Nigerian Institute of Town Planners. Ministry of Information and Culture Printing Department, Lagos Nigeria, 1992.

Ogunsesan. D. K. (2004). The process and problems of development control and some remedial solutions. *A Multidisciplinary Journal Environscope*, 1(1).

Pidgeon, N., & Butler, C. (2009). Risk analysis and climate change. *Environmental Politics*, 18, 670–688.

Price, R. K., & Vojinovic, Z. (2008). Urban flood disaster management. *Urban Water Journal*, 5(3), 259–276.

Van Heezik, A. (2008). *Battle over the rivers: Two hundred years of river policy in the Netherlands*. The Hague, the Netherlands: Van Heezik Beleidsresearch in Association with the Dutch Ministry of Transport, Public Works and Water Management.

White, I. (2010). *Water and the city: Risk, resilience, and planning for a sustainable future*. Abingdon: Routledge.

White, I., & Richards, J. (2007). Planning policy and flood risk: The translation of national guidance into local policy. *Planning Practice and Research*, 22(4), 513–534.

Williamson, I. P., Rajabifard, A., & Feeney, M.-E. F. (Eds.) (2003). *Developing spatial data infrastructures. From concept to reality*. London/New York: Taylor & Francis.

8

Understanding engineering approaches to urban flood management in Indonesia

Rian Mantasa Salve Prastica, Wakhidatik Nurfaida, Amalia Wijayanti, Abdunnavi Alfath and Muhammad Sulaiman

Introduction

Flooding is one of the most common natural disasters in Indonesia. Records from the National Agency for Disaster Countermeasure (Indonesian: Badan Nasional Penanggulangan Bencana [BNPB]) database have shown that approximately 30%–40% of the natural disasters in the past decade were caused by flooding, with an increasing trend. Several cities in Java Island, including Jakarta, Ambon, Semarang, and Bogor, are always on the top list of urban flood-prone areas. In Jakarta, the capital city of Indonesia, approximately 80%–90% of disasters are urban flooding. With a high population density of approximately 13,000 people per km^2 (Martinez & Masron, 2020), floods are expected to impact many residents. BNPB reported that more than 100,000 people in Jakarta were affected by floods in the first two months of 2020 (BNPB, 2021). Local news suggested that floods might have caused more than hundreds of billions USD of infrastructure losses. Although the infrastructure for urban flood management is being developed, the problem is unlikely to vanish in the near future, especially with the rapid urbanization in the city.

Urban flooding in Jakarta is influenced by at least three physical influencing factors: increasing frequency of extreme rainfall events, land subsidence, and sea level rise. A climate study has captured a significant increase of heavy rainfall in Jakarta over the period of 1961–2010, followed by a decrease in the return period of 200-mm rainfall intensity (Siswanto, van Oldenborgh, van der Schrier, Jilderda, & van den Hurk, 2016). Located in a low-lying area, the topography of Jakarta is relatively low, with some areas below the sea water level. This condition is worsened by the land subsidence of approximately 1–15 mm per year and 20–28 mm per year in a few locations (Abidin et al., 2011). Adaptation to land subsidence is becoming increasingly difficult due to the excessive groundwater extraction for domestic and industrial use. The Intergovernmental Panel on Climate Change has projected a global sea level rise between 0.48 and 0.84 (Oppenheimer et al., 2019). Furthermore, a coastal hydrodynamic modeling study using the finite-volume coastal ocean model (FVCOM) investigated the tidal characteristics of Jakarta Bay and estimated increases in the local sea level rise under different scenarios due to the increasing tidal force (Yahya Surya, He, Xia, & Li, 2019). This condition could lead to an even higher risk of coastal flooding in Jakarta.

Flood disaster management strategies in Jakarta have been undertaken. During the 1600s–1800s, the Dutch government carried out channel straightening to increase flow velocities. After a large flood in 1918, the West Flood Canal was constructed. This flood diversion was installed to reduce discharges from one of the biggest rivers that flows to Jakarta, the Ciliwung River. After the declaration of the independence of Indonesia, many infrastructures were developed that altered watersheds and reduced water infiltration. Then several retention ponds and the East Flood Canal were also constructed, followed by the implementation of other strategies, such as river canalization, channel sediment dredging, river capacity improvement, revitalization of retention ponds, and construction of sea dikes. Further disaster management strategies are still being considered for the near future to anticipate increasing flood risks, such as the outer sea wall or giant sea wall.

Understanding the complexities of urban flood management approaches in Jakarta would be useful for other cities. Accordingly, this chapter will try to cover several engineering approaches of flood management in Jakarta. In addition, it will describe another case of urban flooding in Wai Batu Merah, a small city with a high population density in Ambon, Maluku Province. Moreover, this chapter will explore the mathematical modeling approaches that are often used in analysis and design.

Overview of the research method: mathematical modeling approaches

Mathematical models are classified as analytical or numerical models based on the form of calculation used. Analytical solutions are solutions in which answers are obtained through a direct application of mathematical expressions or equations that represent a physical phenomenon. Analytical models can usually be completed at various locations manually or using a computer program on a channel network with floodplains.

Many alternatives can be used to create river modeling with the provided software, which is accessible worldwide. Flood modeling can be classified according to the model dimension and the availability, where it is open source or commercially used. Table 8.1 shows a brief description of several software options for flood modeling. Based on the practices in the field, this chapter elaborates on flood modeling using HEC-RAS and Personal Computer Stormwater Management Model (PC-SWMM) for urban flooding in Jakarta and Ambon, respectively.

Results and discussion

Flood mitigation strategies in urban cities (study case: Jakarta)

Located in the downstream area of several rivers, Jakarta receives a huge amount of runoff water from the watershed. A total of 13 main rivers flow across the city, including the Ciliwung River, with a basin area of more than 430 km^2 (Figure 8.1). The river discharge from the Ciliwung River is actually one of the two core problems of urban flooding in Jakarta, with coastal flooding as the second. Therefore, studies on the Ciliwung River have drawn the attention of engineers, policy makers, and researchers. Governments have also tried several approaches to tackle the disaster (Asdak, Supian, & Subiyanto, 2018; Lin, Shaad, & Girot, 2016). To obtain a better understanding to this approach, the Ciliwung River in Jakarta is used as an area of study.

Many scenarios have been proposed from the government or researchers in the form of regulations or policies and scientific papers, which, however, actually received less attention from

Table 8.1 Flood modeling software characteristics

Software	Mathematical equation used in the model	Modeling purposes	Benefits using the model	Limitation or boundary condition	Dimension of the model	Access to the software	Correlated research
HEC-RAS	One-dimensional energy equation	Shallow water, steady flow, unsteady flow	Wide range of parameters used in the model, user friendly, good flexibility	Certain conditions need multidimensional modeling	1D, 2D	Open source	(Ezzine et al., 2020; Geravand, Hosseini, & Ataie-Ashtiani, 2020; Prastica et al., 2018; Prastica & Wicaksono, 2019; Quirogaa, Kurea, Udoa, & Manoa, 2016; Zellou & Rahali, 2017)
LISFLOOD-FP	1D kinematic equation 2D diffusive wave equation	Hydrological analysis in geographic information systems environments	User friendly, spatial-based visualization, extensive documentation	Spatial and topographic data for modeling need high resolution to obtain better results	1D, 2D	Open source	(Luke et al., 2015)
HEC-HMS	Continuity and momentum equation	Simulation or analysis for precipitation and runoff in certain catchments and hydrodynamic modeling	Hydrological application for a number of alternatives; the software can be integrated with other software, and the computer tools have extensive documentation features	The model is not suggested for flood simulation	1D, 2D	Open source	(Rangari, Sridhar, Umamahesh, & Patel, 2019)
MIKE-11	1D Saint Venant equation, momentum equation, and continuity equation	Flow simulation, water level analysis, water quality modeling, sediment transport in water bodies, irrigation engineering, and reservoir modeling	Extensive documentation, river modeling with comprehensive features	Limited to river flooding events; 2D modeling is not recommended	1D	Commercial use	(Billa, Mansor, & Mahmud, 2004; Bisht et al., 2016; Jamali, Bach, & Deletic, 2020)

(Continued)

Table 8.1 (Continued)

Software	Mathematical equation used in the model	Modeling purposes	Benefits using the model	Limitation or boundary condition	Dimension of the model	Access to the software	Correlated research
MIKE-21	2D shallow water equation	Flow simulation, wave simulation, sediment transport analysis, ecology in rivers, lake analysis, estuary modeling, estuary simulation and sea surface modeling	Hydrodynamic modeling, flow velocity in various directions of flow	Calibration is needed	2D	Commercial use	(Khadr & Ahmed, 2016; Panda, Mahanty, Ranga Rao, Patra, & Mishra, 2015)
SWMM	Finite differential equation	Modeling for surface runoff, groundwater analysis, contaminant modeling	Extensive documentation, adaptive to hydrodynamic operation	Spatial and topographic data for modeling need high resolution to obtain better results	Generic	Open source	(Baek, Ligaray, Pachepsky, Ahn, & Yoon, 2020; Kourtis, Tsihrintzis, & Baltas, 2018; Li et al., 2019; Xu, Jia, Wang, Mao, & Xu, 2017; Zhao et al., 2019)
FLO-2D	1D, 2D shallow water equation	1D flow in open channels, 2D flow in the floodplain, solving the dynamic equation	Combination of hydrological and hydraulics modeling for urban and river flooding	Limited analysis for bridge and culvert computation, which need external data for National Flood Insurance Program usage	2D	Open source	(Erena, Worku, & De Paola, 2018)

Engineering approaches to urban flood management in Indonesia

Figure 8.1 Jakarta area and the river basins flowing to the area

policy makers. Unacceptable flooding scenarios are caused by many factors: funding (Yildirim & Demir, 2021), political views, and vertical or horizontal conflicts in the area (Prastica & Wicaksono, 2019). The proposed alternatives of flood mitigation strategies in the Ciliwung River are normalization or canalization, a green infrastructure (GI) scenario, and a partial normalization scenario.

In the engineering approach, two compulsory components are used to make the model work. The first component is hydrological aspects, such as rainfall data in climate or geophysics station. Hydrologic analysis aims to quantify the water volume in the watershed or coverage study area. The amount of water needs to be balanced from the perspective of water resources management, which defines that water is practically constant in Earth in a certain time period. Hence, hydrological components, such as water balance equations, are necessary to create the water balance. The main elements of the water balance system are as follows (Bogardi, 2021): precipitation, evaporation, transpiration, sublimation, percolation or infiltration, surface and subsurface runoff, and vapor transport.

In this chapter, flood volume is analyzed using the Nakayasu synthetic unit hydrograph (Kang et al., 2009; Prastica et al., 2018). In brief, flood discharge data that have been obtained will become input data to the HEC-RAS modeling software. The HEC-RAS modeling in this study constructed a model of the river section in STA 0 + 00 or downstream (located in Matraman District) to the upstream point with STA 23 + 618 (located in Pasar Rebo District).

The next procedure is determining the hydraulic geometry of the river, where the coordinates can be extracted as the drawing points of the model in the HEC-RAS software. The

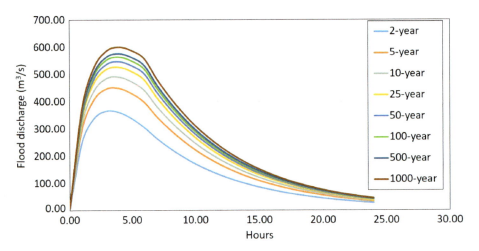

Figure 8.2 Flood hydrograph in Ciliwung watershed based on the SCS unit hydrograph

coordinates of the existing cross-sectional image of the Ciliwung River range from STA 0 + 00 to STA 0 + 050 in the section of the Pasar Rebo–Matraman District based on the existing cross-sectional image from River Basin Agency, locally called as BBWS, of Ciliwung Cisadane. The next data needed are the flood discharge, which can be obtained from the Nakayasu synthetic unit hydrograph, as depicted in Figure 8.2.

After entering the station and elevation points, the next step is to fill in the parameters and then run the existing cross-sectional design with a steady flow status. We fill in the steady flow edit section with the return period discharge or hydrological data from Figure 8.2. In HEC-RAS, the flood discharge status is the maximum value in each return period. The ready status HEC-RAS from this step can be automatically generated from the software. When the software successfully reads the data parameters, the program can visualize the existing cross section with the available flood discharge data. In brief, the existing hydraulic condition of the river cannot accommodate the flood volume occurring in the watershed. Hence, along with the policy-making process, mitigation strategies for this study area are needed. As a well-known structural mitigation strategy in Indonesia and other developing countries, river normalization or canalization is simulated using HEC-RAS to obtain more understanding about the normalization strategy toward rivers' existing conditions.

Furthermore, Table 8.2 identifies the flood height estimation extracted from the HEC-RAS running result. The results will help decision makers easily formulate mitigation strategies in the study area.

In the two-dimensional modeling of the Ciliwung River runoff, an unsteady flow simulation is used to make it more suitable for natural flood events. Unlike the steady flow simulation, unsteady flow takes into account the parameters that affect the flow of discharge in the river, such as elevation, river shape, and energy. In this unsteady flow simulation, the model of the river starts from STA 6 + 500 in the Jatinegara–Matraman sub-district. The complicated shape of the river flow makes it difficult to determine the image of the river flow according to the cross-sectional axis of the river. This unsteady flow simulation is simulated with a time span of 2 days or 48 hours (June 22, 2021–June 24, 2021) with a return period of 100 years. The initial program is depicted in Figure 8.3. The peak of the flood occurred at 2:00 PM on June 22, 2021. At 00.00 AM, June 23, 2021, the flood discharge began to decrease, and the water level began to fall.

Engineering approaches to urban flood management in Indonesia

Table 8.2 Flood height estimation

STA	Left border (meter)	Right border (meter)
23+618	—	0.15
19+600	0.72	2
12+800	1.89	1.85
12+300	1.18	1.48
9+150	0.94	1.85
4+050	2.21	2.21
0+550	1.11	1.04

Figure 8.3 Unsteady flow simulation at 00:00 AM, June 22, 2021

Flood disaster mitigation

The HEC-RAS simulation results show several flood points on the Ciliwung River. This phenomenon occurred because the Ciliwung River cannot accommodate the design flood discharge. Hence, efforts are needed so that the flowing water does not run out of the river cross section and flood the area around the river. One of the mitigation efforts that can be done is river normalization. River normalization is a method to form a cross section of the river to meet the flow of the river water so that it does not overflow. Normalization, also known as canalization, is also performed because the capacity of the river is gradually shrinking, caused by the silting and narrowing of the river body by settlements. In this chapter, the design of the cross-sectional shape of the river channel is as follows: (a) The river's cross-sectional depth is as high as 8 m from the ground, (b) the channel edge or side slope (Z) is made with a 1:2 slope, and (c) the bottom of the river is 20 m wide. Figure 8.4, which consists of three images, is a simulation of the cross-sectional water level comparison of the Ciliwung River between the existing condition and normalization scenario.

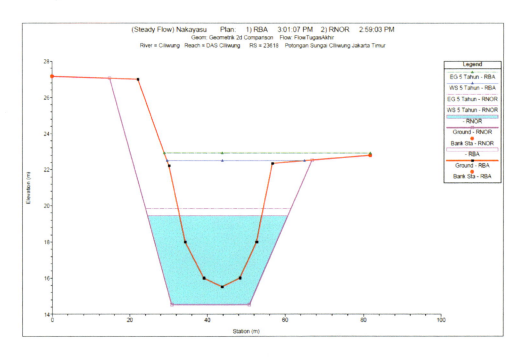

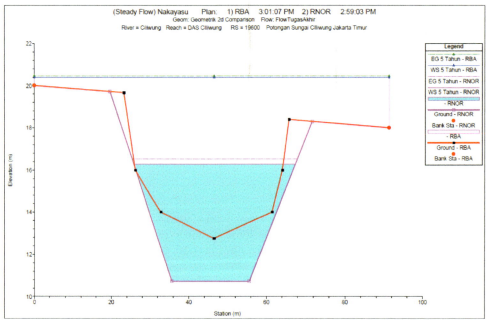

Figure 8.4 Comparison of the canalization scenario and existing condition of the Ciliwung River

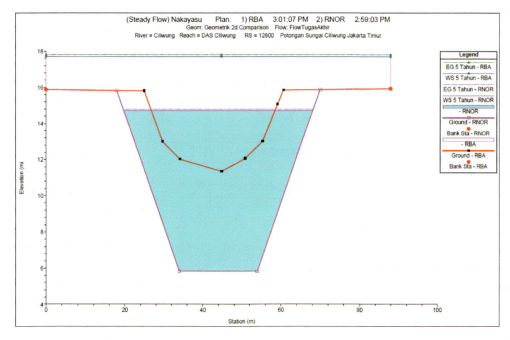

Figure 8.4 (Continued)

Flood disaster mitigation strategies using PC-SWMM modeling (case study: Wai Batu Merah, Ambon)

Instead of the gray infrastructure approach, GI offers beneficial results and hence is a better alternative for flood mitigation in urban areas. GI is also referred as low-impact development (LID), water-sensitive urban design, integrated urban water management, best management practices, and sustainable urban drainage system. The concept is widely used worldwide due to its cost-effective way of reducing water contamination and runoff in urban landscapes.

To understand how GI can help drainage systems tackle runoff and contamination problems, an approach for a flood mitigation scenario is employed, such as PC-SWMM. It is a water resource management modeling software developed by Computational Hydraulics International. PC-SWMM can be used for clean water supply solutions, drainage design and GI, floodplain delineation, overflow mitigation from sewers, and water quality and watershed analyses in 1D and 2D modeling. In modeling the PC-SWMM approach, a study area of the Wai Batu Merah watershed is introduced, as depicted in Figure 8.5.

Drainage analysis

The Wai Batu Merah watershed has an area of 698 hectares, with the main river called the Batu Merah River. The center of the population density is in the downstream part of the watershed, which is also a coastal area. There are 13 sub-watersheds, with the area and length of the river shown in Figure 8.5. In this flooding mitigation approach, the sub-watershed is divided into 13 catchment areas for model construction and mitigation strategies implementation, symbolized with codes A, B, C, D, E, F, G, H, I, J, K, L, and M.

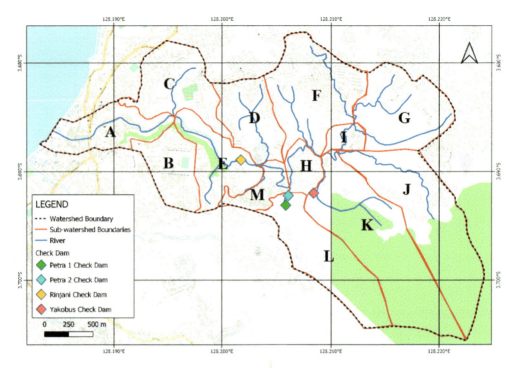

Figure 8.5 Subcatchment division analysis in the Wai Batu Merah watershed

After understanding the watershed system and river system and observing land use in the Wai Batu Merah watershed, the areas that will be served by drainage or in the PC-SWMM program called subcatchment can be determined. The subcatchment area is determined based on areas that have land cover, such as residential areas, offices, economic centers, and industries. Observations of land cover areas were performed through the Google Earth Engine. Figure 8.5 also shows the subcatchment area to be served by the drainage plan. As shown in Figure 8.5, the subcatchment is divided into several zones, with reference to the sub-watershed system and other connecting infrastructure, namely the road system. This zoning is performed to facilitate the planning process, construction work, and drainage maintenance in the future.

Modeling

The PC-SWMM software helps visualize the drainage network plan and ensures that the capacity dimensions of the planned drainage can accommodate the discharge that occurs in the channel. The drainage master plan modeling is divided into several alternative scenarios, as presented in Table 8.3.

The planned GI component is added to reduce the runoff by collecting and absorbing rainwater. The concept of *tampung, resapkan, alirkan, pelihara* (Indonesian term) or accommodate, absorb, flow, and maintain is implemented through the application of LID, where the technologies applied are permeable pavement, rain barrel, infiltration well, and biopore infiltration holes. Among the running codes, Figure 8.6 presents a representative sample to visualize the scenario modeling results. Furthermore, the overall results can be observed in Table 8.4 to distinguish the modeling results among scenarios.

Table 8.3 Scenario of GI for flood mitigation

No	Scenario	GI condition				Flooding return period discharge (Q)
		Permeable pavement	Rain barrel	Infiltration well	Biopore infiltration holes	
1	Scenario 1					Q2, Q5, Q25, Q100
2	Scenario 2	√				Q2, Q5, Q25, Q100
3	Scenario 3		√			Q2, Q5, Q25, Q100
4	Scenario 4			√		Q2, Q5, Q25, Q100
5	Scenario 5				√	Q2, Q5, Q25, Q100
6	Scenario 6		√	√		Q2, Q5, Q25, Q100
7	Scenario 7	√	√	√		Q2, Q5, Q25, Q100
8	Scenario 8	√	√	√	√	Q2, Q5, Q25, Q100

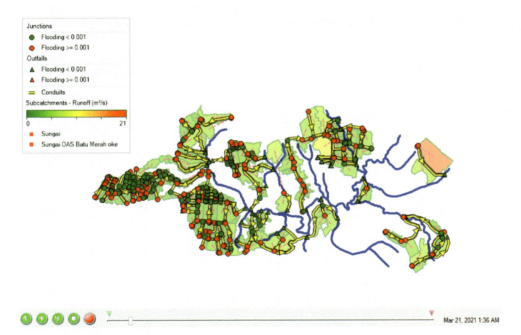

Figure 8.6 GI modeling result for flood mitigation with the implementation of Scenario 6 with flooding characteristics provided by the return period of 100-year discharge (Q100)

When there is extreme rain or rain with a long duration, the LID infrastructure will experience a decrease in capacity. The condition when these infrastructures are fully filled with rainwater is called the saturated condition. The modeling results in the PC-SWMM software show differences in the reduction of runoff that occurs when the LID is in the initial condition and the LID is in the saturated condition. Table 8.4 shows the difference in the reduction of runoff when the LID is in the initial condition compared to the LID in the saturated condition.

Table 8.4 Flood reduction difference between the initial and saturated conditions

No	Scenario	Runoff reduction percentage in initial condition	Runoff reduction percentage in saturated condition	LID performance degradation
1.	Scenario 1	0%	0%	0%
2.	Scenario 2	8.2%	7.5%	8.5%
3.	Scenario 3	7.9%	0%	100%
4.	Scenario 4	3.6%	0.02%	99.4%
5.	Scenario 5	1.5%	0%	100%
6.	Scenario 6	8.1%	0.02%	99.8%
7.	Scenario 7	15.9%	7.52%	52.7%
8.	Scenario 8	16.9%	7.52%	55.5%

As shown in Table 8.4, in Scenarios 3 and 5, the LID performance decreases to 100%, which means that when extreme rains of long duration or intensity occur, the LID infrastructure is no longer effective in reducing runoff. In Scenarios 4 and 6, the decrease in the LID performance is also very large, reaching more than 99%, which indicates that the effectiveness of LID performance in reducing runoff when saturated is very small. Meanwhile, in Scenarios 7 and 8, the effectiveness of the LID performance in the saturated condition is close to half of the normal condition performance. Meanwhile, the smallest decrease in performance occurred in Scenario 2, where the decrease in performance was only 8.5% from the initial condition.

Conclusions and future practices

In this chapter, we propose and demonstrate flood modeling mitigation strategies using the computational model approach. The research considered case studies in Jakarta and Wai Batu Merah, Ambon, to investigate the response of river and watershed systems toward the structural and non-structural interventions in terms of engineering approaches.

HEC-RAS provides a flood inundation simulation and can visualize the water profile when the canalization scenario is implemented. It gives flexibility to decision makers to modify the hydraulic characteristics of the river so that the flood disaster will be fully overcame. The study was supported by previous research in the same study area (Lin, Shaad, & Girot, 2016) in the discussion part. Full normalization affects the flooding volume and inundation area in the impacted river system. However, the GI scenario was suggested to be implemented. This recommendation then was brought into the Wai Batu Merah case study to implement the GI application in the area to control flooding and, even more, water pollution in urban waters. The models were captured using PC-SWMM, and several scenarios were simulated to bring certain GI to be best constructed to tackle urban flooding. The combination permeable pavement, rain barrel, infiltration well, and biopore infiltration holes contribute the most to reducing flooding in urban settings.

Overall, the engineering approach for flood mitigation strategies can be performed in structural and non-structural ways. Currently, the combination of both scenarios may compromise the characteristics of the watershed and society in the study area. HEC-RAS and PC-SWMM provide understanding of the response of water systems in various scenarios. However, social factors may become necessary due to their daily involvement in water management in the watershed. In the future, engineering and social engagements need to be developed (Webber et al., 2020) to obtain better solutions to solve flood disasters in urban areas.

References

Abidin, H. Z., Andreas, H., Gumilar, I., Fukuda, Y., Pohan, Y. E., & Deguchi, T. (2011). Land subsidence of Jakarta (Indonesia) and its relation with urban development. *Natural Hazards*, *59*(3), 1753–1771. https://doi.org/10.1007/s11069-011-9866-9

Asdak, C., Supian, S., & Subiyanto (2018). Watershed management strategies for flood mitigation: A case study of Jakarta's flooding. *Weather and Climate Extremes*, *21*(August), 117–122. https://doi.org/10.1016/j.wace.2018.08.002

Baek, S., Ligaray, M., Pachepsky, Y., Chun, J. A., Yoon, K. S., Park, Y., & Cho, K. H. (2020). Assessment of a green roof practice using the coupled SWMM and HYDRUS models. *Journal of Environmental Management*, *261*(January), 109920. https://doi.org/10.1016/j.jenvman.2019.109920

Billa, L., Mansor, S., & Mahmud, A. R. (2004). Spatial information technology in flood early warning systems: An overview of theory, application and latest developments in Malaysia. *Disaster Prevention and Management: An International Journal*, *13*(5), 356–363. https://doi.org/10.1108/09653560410568471

Bisht, D. S., Chatterjee, C., Kalakoti, S., Upadhyay, P., Sahoo, M., & Panda, A. (2016). Modeling urban floods and drainage using SWMM and MIKE URBAN: A case study. *Natural Hazards*, *84*(2), 749–776. https://doi.org/10.1007/s11069-016-2455-1

BNPB. (2021). *Data Informasi Bencana Indonesia (Indonesian disaster information database)*. Retrieved from http://dibi.bnpb.go.id/

Bogardi, J. J. (2021). *Handbook of water resources management: Discourses, concepts and examples*. Springer. https://doi.org/10.1007/978-3-030-60147-8

Erena, S. H., Worku, H., & De Paola, F. (2018). Flood hazard mapping using FLO-2D and local management strategies of Dire Dawa city, Ethiopia. *Journal of Hydrology: Regional Studies*, *19*(June), 224–239. https://doi.org/10.1016/j.ejrh.2018.09.005

Ezzine, A., Saidi, S., Hermassi, T., Kammessi, I., Darragi, F., & Rajhi, H. (2020). Flood mapping using hydraulic modeling and Sentinel-1 image: Case study of Medjerda Basin, northern Tunisia. *Egyptian Journal of Remote Sensing and Space Science*, *23*(3), 303–310. https://doi.org/10.1016/j.ejrs.2020.03.001

Geravand, F., Hosseini, S. M., & Ataie-Ashtiani, B. (2020). Influence of river cross-section data resolution on flood inundation modeling: Case study of Kashkan river basin in western Iran. *Journal of Hydrology*, *584*, 124743. https://doi.org/10.1016/j.jhydrol.2020.124743

Jamali, B., Bach, P. M., & Deletic, A. (2020). Rainwater harvesting for urban flood management – an integrated modelling framework. *Water Research*, *171*, 115372. https://doi.org/10.1016/j.watres.2019.115372

Kang, M. S., Koo, J. H., Chun, J. A., Her, Y. G., Park, S. W., & Yoo, K. (2009). Design of drainage culverts considering critical storm duration. *Biosystems Engineering*, *104*(3), 425–434. https://doi.org/10.1016/j.biosystemseng.2009.07.004

Khadr, M. E. M., & Ahmed, Y. A. A. (2016). Hydrodynamic and water quality modeling of Lake Manzala (Egypt) under data scarcity. *Environmental Earth Sciences*, *75*(19), 1–13. https://doi.org/10.1007/s12665-016-6136-x

Kourtis, I. M., Tsihrintzis, V. A., & Baltas, E. (2018). Simulation of low impact development (LID) practices and comparison with conventional drainage solutions. *Proceedings*, *2*(11), 640. https://doi.org/10.3390/proceedings2110640

Li, Q., Wang, F., Yu, Y., Huang, Z., Li, M., & Guan, Y. (2019). Comprehensive performance evaluation of LID practices for the sponge city construction: A case study in Guangxi, China. *Journal of Environmental Management*, *231*(October), 10–20. https://doi.org/10.1016/j.jenvman.2018.10.024

Lin, E., Shaad, K., & Girot, C. (2016). Developing river rehabilitation scenarios by integrating landscape and hydrodynamic modeling for the Ciliwung River in Jakarta, Indonesia. *Sustainable Cities and Society*, *20*, 180–198. https://doi.org/10.1016/j.scs.2015.09.011

Luke, A., Kaplan, B., Neal, J., Lant, J., Sanders, B., Bates, P., & Alsdorf, D. (2015). Hydraulic modeling of the 2011 New Madrid floodway activation: A case study on floodway activation controls. *Natural Hazards*, *77*(3), 1863–1887. https://doi.org/10.1007/s11069-015-1680-3

Martinez, R., & Masron, I. N. (2020). Jakarta: A city of cities. *Cities*, *106*, 102868. https://doi.org/10.1016/j.cities.2020.102868

Oppenheimer, M., Glavovic, B. C., Hinkel, J., Wal, van de, Magnan, A. K., Abd-Elgawad, A., . . . Sebesvari, Z. (2019). Sea level rise and implications for low-lying islands, coasts and communities. In H. O. Pörtner, D. C. Roberts, V. Masson-Delmotte, P. Zhai, M. Tignor, E. Poloczanska, K. Mintenbeck, A. Alegría, M. Nicolai, A. Okem, J. Petzold, B. Rama, & N. M. Weyer (Eds.), *IPCC special report on the ocean and cryosphere in a changing climate*. Retrieved from www.ipcc.ch/srocc/

Panda, U. S., Mahanty, M. M., Ranga Rao, V., Patra, S., & Mishra, P. (2015). Hydrodynamics and water quality in Chilika Lagoon-A modelling approach. *Procedia Engineering*, *116*(1), 639–646. https://doi.org/10.1016/j.proeng.2015.08.337

Prastica, R. M. S., Maitri, C., Hermawan, A., Nugroho, P. C., Sutjiningsih, D., & Anggraheni, E. (2018). Estimating design flood and HEC-RAS modelling approach for flood analysis in Bojonegoro city. *IOP Conference Series: Materials Science and Engineering*, *316*(1). https://doi.org/10.1088/1757-899X/316/1/012042

Prastica, R. M. S., & Wicaksono, D. (2019). Integrated multimodal disaster mitigation management for urban areas: A preliminary study for 2-d flood modeling. *IOP Conference Series: Materials Science and Engineering*, *650*(1). https://doi.org/10.1088/1757-899X/650/1/012056

Quirogaa, V. M., Kurea, S., Udoa, K., & Manoa, A. (2016). Application of 2D numerical simulation for the analysis of the February 2014 Bolivian Amazonia flood: Application of the new HEC-RAS version 5. *Ribagua*, *3*(1), 25–33. https://doi.org/10.1016/j.riba.2015.12.001

Rangari, V. A., Sridhar, V., Umamahesh, N. V., & Patel, A. K. (2019). Floodplain mapping and management of urban catchment using HEC-RAS: A case study of Hyderabad city. *Journal of the Institution of Engineers (India): Series A*, *100*(1), 49–63. https://doi.org/10.1007/s40030-018-0345-0

Siswanto, S., van Oldenborgh, G. J., van der Schrier, G., Jilderda, R., & van den Hurk, B. (2016). Temperature, extreme precipitation, and diurnal rainfall changes in the urbanized Jakarta city during the past 130 years: Precipitation characteristics changes in the urbanized city. *International Journal of Climatology*, *36*(9), 3207–3225. https://doi.org/10.1002/joc.4548

Webber, J. L., Fletcher, T. D., Cunningham, L., Fu, G., Butler, D., Burns, M. J., . . . Burns, M. J. (2020). Is green infrastructure a viable strategy for managing urban surface water flooding? *Urban Water Journal*, *17*(7), 598–608. https://doi.org/10.1080/1573062X.2019.1700286

Xu, T., Jia, H., Wang, Z., Mao, X., & Xu, C. (2017). SWMM-based methodology for block-scale LID-BMPs planning based on site-scale multi-objective optimization : A case study in Tianjin. *Frontiers of Environmental Science and Engineering*, *11*(4), 1–12. https://doi.org/10.1007/s11783-017-0934-6

Yahya Surya, M., He, Z., Xia, Y., & Li, L. (2019). Impacts of sea level rise and river discharge on the hydrodynamics characteristics of Jakarta Bay (Indonesia). *Water*, *11*(7), 1384. https://doi.org/10.3390/w11071384

Yildirim, E., & Demir, I. (2021). An integrated flood risk assessment and mitigation framework: A case study for Middle Cedar River Basin, Iowa, US. *International Journal of Disaster Risk Reduction*, *56*, 102113. https://doi.org/10.1016/j.ijdrr.2021.102113

Zellou, B., & Rahali, H. (2017). Assessment of reduced-complexity landscape evolution model suitability to adequately simulate flood events in complex flow conditions. *Natural Hazards*, *86*(1), 1–29. https://doi.org/10.1007/s11069-016-2671-8

Zhao, G., Xu, Z., Pang, B., Tu, T., Xu, L., & Du, L. (2019). An enhanced inundation method for urban flood hazard mapping at the large catchment scale. *Journal of Hydrology*, *571*, 873–882. https://doi.org/10.1016/j.jhydrol.2019.02.008

9
Restoration and recovery of flood-affected communities

Anoradha Chacowry

Introduction

Restoration and recovery in the context of a flood disaster

Restoration and recovery in the aftermath of a flood disaster are generally understood as the process of cleaning up and getting communities' activities back to normal. In real situations, they go far beyond these general activities. The associated processes include the restoration of their environmental and economic assets such as damaged infrastructures and disrupted services. In addition to this, they involve the rebuilding of social capital and support to affected communities in enhancing their capacities to cope with future flood events.

In developing countries, these processes present challenging issues for sustainable recovery in flood-prone communities, as they are influenced by a host of complex factors at local and national levels. The purpose of this chapter is to explore these issues through the lenses of vulnerability, resilience and environmental justice, attributes which are deeply rooted in the social capital aspect of a community. A case study illustrates how a flood-prone community in a small developing island state used these attributes and planned to 'live with floods'.

Recovery in the flood disaster management cycle

In a flood disaster cycle, recovery is seen as a part of a continuous process that covers all four phases of activities described in Figure 9.1. In the event of a severe flood disaster, the four phases may overlap over a given span of time. For example, the activities undertaken in favour of flood victims during the relief and assistance phase can be pursued during the rehabilitation/reconstruction phase.

The activities carried out as part of the restoration and recovery process when a flood disaster occurs involve the following phases:

Emergency and rescue phase (phase 1)

This phase takes place during or immediately after a disaster. It involves helping or evacuating trapped victims. In most cases the first responders are family members, neighbours, volunteers, firefighters, police and the medical team.

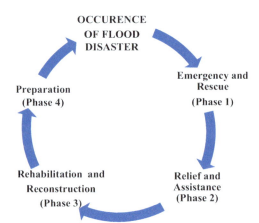

Figure 9.1 Four phases of recovery process in a flood risk management cycle, starting with the occurrence of a flood disaster

Relief and assistance phase (phase 2)

This phase involves cleaning up of debris and evacuation of accumulated floodwater from homes and the neighbourhood over an extended period of time. Financial and other forms of assistance such as building materials and necessities for sustenance are provided to vulnerable households.

Rehabilitation and reconstruction phase (phase 3)

This is a medium- to long-term phase related to the rebuilding, restoration of damaged infrastructure and drainage systems, provision of assistance to vulnerable groups and relocation of severely affected households to safer places. In some developing countries, health officers undertake occasional visits to check for flood-borne diseases in affected communities.

Preparation phase (phase 4)

This phase involves contingency planning and construction of flood drains. Campaigns of awareness-raising through informal education, the media and participatory activities of communities for effective response are also undertaken.

Each of the four phases has a time frame of restoration in the aftermath of flood disasters and it depends on several factors, including (i) the severity of the damage, (ii) the livelihood situation, (iii) the availability of resources to the disaster-affected communities and (iv) the efforts that various stakeholders involved in the recovery process are able to deploy.

Restoration and recovery of the community may be a short- or long-term process. Short-term recovery generally focuses on the immediate tasks of securing the impacted area and restoring basic living conditions and the essential local infrastructure. Disrupted transport systems and essential services are established. Long-term recovery focuses on planning strategies for reconstruction processes for sustainability and resilience aligning with the Building Back Better (BBB) principles of the Sendai Framework for Disaster Risk Reduction 2015–2030 (UNISDR, 2015).

Why recovery?

Developing countries including small island states are characterised by rapid urbanisation, population growth, fragile economies and the gradual disintegration of self-supporting traditional social structures. These factors and the lack of robust institutional capacity hinder initiatives and recovery programmes and result in considerable economic, social and environmental costs (IMF, 2019). As an example, the damages incurred from the 2015 flood in Malawi exceeded US$330 million, while recovery and reconstruction amounted to about US$500 million (Kita, 2017).

Intense human activities and climate change are more likely to aggravate localised flooding, exacerbate vulnerability and foster social inequalities (Muñoz and Tate, 2016). In order to address the inequalities and ensure effective recovery of all community members, ways and means should be planned and deployed to reduce vulnerability and enhance resilience in a holistic manner.

Understanding social vulnerability in recovery process

The concept of social aspect of vulnerability became prominent with the increasing number of disasters affecting communities and the unsuccessful mitigation efforts in reducing risk (Cardona, 2004; UNISDR, 2015). In hazard studies, vulnerability is broadly defined as the social conditions of exposure and the degree of resilience to cope with and recover from the impact of disasters. Factors that increase vulnerability are predominantly determined by the socio-economic conditions and the safety of the environment in which an individual lives (Wisner et al., 2006).

Socio-economic conditions

Households with low economic resources or with unsustainable means of survival are often driven to occupy low market-value environment such as wetlands, riverbanks and overcrowded settlements which are prone to frequent flooding in times of intense rainfall. Wealthier people build on hillsides or near the coast for the view and are able reconstruct and recover quickly in the aftermath of flood disasters, whereas lower-income groups are often left to fend for themselves, with the result that they are unable to cope and recover from successive events of localised flooding.

After the tsunami disaster of 2004 in Sri Lanka and post-flooding, disparity in the reconstruction activities among those who were affected was noted (Ingram et al., 2006). Poor fishing communities were forced to relocate, causing lasting disruption of their social networks. The change adversely affected their economic subsistence and contributed to an increase in vulnerability, and it took them several years to recover.

Exposure to unsafe environment

Health concerns in the recovery process from flood disasters are often neglected once the emergency and relief stages are over. Exposure to flood conditions arising from torrential rains and flash floods leads to the proliferation of water-borne diseases and contamination of drinking water. In the longer term, the propagation of vector diseases such as malaria, dengue and chikungunya (Olanrewaju et al., 2019) amplify the growing risk, vulnerability and disruption of family life.

The emergence of psychological and mental health issues in many developing countries is associated with the trauma of living with flood (Mekkodathil et al., 2019). These problems are seldom visible in the long term and need to be addressed in the national health care services (Shaw, 2019).

Many countries have been experiencing the double hazards of flood and the COVID-19 pandemic (Simonovic et al., 2021). As such, the pandemic will further adversely impact socio-economic recovery of vulnerable communities living in overcrowded settlements, especially in developing countries. Measures such as social distancing are expected to impact humanitarian assistance, coordination measures and cultural alienation that may further affect the recovery of communities. Such pandemics are predicted to recur. In developing countries with inadequate health facilities, the recovery of communities from the dual impact of flood and health disasters has to be accounted for in future health plans and strategies.

Social equity and environmental justice in the framework for sustainable recovery

Inequity in resilience building for sustainable recovery

The efforts and resources required in building the resilience of community groups and achieving recovery are determined by the extent of social injustice and exposure to unsafe environment. The government often uses 'one tone' to communicate to the population on resource allocation, often ignoring the cultures, attitudes and specific needs of underprivileged and minority communities. Local elites involved in relief and aid programmes are often politically connected and exclude vulnerable sectors of communities in decision-making processes. Post-relief assistance handed down to disadvantaged communities by donor institutions seldom goes towards their desired needs in welfare and reconstruction.

Environmental justice in sustainable recovery policy agenda

The prevalence of geographically marginalised communities to natural disasters springboarded into the idea of environmental justice (EJ) in recent years (Cutter, 2006; Flores et al., 2021). It fundamentally addresses the right of an individual to a healthy environment and a good quality of life. In disaster studies, its application has been diversified to explore inequality in social groups experiencing recovery disparities.

Research methods that apply tools such as geographic information systems and land survey are used to acquire quantitative measures. However, these techniques do not take into account all dimensions of situational inequality that entrenches and perpetuates environmental injustice. Qualitative approaches, such as focus groups and participatory methods that draw on people's local knowledge to identify sources of hazard risks, could help in identifying patterns of exposure to such risks and improving national policy.

EJ gained recognition as policy priority worldwide in the last decades. These attributes were initially enshrined in Sustainable Agenda 21 (UN, 1992) as a global concern in formulating policies to achieve sustainable development. In developing countries, the emphasis is protection of underprivileged groups and their involvement in decision-making (Ako, 2013), hence the importance of public participation and exchange of local and scientific knowledge to enhance resilience and sustainable recovery with support from the government, NGOs, private sector and institutional agencies as stakeholders.

Building community resilience in recovery strategies

The need for communities of developing states to ensure quick and effective recovery as part of the resilience-building measures against disasters was particularly emphasised in the Sendai Framework for Disaster Risk Reduction 2015–2030 (UNISDR, 2015). To reduce risk and improve recovery and living conditions of disaster-impacted communities, it called for the strengthening of community resilience by building back better. Community resilience is built on several premises, as given in Box A.

Box A. Basic premises related to strengthening community resilience

The guiding principles as laid out in the Sendai Framework and UN sustainable development goals include:

a) the need and the capacity of communities to build back better in the restoration and recovery phases of disasters
b) tackling problems of poverty, inequality, environment changes and climate change
c) strengthening good governance in disaster risk reduction strategies at the national, regional and global levels
d) improving preparedness and national coordination for disaster response, rehabilitation, reconstruction and the use of post-disaster recovery and reconstruction to build back better
e) engagement of government institutions with civil society as stakeholders at all levels of the recovery process.

The principle of close interaction between the community and other stakeholders are stressed but more often in practice the quick fit top-down solution is often applied.

Social capital aspect of community resilience and recovery

In post-flood disasters, governments, for several reasons, often appear to be mostly concerned with economic and infrastructural restoration, leaving communities to fend for themselves once the flood is over. Communities are thus compelled to rely on their social capital as resources in long-term recovery. Social capital involves local knowledge, experience and the social network of collaboration as pressure group and of helping each other with basic needs and support in times of distress.

Participatory activities of relevant stakeholders encourage social learning of affected communities using various techniques such as 'mapping out' and 'visual simulation' of flooding in their areas. Mutual support among community members, the use of situational knowledge and collaboration among community members are found to enhance community resilience and enable faster and more effective recovery (Lopez-Marrero and Tschakert, 2011).

Over centuries, traditional and riparian communities in developing countries have developed flood resilience and coping strategies by applying their local knowledge and experience. Social capital in the form of civic action is found to be useful in disaster relief efforts of vulnerable individuals such as the elderly, disabled persons and children. It can hasten the psychological recovery of older adults who are affected by the traumatic experience of disaster events (Phillips and Cherry, 2019).

Bridging the gap between centralised institutions and local communities in decision-making processes through networking of social capital, including the sharing of local communities' knowledge with experts, could contribute to multi-stakeholder participation in resolving the problems of social inequity and environmental justice. The concerted efforts of all stakeholders in a community, including neighbours, relatives, the private sector and NGOs, as well as government authorities involved in the reconstruction process, offer opportunities to reduce vulnerability, accelerate recovery and promote community well-being.

Managing and living with floods

As flood disasters cannot be prevented, a paradigm shift from traditional flood risk management to flood resilience has currently been focused on the principle of 'living with floods' and building above flood level (Adebimpe et al., 2021). It has mostly been studied in urban environments but nevertheless offers the added value of improving disaster knowledge and raising awareness of flood-affected communities in coping with flood disasters and thus facilitating recovery.

Strengthening social capital is a crucial asset in hastening community recovery (Storr et al., 2016) as affected communities rebuild their own homes and live with floods as a 'way of life'. Figure 9.2 illustrates the technique of building houses on stilts, an adaptation strategy of flood-prone communities in Cambodia.

Figure 9.2 House on stilts that enable the community to adapt to floods in Cambodia
Source: Author (2018)

Considering risk perception of communities in recovery processes

Perceived risk is defined as 'a set of mental strategies or heuristics that people employ to make sense of the uncertain world' (Slovic, 1987). Many of the decisions people take are based on their perception and understanding of risk. An individual's interpretation of risks is also shaped by their own experience, personal values and cultural beliefs and by a changing social environment. Risk perception of communities exposed to flood disaster is an added form of social capital that resides on the intuitive characteristics of individuals on how to cope with flood impacts, build resilience and recover on their own.

Case studies carried out in Fiji and Indonesia between 2018 and 2020 showed promising results of how flood-prone communities can manage their risk and recover effectively from recurrent flood disasters. The participatory activities involved the recording of the levels of flood during raining seasons and communicating the visual images to other members of the community on social media. A database was established for interpretation by each member of the community (Wolff, 2021). It enabled the community to be better prepared to manage hazard risk.

Case study applying the concepts of vulnerability and resilience for recovery in a flood-prone community

These issues are framed in a case study to illustrate how flood-affected communities cope with flood recovery (Chacowry et al., 2018). The study area is located in the suburban town of Port Louis, the capital of Mauritius. The area is characterised by a river that drains through a wetland and by other rivulets that overflow in times of torrential rain and flash floods and lead to water accumulation in the surrounding neighbourhood. The study is focused on the perception of households on their vulnerability, resilience and recovery from successive floods.

Methodology

A mixed method of quantitative and qualitative approaches consisting of a questionnaire survey, focus group interviews and a participative exercise were used to examine the vulnerability, resilience and recovery of the community at household level. The questionnaire survey was carried out in some 236 households using the variables whose details are given in Table 9.1.

Results of survey

The results from quantitative analysis showed significant associations at $p \leq 0.05$ between variables relating to recovery and those of income level, literacy level and household size with children and/or elderly persons.

The survey results further indicated that low-income groups within the community were more likely to be exposed to frequent flooding, as they occupied hazardous wetland areas and poorly drained marginal lands along riverbanks. The extent of impacts from a flood event was felt differently by groups of people with varying levels of preparedness, resilience and capacity to recover. Some households took a long time to recover, if at all. Households with lower socio-economic backgrounds very often did not recover by the time the next flood event occurred, thus further entrenching their vulnerability. Some low-income families remained in damp conditions for many days and suffered social disruption and economic stress.

The psychological impact of flooding was mostly associated with households with large families and those with children. Concern about diseases and chronic stress have been found to

Table 9.1 Variables of vulnerable households in the study of recovery

Variables		Focus group (n* = 7)	Participatory (n* = 15)
Household characteristics: Small: ≤ 4 members Large: ≥ 5	i) Age groups ii) Literacy iii) Occupation, income	**Themes:** i) Perception of living conditions ii) Coping strategies during floods iii) Recovery from floods iv) Resilience building v) Assistance from local government	**Themes:** i) Identification of the flood problem ii) Discussion on addressing the flood problem iii) Proposal of solutions for reducing vulnerability and resilience building of the community
Vulnerability to flood	i) Type and frequency of floods experienced ii) Tangible and intangible impact iii) Socio-economic situation iv) Environmental factors		
Recovery from flood hazards	i) Getting back to 'normal' after the flood ii) Relocation after the flood		
Resilience to flood hazards	i) Precautions taken before a flood ii) Adaptation measures taken iii) Collaboration with the community iv) Reliance on building resilience from flood protection v) Knowledge on flood increase vi) Respondents' opinion on measures on enhancing resilience		

n*: Number of participants

be factors that have led to lowering of resilience, increased vulnerability and reduced capacity to recover. Greater awareness and education on health issues, especially of the risk of catching flood-borne diseases, could be enforced in the agenda of community health programmes.

Effective coping strategies, sound resilience-building measures, local experiential knowledge and community solidarity were essential in recovery. In terms of enhancing environmental resilience and of reducing their exposure to associated risks, most of the households took pre-emptive measures ahead of flood events like making furrows to divert water during and after a flood hazard (Figure 9.3).

Helping neighbours and liaising with local authorities to clean up soon after flood hazards was a common sign of solidarity. Figure 9.4 shows, in percentage, the source of support in recovery and during resilience-building. Figure 9.5 shows the source and nature of assistance received by households in flood mitigation. Most of the assistance came from family members and neighbours.

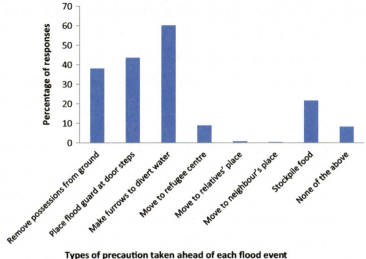

Figure 9.3   Precautions taken prior to flood events
Source: Author's survey

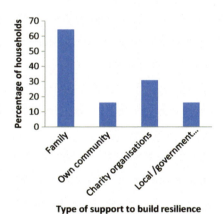

Figure 9.4   Assistance during recovery

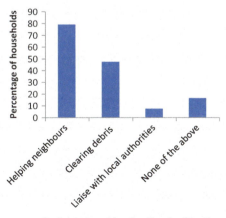

Figure 9.5   Community solidarity network
Source: Author's survey

Results from focus group interviews and participatory exercise

Qualitative results from focus group interviews (Table 9.1) indicated that social inequity and environmental injustice hindered recovery among low-income households. However, some resilience was present through social capital solidarity in times of adversity amongst some community subgroups. Outcomes from a participatory exercise (Table 9.1) showed that local and experiential knowledge of how to cope with floods was crucial in resilience-building strategies of households and communities. Developing civic action and awareness among the participants such as not throwing garbage on the road and of keeping the area and the drains clean were seen to reinforce community ties. These were seen as essential elements in the case study for recovery during each flood event and in long-term vulnerability reduction and resilience-building to flood disasters.

This research has highlighted the importance of understanding how the recovery and rehabilitation phase within the disaster cycle plays out locally in unravelling factors affecting community resilience in an exposed, flood-prone, suburban area of Port-Louis. Many of the challenges faced by affected community groups in reducing their vulnerability and in building resilience in the recovery phase have yet to be addressed.

Marginalisation and environmental justice

Findings from the questionnaire survey showed that there was a feeling of 'being left out' and abandonment among low-income sectors of the community. Both small and large households with children perceived a lack of support from the authorities to assist them in the recovery phase. Marginalised and underprivileged groups, mainly households with children and elderly persons within the communities, who were differentially exposed to flood risk, were threatened by higher levels of health hazards.

The perception of marginalisation was profound among the low-income groups, who perceived that there were strong disparities in the way they were treated regarding land allocation and the construction of flood-proof housing during rehabilitation and recovery phases. Any complaint about environmental problems and poor living conditions during flooding was often disregarded by the authorities. The feeling of environmental injustice seemed to grow with time, leading the underprivileged group to adopt a fatalist attitude of acceptance of impacts, while its vulnerability increased from one flood to the next, leading to a gradual decline in households' resilience. The factors that influenced the recovery patterns in the case study could be used as conceptual tools in a framework or model of sustainable recovery programme of a flood-affected community.

An overview of the concept of recovery and its linkages to resilience and vulnerability as discussed is given in Figure 9.6. The capacity to recover depends largely on the resilience of the community. In turn, aspects of resilience that need strengthening can be derived from an understanding of the factors that lead to community vulnerability. The major variables that determine vulnerability and those that could be used to assess resilience are highlighted. A few extraneous issues that aid in recovery are also reflected in the figure. These factors determine to varying degrees the timeframe, capacity and sustainability of recovery.

Challenging issues related to recovery of affected communities

Social disparities and uncertainties

The case study shows that recovery of affected households and communities is a complex and multifaceted process that is entwined with a number of concepts that act as precursors in

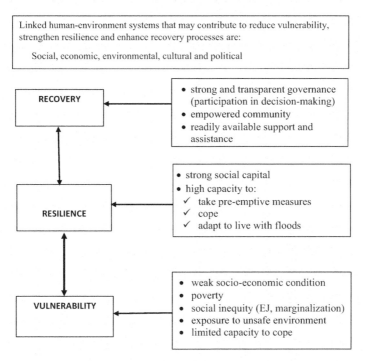

Figure 9.6 Schematic representation of the interplay between vulnerability, resilience and recovery

Source: Author from survey results

mitigating risk and restoring the livelihood of communities. At the same time, approaches in solving social disparity in recovery go far beyond the normal practices of cleaning up. Unexpected challenges such as those arising from COVD-19 have driven the world and in particular developing countries and small island states into an environment of uncertainties. The tools available and the known practices of dealing with uncertainties and in the complex activities of restoration and recovery are not well understood (McCreight and Harrop, 2019). Consequently, they pose major challenges for sustainable recovery of affected communities.

Eradicating poverty and social exclusion and restoring social justice

Sustainable recovery from flood disasters has become a challenging task due to social inequality and differential exposure to flood conditions. Poverty drives people to precarious living conditions in both urban and rural areas, often resulting in environmental injustice. Critical views of traditional methods in environmental disaster risk assessment point out the deficiency of centralised technical 'expert-only' models of analysis, missing out the knowledge of the many interacting complexities experienced by residents.

Governance approach in recovery processes

Many developing and small island states are governed by weak institutional capacity, limited resources and unsustainable development patterns (Horn and Elagib, 2018) that hinder effective recovery in flood-prone communities. Partisan politics where assistance is given to selected communities which benefit from flood preventive structures built on a priority basis excludes the needy communities, thereby entrenching social marginalisation and vulnerability, leading to ineffective recovery from the impact flood disasters for deserving communities.

A prerequisite for effective recovery is to recognise and consider the needs of the vulnerable community and its experiential flood knowledge. Often, for the sake of expediency, its needs and priorities, in the aftermath of a flood disaster, are overlooked. Local NGOs that work at the grassroots level offer a pool of local volunteers that can effectively contribute to recovery and resilience-building. The vulnerable communities and local NGOs should be empowered and be involved in the formulation and implementation of resilience-building strategies. In this way, the key challenges and difficulties often encountered in the planning and flood disaster management for recovery could be effectively addressed.

'Building back better' – a window of opportunity for greater focus on recovery processes?

The mantra of 'building back better' enshrined in the Sendai Framework for Disaster Risk Reduction that is meant to increase resilience of nations and communities is not well understood. Implementing Building Back Better (BBB) in a disaster reduction framework as a way of increasing resilience of communities has been mainly driven by top-down centralised approaches, overlooking the needs and priorities of affected communities. Furthermore, BBB concepts are extensively used by policy makers, scholars and humanitarian agencies and funding agencies that do not account for a holistic approach to development (Su and Le Dé, 2020). Such wholesale approaches to BBB may hinder recovery as the focus is shifted to rehabilitation.

The central premise of the framework of community working together with all other stakeholders' agencies is often hampered by numerous difficulties in the implementation phase. On the other hand, institutional challenges, lack of technical expertise and conflicts could exclude stakeholders from flood recovery processes and management. One way to overcome the handicap is for the leadership to ensure that the focus of any activity is on building the resilience and well-being of a community.

Psychological support underestimated as a tool for recovery

The psychological trauma of living with floods from one event to the next further amplifies the vulnerability of the affected victims and leads to decline in the early recovery of affected communities. This aspect of vulnerability is often not accounted for in developing countries. The sudden upsurge of COVID-19 pandemic and its impact on the health of vulnerable communities and the uncertainties associated with climate change are also unknown and are sources of trauma. How to address the psychological trauma associated with these uncertainties in addition to those associated with daily life for the poor remains an enormous challenge to policymakers in developing countries.

Recommendations for an effective recovery

A few recommendations for an effective recovery are listed in the following:

a) Integration by policymakers of the social dimension of vulnerability in national disaster risk management strategies in view of the increasing number of natural disasters affecting a larger number of people worldwide.

b) Empowering flood-affected communities to integrate local and experiential knowledge in decision-making through participatory activities with other stakeholders into flood risk management policies. A shift from reactive to proactive flood management requires thorough involvement of the community and stakeholders.
c) Introduction of new approaches to respond to disasters while managing uncertainties like the world pandemic of COVID-19 and climate change.
d) Establishment of a Marshall-like plan at the national level in the restoration and enhancement of flood-affected communities, which would include infrastructure as well as social capital.

Conclusion

Many of the challenges faced by flood-affected communities in reducing their vulnerability and in building resilience in the recovery phase have remained unanswered. A prerequisite to addressing them is to recognise and consider the needs of the vulnerable and their experiential flood knowledge and ensure their involvement in the implementation of resilience-building strategies.

Further research needs to focus initially on ways of co-working with all sectors of the community to plan and implement strategies for resilience-building that place emphasis on participation and developing and sharing community-based knowledge. The restoration and recovery of flood-affected communities, however, require a solid integrated long-term effort at individual, household, local community, regional, national, institutional and political levels. A sound policy framework and science-based forward planning and sustained implementation strategy, combined with determined and agreed actions through collaborative governance, are essential to building and ensuring robust and long-lasting community resilience. Restoration and recovery initiatives should not be a one-off process but an ongoing one that includes monitoring and assessment approaches that should be accountable to flood-prone communities.

References

Adebimpe, O., Proverbs, D., and Oladokun, V. (2021). A fuzzy-analytic hierarchy process approach for measuring flood resilience at the individual property level. *International Journal of Building Pathology and Adaptation*, 39 (2), pp. 197–217.

Ako, R. T (2013). *Environmental justice in developing countries: Perspectives from Africa and Asia-Pacific*. London: Routledge, p. 168. ISBN 9781138686847 Published May 11, 2016.

Cardona, O. (2004). The need for rethinking the concepts of vulnerability and risk from a holistic perspective: A necessary review and criticism for effective risk management. In G. Bankoff et al. (eds.), *Mapping vulnerability: Disasters, development and people*. London: Earthscan Publishers, pp. 1–17.

Chacowry, A., McEwen, L., and Lynch, K. (2018). *Recovery* and resilience of communities in flood risk zones in a small island developing state: A case study from a suburban settlement of Port Louis, Mauritius. *International Journal of Disaster Risk Reduction*, 28, pp. 826–838.

Cutter, S. L. (2006). *Hazard, vulnerability and environmental justice*. New York: Earthscan, p. 407. ISBN 13:978-1-84407-310-8.

Flores, A. B., Collins, T. W., Grineski, S. E., Griego, A. L., Mullen, C., Nadybal, S. M., Renteria, R., Rubio, R., Shaker, Y., and Trego, S. A. (2021). Environmental injustice in the disaster cycle: Hurricane Harvey and the Texas Gulf Coast. *Environmental Justice*, 14 (2), pp. 146–158, April. http://doi.org/10.1089/env.2020.0039

Horn, F., and Elagib, N. A. (2018). Building socio-hydrological resilient cities against flash floods: Key challenges and a practical plan for arid regions. *Journal of Hydrology*, 564, pp. 125–132. ISSN 0022-1694. https://doi.org/10.1016/j.jhydrol.2018.07.001 [Accessed 12 June 2021].

IMF. (2019). *Building resilience in developing countries vulnerable to large natural disasters*. ISBN/ISSN: 9781498321020/2663-3493. www.imf.org/ [Accessed 16 July 2021].

Ingram, J., Franco, G., Rumbaitis-del Rio, C., and Khazai, B. (2006). Post-disaster recovery dilemmas: Challenges in balancing short-term and long-term needs for vulnerability reduction. *Environmental Science and Policy*, 9, pp. 607–613.

Kita, S. M. (2017). *Can we 'build back better'? Lessons from floods recovery framework development and implementation in Malawi*. African Perspectives on Disaster Risk Reduction Issue, June. Gravitazz Institute for Disaster Reduction and Emergency Management. www.gravitazzcontinental.com [Accessed 29 September 2021].

Lopez-Marrero, T., and Tschakert, P. (2011). From theory to practice: Building more resilient communities in flood-prone areas. *Environment & Urbanization Institute for Environment and Development (IIED)*, 23 (1), pp. 229–249. https://doi.org/10.1177/0956247810396055.

McCreight, R., and Harrop, W. (2019). Uncovering the real recovery challenge: What emergency management must do. *Journal of Homeland Security and Emergency Management*, 16 (3).

Mekkodathil, A. M., Sathian, B., Elayedath, R., Simkhada, P., & van Teijlingen, E. (2019). Post-traumatic stress disorder among the flood affected population in Indian subcontinent. *Nepal Journal of Epidemiology*, 9 (1), pp. 755–758.

Muñoz, C. E., and Tate, E. (2016). Unequal recovery? Federal resource distribution after a midwest flood disaster. *International Journal of Environmental Research and Public Health*, 13 (5), p. 507. Published online 2016 May 17. doi:10.3390/ijerph13050507.

Olanrewaju, C., Chitakira, M., Olanrewaju, O., and Louw, E. (2019). Impacts of flood disasters in Nigeria: A critical evaluation of health implications and management. *Jàmbá: Journal of Disaster Risk Studies*, 11 (1).

Phillips, J. R., and Cherry, K. (2019). Interest group session – disasters and older adults: Harnessing the power of social networks during disasters to facilitate preparation and recovery. *Innovation in Aging*, 3 (Supp 1), p. S390. https://doi.org/10.1093/geroni/igz038.1434 [Accessed 23 June 2021].

Shaw, R. (2019). Disaster recovery: Community-based psychological support in the aftermath. *Progress in Disaster Science*, 2.

Simonovic, S., Rudzewicz, Z., and Wright, N. (2021). Floods and the COVID-19 pandemic – a new double hazard problem. *Wires, Water*, 8 (2).

Slovic, P. (1987). Perception of risk. *Science*, 236 (4799), pp. 280–285.

Storr, V., Haeffele-Balch, S., and Grube, L. (2016). Social capital and social learning after Hurricane Sandy. *The Review of Austrian Economics*, 30 (4), pp. 447–467.

Su, Y., and Le Dé, L. (2020). Whose views matter in post-disaster recovery? A case study of "build back better" in Tacloban city after Typhoon Haiyan. *International Journal of Disaster Risk Reduction*, 51 (101786). ISSN 2212–4209. https://doi.org/10.1016/j.ijdrr.2020.101786. [Accessed 10 July 2021].

UN. (1992). *United Nations conference on environment & development Rio de Janeiro, Brazil, 3 to 14 June 1992 AGENDA 21*. https://sustainabledevelopment.un.org › Agenda21 [Accessed 27 September 2021].

UNISDR. (2015). *Sendai framework for disaster risk reduction 2015–2030*. www.preventionweb.net/files/43291_sendaiframeworkfordrren.pdf [Accessed 24 March 2021].

Wisner, B., Blaikie, P., Cannon, T., and Davis, I. (2006). Natural hazards, people's vulnerability and disasters. In *At risk*, 2nd ed. London: Routledge.

Wolff, E. (2021). The promise of a "people-centred" approach to *floods*: Types of participation in the global literature of citizen science and community-based *flood* risk reduction in the context of the Sendai framework. *Progress in Disaster Science*, 10 (100171).

10
Preparedness and management of (flood) disaster amid a pandemic in a developing country

Lessons from Cyclone Amphan in southwestern Bangladesh

Sabiha Lageard and Namrata Bhattacharya-Mis

Introduction

The challenge of managing dual disasters is not new for developing nations, and a number of examples can be cited where multiple hazards were encountered in disaster management (e.g. natural disasters during Zika virus or Ebola outbreaks). However, circumstances are very different when 'frequent' disasters such as flooding and cyclones coincide with a prolonged global pandemic such as COVID-19, as significant differences in preparedness challenges and response trade-offs compound the vulnerabilities of the most at-risk groups (Peleg et al. 2021). Efforts of frontline workers to coordinate institutional response within civil society are hindered by additional mobility restrictions and social distancing requirements, resulting in overwhelming pressures on already over-stretched resources. Literature on managing dual disasters is currently limited, and the need for coordinated action to combat such events has only become compelling since the outbreak of COVID-19 (Simonovic et al. 2021). Since the start of the pandemic, a number of developing nations have been affected by climate-related hazards (EMDAT 2020; Izumi & Shaw 2022), providing lessons and establishing the need for developing a multi-hazard, integrated disaster risk management approach (UNDRR 2020) to tackle 'dual' disasters.

On 20 May 2020, the southwestern part of Bangladesh was hit by Cyclone Amphan. Although the eye predominantly traversed West Bengal (India), parts of Bangladesh close to the border with India, particularly Satkhira District, endured a heavy blow. The impact of the cyclone was worsened by the pre-Amphan lockdown and social distancing measures (commenced 26 March 2020) adopted in response to COVID-19, already seriously impacting the country's economy. Most people in the cyclone-affected area living on daily subsistence wages were already experiencing extreme hardship due to the health emergency, and Amphan imposed additional stresses, taking away livelihood capital, that is, health, assets, livestock, farms, and income (Momtaz & Shameem 2015; reliefweb 2020).

Given the COVID-19 background to Cyclone Amphan, this research was undertaken to understand how the coastal population prepared themselves in advance of the disaster and assess the effectiveness of measures put in place to cope with the dual disasters of a cyclone and a global pandemic. In the case of a poorly coordinated response to Amphan, COVID-19 would be expected to spread more rapidly, posing greater risk to lives and livelihoods, especially to already resource-stretched areas. This chapter uses a case study approach to identify whether, as a signatory to the Sendai Framework for Disaster Risk Reduction (SFDRR), Bangladesh adopted appropriate preparedness measures for all the hazards associated with the dual/concurrent disasters.

Background research

Disaster preparedness involves prediction, prevention, and mitigation to reduce disaster impacts on vulnerable populations, and enables them to effectively respond and cope with the consequences of a disaster (Rodríguez-Espíndola et al. 2018; UNDRR 2020; IFRC 2021).

Preparedness is one of the key stages in the disaster management cycle and involves risk awareness, advanced warning, risk communication, and capacity building (Skinner & Rampersad 2014; Oloruntoba et al. 2018; Erbeyoğlu & Bilge 2020). It is acknowledged that accurate and timely prediction, together with effective preparation, can reduce negative disaster experiences (Sadeka et al. 2020). The focus of disaster preparedness has, however, shifted from saving lives through infrastructure provision to effective risk communication to 'the last mile' to enhance the capacity of a community. The consensus is that capacity enhancement based on local needs and concerns will allow preparedness measures to be most effective (Ginige et al. 2010). There has also been a change in the global institutional approach from top down to bottom up, where importance is placed on 'the last mile' by considering it the 'first mile', putting the community at the centre of the disaster risk management system (Sarabia et al. 2020; Haklay 2013).

Dual disaster management requires a combination of DRR measures and emergency relief to cater to all resultant hazards. Cyclone Amphan and COVID-19 both posed health risks, whether through water contamination or the unavailability of water sanitation and hygiene (WaSH) facilities. Risk reduction (RR) measures for both therefore required preparedness to reduce health risks, as well as risks to life. Some have indicated the suitability of SFDRR during the COVID-19 pandemic, as it recognises health at the heart of disaster risk management (DRM), and it can also act as a catalyst for health emergency and disaster risk management (HEDRM) frameworks (Djalante et al. 2020; Ishiwatari et al. 2020; Wright et al. 2020; Dariagan et al. 2021).

Capacity to respond

The effectiveness of disaster preparedness (single or multiple disasters) depends mainly on the capacity of the community to carry out measures required to reduce risks. The capacity of a community or individual to respond to a disaster often reflects their own degree of vulnerability. Hamidi et al. (2020) identified building type and location, dependent population, illiteracy, unemployment, and weak economic capacity as key contributory factors to human vulnerability. Blaikie et al. (1994: 9), on the other hand, defined personal/group vulnerability as the 'capacity to anticipate, cope with, resist, and recover from the impact of a natural hazard'. The role of social networks/social capital as a safety net for vulnerable groups is also key to enhancing their capacity, since these groups rely on social networks to enhance their capacity to cope. The general tendency amongst the most vulnerable individuals, in the absence of formal institutional support, is to turn to informal social networks such as relatives, neighbours, and friends to prepare or recover from disasters (Haque et al. 2012; Soloman 2014; Masud-All-Kamal & Monirul Hassan 2018; Li & Tan 2019).

Disaster preparedness: Bangladesh

Bangladesh is prone to cyclones, and the frequency and severity of these are increasing due to climate change. The country's vulnerability to these natural hazards has attracted support from organisations such as the World Bank, UNDP (multilateral), USAID, EU (bilateral), and the Red Crescent (international NGO) to improve disaster preparedness and foster enhanced resilience in people/communities.

Cyclones pose challenges to the lives and livelihoods of around 30% of Bangladeshis inhabiting coastal areas, who are also identified as vulnerable due to poor ability to access infrastructural facilities and poor socioeconomic status (Mazumdar et al. 2014). It is these vulnerable groups who share disproportionately the negative consequences of cyclones, and enhancing their capacity to withstand such events is non-negotiable. Understanding the risks and disruption (storm surges, flooding) posed by cyclones has received considerable attention in Bangladesh's disaster preparedness measures (Mallick 2014; Saha 2015; Wedawatta et al. 2016; Quader et al. 2017; Choudhury et al. 2019; Islam et al. 2021).

Risk awareness: Bangladesh

The Ministry of Disaster Management and Relief (MoDMR) is responsible for coordinating national disaster management efforts. The main goal of the National Plan for Disaster Management (NPDM) 2016–2020 (updated to 2025 in March 2020) is to save lives, protect investment, and facilitate effective recovery. Common health concerns stem from cyclonic hazards resulting from the destruction of water sources and sanitation facilities. Failure to put appropriate recovery measures in place promptly, regarding water supply and sanitation, can threaten lives. The coastal population has a reasonable understanding of cyclone risks and ways of reducing their impacts. They know to secure their houses with ropes, bury their non-perishable assets, seek shelter in designated buildings (storm shelters), and safeguard animals (USAID 2013). However, knowledge does not always translate into action, due to fatalism, indecision, not realising when to act, and unreliable shelters (Islam et al. 2021).

Preparedness measures: Bangladesh

In 2014, Bangladesh's government pledged to build preventative infrastructure, including multi-purpose cyclone shelters and permanent embankments and to improve transport networks for 14 million of the coastal population. This resulted in 552 new shelters, rehabilitation of 450 existing shelters (total 4,760), building killas (raised ground for livestock), improving housing, afforestation, and constructing 550 km of rural roads to better access cyclone shelters (The World Bank 2019).

Such projects reflected a commitment to remove potential transport bottlenecks that can hamper rescue operations. The location of cyclone shelters was, however, a more contentious issue, as this is influenced by social power structures. According to Mallick (2014), socially vulnerable households (defined by income or housing conditions) were given disproportionately less access to shelters compared to less socially vulnerable households. Shelter location is often determined not by areas of greatest need, but by local authority preferences, and they are often poorly equipped with facilities (latrines, disabled facilities, and drinking water). Seventy-five percent lacked an adequate water supply, and there were too few shelters overall (Islam et al. 2021).

Bangladesh has developed a National Preparedness and Response Plan (NPRP) to prevent and contain COVID-19, but this has not been fully coordinated with pre-existing DRR for cyclone mitigation (UNDRR 2020). The NPRP includes limiting human-to-human transmission, communicating critical risk and providing information to communities, and minimising social and economic impacts through multi-sectoral partnerships. During Amphan, COVID-19–specific preparedness measures included the use of additional shelters to reduce over-crowding, making mask wearing compulsory, providing soap/sanitisers, and a room reserved for isolation cases in each shelter. Existing public health delivery infrastructure had been mobilised to reach every citizen in every village (Hasina & Verkooijen 2020). Despite a strong institutional DM setup and inclusive plans for DRR, a number of constraints, including poor institutional capacity building, resource mobilisation hurdles, poor coordination, and poor policy-operational linkages, meant a fully co-ordinated plan had not been implemented (Ahmed 2019).

Advance warning and risk communication: Bangladesh

Bangladesh has made the Storm Warning Centre (SWC), Bangladesh Meteorological Department (BMD), prepare all weather forecasts and hazard warnings, resulting in significant reductions in cyclone-related deaths through modernising its early warning system (Haque et al. 2012; Syed et al. 2021). Once the cyclone development and path are identified, BMD disseminates this information through the media, local authorities, and a Cyclone Preparedness Programme (CPP) Head Office. NGO and INGO volunteers are responsible for passing on forecasts locally (Paul & Dutt 2010; Paul 2012; Ahsan et al. 2016). These warnings are, however, often criticised for inaccuracies and inability to reach remote areas: 86% of households received warning of Cyclone Sidr, but no communications were received at sea (Roy et al. 2015).

Risk communication for cyclones has a long history in Bangladesh and takes into consideration past limitations. Responsibility for 'last mile' warnings is decentralised to locally recruited CPP volunteers, who inform households/communities, using basic techniques such as handheld sirens, megaphones, bicycle-mounted loudspeakers, knocking on doors, danger signal numbers (severity level), and advice to evacuate to safe places (Paul 2012; Ahsan et al. 2016). CPP volunteers have made a significant difference to at-risk communities, and their involvement reflects the authority's commitment to saving lives (reliefweb 2019). CPP responses are, however, often criticised, as last minute and under-resourced, failing to assist vulnerable households with embankment protection and securing homes (Ahsan 2016).

Most people in coastal areas understood the risks associated with COVID-19 prior to Cyclone Amphan, as DRR information was disseminated using audio, video, and printed materials (BBC Media Action, WHO, IOM, Bengal Creative Media – Shongjog 2020). Despite this, the participation of at-risk groups in protective action was hampered by misconceptions, superstitions, and religious conservatism (Communication with Communities Working Group 2020). The first cases of COVID-19 were identified in Bangladesh on 8 March 2020, and the whole country was placed in lockdown until 30 May 2020. When Cyclone Amphan hit the coastal region, movement restrictions due to COVID-19 were fully operational, making disaster preparedness and recovery work difficult.

Capacity to respond: Bangladesh

With World Bank support, Bangladesh has developed a Coastal Area Rehabilitation Project involving the construction of shelters, strengthening early warning systems, and supporting communities through investment in climate-resilient infrastructure and innovative technologies. Unfortunately, project implementation depends on limited multilateral and bilateral funds, and restricted priorities imposed by donors are often at odds with local needs. The effectiveness of this support is further hampered by fund misappropriation (Islam et al. 2017; UN Women 2020).

Sultana et al. (2020) suggests the coastal population's capacity to cope is often dictated not only by infrastructure, but also by social capital and urban opportunities. Embankments are highly valued for their protective capacity and for resilience offered to livelihoods. It is however important to note that coastal embankments also serve as part of the road network and at the time of Amphan were in general disrepair due to repeated storm damage, unsanctioned attempts by farmers to access saline water (shrimp farming), and overuse in transportation.

For local people, community organisations (social capital) are also important, since they offer support in accessing relief and rehabilitation services. Similarly, seasonal migration to urban areas is an important long-term coping and resilience strategy, helping to diversify livelihoods (70% rural households).

The adequacy of DRR measures to enhance the coastal populations' capacity to cope was significantly tested during Cyclone Amphan, due to the parallel threat of COVID-19. The pandemic increased the vulnerability of the coastal population and reduced its capacity to cope with natural hazards due to declining incomes (63% reduction) and unemployment rising to 55% (Rhaman & Matin 2020).

Data and methods

Site selection

To facilitate the research conducted remotely (due to COVID-19), the Executive Director of the Institute of Water Modelling (IWM, Bangladesh) was contacted initially to identify a suitable study area. Shyamnagar, a sub-district of Satkhira District, was subsequently chosen, as it had experienced the full brunt of Cyclone Amphan (Figure 10.1).

Data collection and analysis

Field research/travelling to Bangladesh was impossible due to COVID-19. Personal contacts were therefore used to identify a reliable primary participant in the study area who assisted in subsequent participant selection, applying a snowballing methodological approach.

Bias was avoided by interviewing a cross-section of the local population. Four specific social groups were represented: i) professionals (physician, police superintendent, politician, and teacher), ii) entrepreneurs (one large and two small business owners), and iii) poor (four daily wage earners).

Respondents	Occupation	Social status
A	Physician	Professional (i)
B	Police	Professional (i)
C	Politician	Professional (i)
D	Teacher	Professional (i)
E	Large business owner	Entrepreneur (ii)
F	Small business owner	Entrepreneur (ii)
G	Daily wage earner (X4)	Manual labourer (iii)

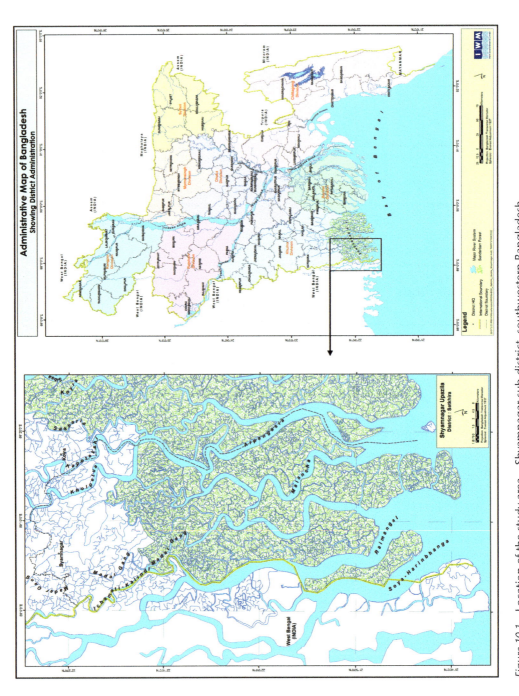

Figure 10.1 Location of the study area – Shyamnagar sub-district, southwestern Bangladesh
Source: Institute of Water Modelling – Bangladesh with permission

The research aimed for equal gender representation (achieved in all groups except entrepreneurs). In groups i–iii, six males and five females were interviewed. The final group (iv) consisted of five male experts, including engineers from IWM (two) and from the Bangladesh Water Development Board, Khulna Division (one), an independent local charity worker from Shyamnagar with first-hand experience of dealing with the impacts of Amphan, and a local government official (responsible for surveying and road maintenance). Group iv was interviewed to gain technical information and insights into issues raised by participants in groups i–iii.

All documents, consent forms, and questionnaires were communicated to most of the participants by email. Those without email gave permission to be interviewed verbally over the phone. Interviews were conducted in Bengali (local language) between 12 and 23 July 2020 using WhatsApp and Viber and were recorded using a Dictaphone. Interviews were transcribed into English and then coded thematically.

Findings were analysed using a conceptual framework developed from the academic literature to explain disaster experiences in relation to disaster preparedness, as well as the capacity to respond/carry out RR measures (Figure 10.2).

The literature revealed that community response to a disaster can be significantly improved by adequate preparedness (Rodríguez-Espíndola et al. 2018; Yadav & Barve 2019). This research used 'disaster preparedness' to embrace all activities undertaken to prepare communities prior to or during a disaster. These include risk awareness, early warning, dissemination of information, and other DRR measures, such as access to cyclone shelters and WaSH facilities. The quality of shelters and availability of WaSH facilities signify whether health risks (including COVID-19) were incorporated in preparedness measures.

Preparedness measures alone cannot guarantee an effective response by all, as they depend on individual capacities to act. Factors influencing an individual's capacity include vulnerability (socio-economic), social relations/capital, institutional support (finance and resources), hazard perception, and overall coping mechanisms. Attempts were made to shed light on these factors to evaluate the participants' response capacity to the dual disasters.

Overall, participant preparedness and capacity to anticipate, cope with, resist, and recover from Cyclone Amphan during the pandemic were assessed against the following preparedness indicators:

1) understanding the risks
2) pre-disaster warnings
3) communication effectiveness
4) capacity to respond and recover (institutional and public)

A semi-structured questionnaire centred around these issues was prepared and used in the interviews. Ethical review of the research was carried out according to University of Chester's research ethics protocol.

Study area

Satkhira is a district in Khulna Division, southwestern Bangladesh. The district consists of two municipalities, seven sub-districts, 79 union porishods, 8 thana (police stations) and 1436 villages.

Shyamnagar (the specific study area) is Satkhira's southernmost sub-district, bordered by India (west) and the Sundarbans (south – 78% is Sundarban forest). Shaymnager is further

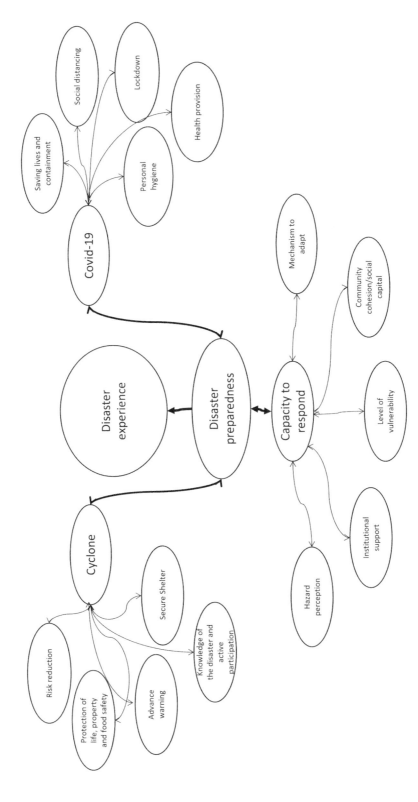

Figure 10.2 Conceptual framework for reducing disaster risk through preparedness measures used in the study (developed by S. Lageard)

divided into 12 administrative unions, the most disaster-prone corner of Bangladesh (Quader et al. 2017; Ahsan & Khatun 2020; Ahsan et al. 2020).

Shyamnagar is highly rural (only 5% urban population – Banglapedia 2015). The area is laced with earth embankments, keeping numerous tidal rivers from flooding agricultural land daily. Embankments have been built as part of state and international donor-funded infrastructural interventions since the 1960s, providing tidal flooding control and boosting the productivity of low-lying farmland. Unfortunately, once these embankments are breached (cyclone or strong tidal surge), the area becomes flooded and waterlogged with seawater. Saline incursions can make farming difficult, jeopardising people's livelihoods (65% in Shyamnagar rely on agriculture to earn a living – Banglapedia 2015).

According to Mudasser et al. (2020), around 35% of Bangladesh's coastal population live in extreme poverty, and their livelihoods are therefore extremely vulnerable to natural hazards. Table 10.1 demonstrates the poor infrastructure and livelihood status of Shyamnagar and the exceptionally low literacy rate (39.6% versus 45.5% – Satkhira District). Literacy clearly limits the population's livelihood options and increases their vulnerability when livelihood capital such as land and water become non-viable.

Prior to 2020, the population of Shyamnagar endured cyclones of a similar strength to Amphan in 2007 (Sdir) and in 2009 (Aila) (Alam & Rahman 2019). Local people therefore had a good understanding of the risks associated with and measures recommended for reducing cyclone hazards. They also constantly adapt their livelihood strategies to mitigate the hazards associated with such disasters, such as the increasing salinity of land and water. Employment accordingly is focused on small-scale shrimp/crab farms, rice cultivation, fishing, and other natural resource extraction (e.g. honey and timber from the Sundarbans).

A rapid needs assessment (RNA) was undertaken in Satkhira District by CBM (UK charity) between 23 and 25 May 2020 among people with disability. They identified food security and livelihood as the two most pressing concerns among this vulnerable group. Interestingly, despite their disability, this group played down the need for healthcare (CBM 2020). Clearly, they believed that if they had sufficient food and a job, health security would then follow, as they must pay for healthcare out of their own pockets. This view is widely held among poor households.

Shyamnagar was battered by Amphan – wind speeds 160–180 kph and a tidal surge exceeding 3 m. An officer in charge of Shyamnagar sub-district stated that the cyclone destroyed 25 km of embankments and 15 km of roads and flooded 30,000 hectares of shrimp farms. At least 5,000 livestock were killed (United News of Bangladesh 2020). In May 2020 the situation for marginal groups in remote parts of Shyamnagar (Munshiganj, Padma Pukur, Gabura) was dire due to COVID-19 lockdown, which halted travel further afield, compromising their primary dry season survival strategy. The arrival of Amphan compounded this problem by destroying their meagre local means (only) of earning livings. Amphan washed away livelihood

Table 10.1 Vulnerability of Shyamnagar residents

Sanitary	Sanitation (% latrines)	Drinking water source (%)	Roads (km)	Land ownership (%)	Source of income (%)	Owned	Landless	Farming	Non-farm labour	Business
44.8	47.5	36	7	51	67	35	811	56.7	43.3	65	6	14.6

Source: Compiled from selected data – en.banglapedia.org

capital such as embankments, crops/trees, and fisheries and also led to the collapse of sanitation infrastructure, clearly demonstrating the need to incorporate the health emergency in DRR preparedness measures.

Results and discussion

Participants represented varying levels of disaster preparedness vulnerability based on education, livelihood strategy, and housing type (Figure 10.3). Group i were professionals (P1–P4), all but one living in and owning brick buildings. Group ii, the poor, all lived in houses built with fragile materials such as mud and bamboo panels. Only one owned their own property, three of four participants were illiterate, and two were female heads of household (P5–P8). Group iii (P9–P11) included two involved in small businesses (a toiletries shop closed at the time of interview due to COVID-19 and a shrimp farm on rented land). The shrimp farmer and the large business owner lived in brick houses, the toiletry shop owner in a mud hut. Those living in mud huts and on rented land were at the greatest cyclone risk and were in most need of a well-functioning support system to enhance their capacity to resist and cope with the dual disasters.

Impact of Cyclone Amphan

Ninety percent of participants felt its impact directly. Direct effects included damage to houses, commercial premises, fruit trees, farm animals, and fishing ponds and loss of the mobile phone network (power cut – DRR measure to prevent electrocutions). Most damage occurred to tangible assets and to physical infrastructure, but no lives were lost (Figure 10.4). Two participants also mentioned stress caused by inability to contact friends and family during the cyclone (no mobile phone network).

Indirect effects were consequences of direct impacts. Loss of farm animals, vegetable patches, fishing ponds, commercial trees, and fruit trees dealt huge blows to livelihoods. Around 50%

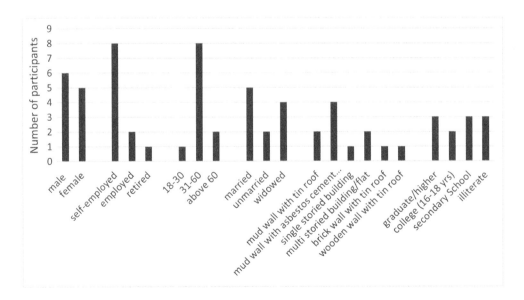

Figure 10.3 Socio-economic characteristics of the participants

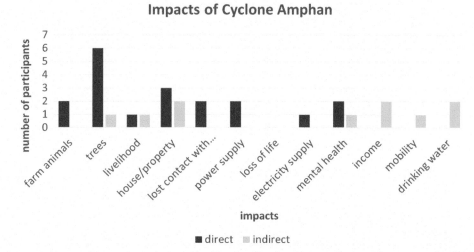

Figure 10.4 Direct and indirect impacts of Cyclone Amphan experienced by participants in Shyamnagar, Satkhira. NB: The storm surge led to the submersion of tube wells and toilets, causing temporary shortages of drinking water

claimed that they were already experiencing reduced incomes and were having to rely on alternative financial sources such as loans from extended family or finding off-farm work. One of the participants mentioned that movement restrictions (COVID-19) hampered their usual coping mechanism – travel to urban areas.

Preparedness for Cyclone Amphan

Advance warning as measure for disaster preparedness

All participants found out about the cyclone well in advance (two days to two weeks – Figure 10.5). According to the experts, information regarding the cyclone development (Bay of Bengal) was relayed to respondents seven days in advance, but the final cyclone track was only confirmed two days ahead. Both the participants and experts therefore agreed that 48 hrs prior to landfall, everyone in the study area had knowledge of the cyclone. The local government in collaboration with the Khulna District Office and BMD announced the disaster risks in Shyamnagar. According to participants and experts, representatives of local NGOs (Red Crescent), CBOs (Uttaron), and volunteers (CPP) informed the population of the severity of the cyclone and advised them on preventative action. In general, people were advised to stay in safe places or move to shelters, stockpile dry food and drinking water, move livestock to safe places (killas), store valuables in a safe place and perishable items on higher ground, tie down houses, and ring the emergency number if required.

Risk communication

Official information (BMD) was relayed using TV and radio bulletins, social media announcements and mobile phone alerts (service providers), and announcements in places of worship. Forty-eight hours before the cyclone arrived, community/disaster management volunteers

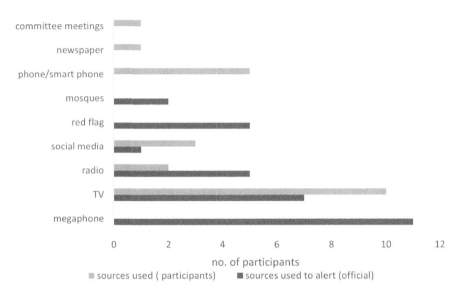

Figure 10.5 Methods of receiving cyclone alerts and gathering cyclone-related information

went round villages with megaphones alerting people to the measures they must undertake to avoid the loss of life and property, and hoisted red warning flags. Local people mainly gathered and updated news from TV and mobile phones. These, however, became redundant due to network outages. Most participants considered TV the most reliable information source, possibly due to its visibility (multiple channels), audibility, and regular updates. Fifty percent of participants (mostly educated and well off) also indicated their reliance on mobile phones for cyclone-related information (Figure 10.5).

Preparedness for COVID-19 containment

All participants were aware of rules and regulations regarding COVID-19 avoidance, and of the lockdown/travel restrictions when Amphan struck, but found these hard to follow because community/CPP volunteers encouraged them to prioritise saving lives and to access cyclone shelters. Participants further claimed that available shelters housed people beyond their intended capacities, making it physically impossible to socially distance.

P6 & P8 state, 'there was a bathroom in the shelter, but was shared between 500 people' (P6); 'no precaution to contain COVID-19. 600 people were packed in one shelter and 30 people in one room. All had to walk to the centre. Volunteers often carry the elderly' (P8).

These clearly indicate limited provisions were made to contain COVID-19 within cyclone shelters, and generally COVID-19 measures took a back seat during the cyclone, contradicting government claims. According to the experts, the local authority did not have sufficient time or funds to prepare shelters to meet required COVID-19 standards. Some shelters were school buildings that had not been used since March 2020 due to the nationwide lockdown. Further, the healthcare system in Bangladesh had limited capacity to provide free masks, soap, and hand sanitisers for all. PPE was made available only to the volunteers who assisted people in reaching the shelters (E2).

As a result, lifesaving DRR measures posed health risks. These were exacerbated by poor WaSH facilities (no separate facilities for the elderly, women, and children) and by shelter locations some distance from homes. To avoid these health risks, more vulnerable participants decided to either stay at home or with neighbours in concrete houses. All participants (Groups i–iii) and the experts agreed that the measures taken to contain COVID-19 were unsatisfactory. These views stem from the lack of universal healthcare provision in Bangladesh. People pay out of their own pockets for masks (TK 40), soap, water, and COVID-19 test certificates (TK4000–5000 requires seeing a doctor), putting considerable strain on dwindling disposable incomes. To make matters worse, water taps in some hard-hit areas ran on meters, making it impossible for some to adhere to the hygiene requirements/prioritise healthcare needs

> A water tap has been installed nearby, but it runs on a meter. It is impossible to maintain hygiene since it is very expensive. Charges range from TK150–200.
>
> *(P6)*

E4 (local government employee) refuted these claims, stating that huge efforts were made by district and sub-district administrative offices with the help of NGOs to set up water taps and distribute masks and hand sanitisers/soap, as well as food (rice, dal, and onions). Official documents also cited efforts made by the Bangladesh Red Crescent Society (BDRCS) to distribute PPE, food, and hygiene materials to every district and sub-district (reliefweb 2020). So why the anomaly in these responses? Further discussion with participants revealed that some assistance was indeed given prior to Amphan, but this was mainly limited to those on a beneficiary list prepared by a local administrative officer. This allegedly left out the names of some in remote areas (extremely poor/no political connections). The nature of their tenancies may also have played a part, as two lived in illegally built houses on an embankment. They claimed that the pandemic had already made them even poorer, as it had affected their livelihoods in a similar manner as a natural hazard.

(P5: 'by abiding [with] the COVID-19 guidelines, many have become poorer, any opportunity of earning a living is lost'.)

The experts viewed preparation for Amphan as 'excellent', with clear guidance given regarding DRR measures. Those who did not comply, but lived near riverbanks were forcefully removed and taken to the nearest shelter. E4 added that each person in the shelter received cash towards food and transport.

It is likely that conflicting recollections regarding preparedness measures can be put down to mistrust and prevalence of selective provision.

Capacity to reduce disaster risks

Capacity to respond during and post-disaster was intertwined with the level of vulnerability of participants. For example, those belonging to the professional and entrepreneur groups did not experience drinking water shortages and avoided temporary shortages by procuring bottled water.

Poorer participants (Group ii) were less fortunate, travelling long distances for drinking water (ponds/tube wells) or relying on charity supplies. The latter faced further disruption due to damaged dirt tracks and mud embankments used as part of the road network. Therefore, water supply, poverty, and lack of institutional support reduced the adaptive capacity of some participants.

Most participants cited inadequate and untimely mobilisation of funds, no urgency in repairing breached/collapsed embankments, and limited capacity of locals/participants in reducing the cyclone-induced risks as hindrances to their capacity. They therefore identified poor institutional support as a reason for their increased vulnerability. In contrast, the experts blamed people's vulnerability on the location of their houses, which were built on mud embankments or by riverbanks, particularly prone to cyclone damage. In addition, these households were apparently also offered financial compensation to move out of the area, which they refused to do. Participants countered that the money offered was insufficient to buy alternative property. This in turn contradicted E1's assertion that the government paid three times the value of the land when acquiring it as compensation (land is generally procured by the District Administrative Officer on behalf of BWDB).

For the participants the consensus was that much-needed assistance was not immediately/promptly available because central government tended to deal with disasters when they arrived. Funds mobilised for relief and rehabilitation had to go through significant red tape before reaching the local authority.

P1 expressed a desire for the local CPP team to have more power and E2 wanted local people to have more capacity. The capacity of most participants was significantly reduced at the time of the cyclone, as it occurred in conjunction with COVID-19 (loss of income/livelihoods and untenable alternatives, e.g. searching for food in the Sundarbans).

> It [government] must make a concerted effort to repair the embankments during the winter season, establish a local disaster management wing under the supervision of grassroot groups, educate people about the risk and increase their capacity to effectively manage the disaster. The disaster funds should be made available to the local team, developing technical knowhow among the local population.
>
> *(P1)*

> Everything is back to normal in his area after two months. Farmers are growing rainy season rice. What can never be recovered is the loss of 30/40 yr old trees.
>
> *(E2)*

Poor participants and small business owners claimed that matters were made worse, as no assistance was provided by the authorities to help them in rehabilitating livelihoods or subsidising incomes. They felt that such issues should receive equal priority to disaster preparedness measures since their lives depend on their livelihoods. They also wanted the authorities to raise the height of the embankments and repair breached/collapsed embankments from previous storms as a matter of urgency, as they viewed embankments as their livelihood capital.

E1 confirmed that there was currently a national plan to address the issues raised by the participants. The plan included building more cyclone shelters, assisting people to build cyclone-resistant houses/houses on killas, raising the height of the embankments, and raising the platforms of tube-wells (avoiding submergence/drinking water shortages). World Bank funds available for such projects had apparently dried up, meaning that these projects were not implemented in Shyamnagar. This area may have been a lesser priority because the Sundarbans

Table 10.2 Prevalence of social networks and local support during Cyclone Amphan

Participant profession	Male/female	Neighbourhood cooperation	Community group involvement	Local government support
Physician	M	yes	yes	Poor
Teacher (primary school)	F	No	yes	Poor
Union Parishad member (Awami League)	F	yes	yes	selective/preferential
Police (deputy superintendent)	M	yes	yes	Poor
Daily wage earner (collecting honey, Sunderbans)	F	no	no	None
Daily wage earner (fisherman, local rivers)	M	yes	no	Selective
Daily wage earner (labourer/road construction)	M	No	no	Poor
Daily wage earner (farm labourer/tenant farmer)	M	No	yes	Poor
Large business owner (trading various)	M	yes	yes	Inadequate
Small business owner (shrimp farming)	M	yes	yes	Inadequate
Small business owner (toiletries/cosmetics shop)	F	yes	yes	selective/preferential

are seen to act as a natural storm barrier, although it still experiences tidal surges, waterlogging, and increased salination. The area is also inaccessible. E5 highlighted that it takes water management engineers half a day to reach the area by water and land, restricting daylight hours available for work. Nevertheless, these hurdles must be overcome, as they affect the livelihood capital of the coastal population.

Participants were asked about the extent to which community cooperation/social capital had played a positive role in enhancing their capacity (Table 10.2). Surprisingly, poor households denied having any support from their neighbours, perhaps due to them all being in a similar situation. This, however, does not explain why they failed to acknowledge support from community groups. Households of two participants were recipients of emergency assistance from a community-based charity, 'Baag Bidhoba' (widowed because of tiger attack). Possibly they failed to fully understand the question or did not want to jeopardise chances of future assistance. The remaining participants acknowledged the contribution made by neighbourhood youth groups, who sought immediate solutions to their problems during and post cyclone. Youth groups went 'above and beyond' to move vulnerable people to safe places, protect and repair the embankments (dumping 'headloads' of soil), remove fallen trees from roads, and supply drinking water and food to those in greatest need (Figure 10.6). All participants agreed that the support provided by the authorities was often inadequate and preferential. Based on these observations, it might be suggested that local welfare groups instead of local authorities were responsible for safe-guarding people during the dual disaster. E3 also asserted the importance of community cohesion:

> [It is] far more important than government as the assistance is given immediately.

(E3)

Figure 10.6 Community support in response to Cyclone Amphan, May 2020. A – distributing drinking water; B – water shortage (submerged tube well); C – cement for repairing embankments; D – distributing clothes and money; E – the rescue operation; F – distributing food and saline

Source: Images courtesy of Dev Prosad Mondal

All participants did, however, acknowledge the role played by NGOs (Red Crescent Society, Uttaran), Boys Scouts/youth groups, community-based women's organisations, and the local Catholic charity Caritas in enhancing participants' capacity to recover through financial and technical assistance (Figure 10.6). E1 also asserted the crucial role played by CBOs and NGOs in meeting basic needs. He further revealed that the authorities are solely responsible for the maintenance of infrastructure. A delay in repairing embankments can therefore be labelled inadequate support by the authorities for enhancing the capacity of vulnerable groups.

All in all, participants blamed poor institutional support/mobilisation of resources at an individual level and loss of livelihood capital as the main reasons for their limited capacity to deal with the dual disasters.

Challenges to preparedness for dual disasters

Bangladesh's global reputation for excellence in DRR was put to the test in the pre-monsoon Cyclone Amphan in 2020 due to the prevalence of another disaster, COVID-19. The antecedent factors such as:

1) lack of coordination between preparedness measures for cyclone DRR and the health emergency,
2) insufficient institutional support for social protection,
3) poor accountability of administrative officials and corruption, and
4) disaster reduction initiatives being non-inclusive
   were perhaps unique in this disaster-prone poor country.

1) Lack of coordination

Two days prior to Amphan, people were alerted to the risk and the measures they should take to reduce its impact, but health security during transit to and occupation of shelters was omitted, demonstrating poor preparedness due to lack of coordination for the dual disasters. Health risks should have been prioritised as part of all hazard preparedness planning. Participants had no knowledge of specific healthcare provision available during Amphan. Normally, participants relied on dispensaries for medical supplies, but these remained shut for 2–7 days depending on location.

To avoid similar challenges in the future, local authorities, as well as other stakeholders, should conduct health need/medical assessments and ensure medical provision in advance of disasters. Also WaSH systems need to form an integral part of the disaster preparedness programme. This must be achieved not only by setting up tube wells and taps in shelters/public venues, but also by clarifying the allocation of responsibility for running and maintaining WaSH facilities, that is, utilising reliable/recognised social groups (e.g. women's groups, school children). This will allow DRR measures to incorporate health aspects of both disasters effectively.

2) Insufficient institutional support for social protection

Participants claimed that reduction in their livelihood capital strained their capacity to adjust and cope. Well-organised and functioning institutional support could effectively improve people's capacity. Experts disclosed that a relief fund was made available for each cyclone-affected household (central government – TK2500). However, participants considered this amount too meagre, as it did not take into consideration loss of livelihood and people's inability to work due to COVID-19 restrictions. These claims are supported by Rahman and Matin (2020), who estimated that the financial support package needed by poor households in rural areas in such circumstances was TK1368 per person per month. They also revealed that rural households experienced a drop in daily incomes due to COVID-19 restrictions and an increase in people living below the poverty line (73%). These are simply the effects of 'disaster poverty'. Additionally, not everyone received relief money, as funds were dispersed via mobile phone accounts to those on beneficiary lists prepared by the local union chairman. Experts considered that the dual disaster prevalence placed a strain on funding required to address cyclone damage (Shammi et al. 2020). Charity organisations/NGOs were not able to fill gaps left in provision as normal, because their funds had dried up due to cutbacks and demands elsewhere (national/global emergencies, e.g. within Bangladesh relief funds are being diverted to the Rohingya refugee camps in Coxes Bazar). Relief, however, can only temporarily support poor households, and they really need long-term solutions through effective socioeconomic recovery programmes.

3) Poor accountability and corruption

Eighty percent of participants blamed mismanagement of funds for them not receiving assistance (P3: 'Government policies are poorly implemented and monitored. Often it isn't fully realised due to corruption and nepotism. Union members/the chairman are responsible for a fair distribution, but that is a rarity'.)

Most participants and some of the experts claimed that commissioners, union chairmen, and village members (political representatives who have administrative power in decision making) were openly practicing nepotism, but nobody could challenge them, as they hold office for four/five years. Some allocated funds and resources to people, not all of whom were badly affected. Allegedly, such practice is common among people with power and not the preserve of one particular political party. Poorer participants were extremely bitter (P6: 'Members are dishonest. They don't distribute funds fairly. The list of affected people often doesn't include the neediest. People favoured by Members mostly received the fund'.)

This view wasn't restricted to poor participants and was supported by E2, who was directly involved in helping the community through distributing relief. He went as far as claiming that corrupt BWDB officials overlooked the damage done to the embankments by the fish farmers and therefore perpetuated the crisis regarding livelihood capital. His assertion has some truth, as the local authority is permitting shrimp aquaculture as part of coastal area's resilience (climate change). Sadly, such industries only benefit wealthy farmers, not those growing food crops for a living (who suffer increasing salinity).

E1 also acknowledged the prevalence of corruption in the system, blaming this on the long supply chain for fund distribution and the lack of a fair distribution mechanism. E2 suggested that if the government knew certain members/chairmen were corrupt, then relief funds should have been distributed through alternative channels such as the army and NGOs. The army especially could be used to repair breached embankments in remote areas.

The involvement of patronage networks and class hierarchy in misappropriation of disaster relief and rehabilitation resources favours local elites and leaves the most needy out of the equation (Masud-All-Kamal & Monirul Hassan 2018). This practice not only erodes community cohesion, but also makes poorer households distrustful of those in charge, making risk prevention harder to achieve.

4) Inclusion

Participants considered disaster preparedness should not only focus on reducing loss of life and property; it should also extend to the security of livelihoods and tenancies. As people live off the land and any damage to embankments disrupts their livelihoods, greater importance should be placed on maintaining this key infrastructure.

There is logic to this as repairing embankments would reduce people's vulnerability to the next disaster. Participants claimed that the local authorities (union/UDMC) often failed to seek people's opinions regarding their needs. This claim was refuted by the experts, who gave assurances that locals were/are always consulted at all stages of initial planning and that embankment development does feature in future coastal development plans. Apparently, there is a plan to design/redesign all the embankments and to raise their height by 1 m, allowing them to withstand sea level rise and extreme weather events (concrete embankments are too costly). Efforts were allegedly being made to reinforce mud embankments through turfing and afforestation [but NB E3: 'areas without embankments recovered quickly, but it took longer in areas where embankments were breached (it won't be before the dry season when the embankments will get repaired)'.] There is a further plan to move 600 inhabitants of Gabura Union, a remote area in Shyamnagar, to more permanent residences elsewhere.

There are two issues with these assertions made by the experts. First, there is a misunderstanding of the term 'inclusive'. The future development plans *are* being formulated by experts in DM, where they consult stakeholders. However, this is not entirely reflective of an inclusive policy. For a plan to be truly inclusive, it must be based on the priorities of vulnerable groups (Chu et al. 2016). Hasan et al. (2019), for example, claimed that little had been done regarding gender-inclusive access to the relief system, ensuring gender security in shelters and access to information and resources. Second, having a plan and its implementation are two different things (Ahmed 2019). Due to poor institutional capacity and varying levels of uptake, plans do not always translate into implementation. Choudhury et al. (2019) asserted that although Bangladesh has decentralised disaster management to local government and disaster management centres (UDMCs – each comprising a chairperson, members, and government officials working at the Union level; NGO representatives; local elites; and community representatives), these bodies lack autonomy and are often poorly governed. Interestingly, participants demonstrated their distrust of some of the individuals involved in their local UDMC.

In summary, the reasons for preparedness measures being ineffective for managing the dual disasters considered in this research are linked to poor governance, poor coordination, inadequate institutional support, and non-inclusive DRR plans. Clearly this is in breach of the SFDRR framework, since it promotes increased resilience through investing in DRR and good governance (UNDRR 2015). Surprisingly, the experts contributing to this research associated the problem with housing location and with livelihood strategies of the local population. Clearly these are the factors that are linked to participants' vulnerability and therefore should have received attention in the DRR resilience measures.

Conclusion and policy recommendations

This research was carried out to ascertain the adequacy of existing preparedness measures for the dual disasters of Cyclone Amphan and COVID-19. It revealed that participants had a good understanding of both hazards and were alerted to Cyclone Amphan 48 hrs in advance.

Participant responses, however, raised questions regarding the effectiveness of the measures put in place to reduce disaster risks. Cyclone shelters were over capacity, there was little emphasis on social distancing, and sanitary hygiene also suffered, as WaSH facilities were often non-existent. This should not have been the case, as the population of Shamynagar, despite their remote location, had had time to adjust to COVID-19 and also had a reasonable understanding of disease containment and how to save lives. These issues reflected a lack of coordination between health and DRR stakeholders, as health risks were not effectively integrated into cyclone preparedness measures.

The local population's capacity to prepare for the dual disaster was hampered by pressure on their livelihoods and incomes during Amphan. Most participants were forced to prioritise needs, making choices between buying soap, masks, and medicine *or* food, reflecting the effects of unequal access to healthcare in the country as a whole.

For the vulnerable, this dual disaster only differed from a single disaster because their capacity was reduced even further. This fact resonated throughout the research, as vulnerable participants specifically reiterated the need to preserve their livelihood capital at the time of any disaster. They played down the need to prioritise health hazards, as they would have to pay for healthcare from their own pockets. Hinderances to their capacity to cope were identified as inadequate support from DRR stakeholders, misappropriation of relief funds and nepotism, poor safeguarding of livelihood capital (e.g. embankments), and non-inclusive development plans.

Policy recommendations – DRR stakeholders

1) Make the preparedness measures effective

Prioritise the 'first mile' and mobilise CPP volunteers, youth and women's groups, and other CBOs to educate vulnerable groups (women, people with disabilities/impairments, the poor, and people generally susceptible to disasters) in possible risks and precautions; they must be empowered with additional knowledge, skills, and resources.

2) Increase the capacity of the vulnerable group to cope with dual disasters

Financial and technical support are needed *before* rather than post-disaster, making protective measures feasible. For example, people at risk from disasters could buy ropes, store bottled water, and carry out necessary maintenance on their houses, as well as purchase medication and hygiene products all in advance. These measures would lessen the demand for emergency relief.

3) Prioritise primary hazards

For Cyclone Amphan, these should have included the health hazards associated with COVID-19. It is a common practice to deliver saline and fresh water to affected households, but this should be expanded to include products to maintain sanitary hygiene, that is, WaSH facilities. These received inadequate attention during Amphan, evidenced in this research by the lack of drinking water, sanitation, and safe spaces for vulnerable groups (children, elderly, women, adolescents) in cyclone shelters. DRR stakeholders/UDMC members must regularly inspect shelters to assess their efficacy and where necessary, promptly repair or install drinking water facilities and latrines with vulnerable groups in mind. Further, medical treatment available at the time of a cyclone should not only cater to physical injuries, but should also make provision for communicable and non-communicable illnesses.

4) Repair breached embankments

This should be undertaken immediately *as a matter of urgency* to remove the bottleneck in maintaining livelihoods. The largest negative effect of the dual disasters was loss of livelihoods, indirectly affecting the health of the vulnerable. Officially the government has already undertaken programmes to rebuild and repair damaged/breached embankments, but due to poor governance, this work is yet to reach remote areas, and this needs scaling up.

5) Ensure good governance

This research highlighted the misappropriation of relief funds due to poor accountability of those in charge, making the impacts of the dual disaster humanmade rather than purely natural. Every effort needs to be made to make the distribution of the relief funds transparent and the supply channels as short as possible.

6) Strive to achieve health security for all

The dual disaster served once again to highlight not only the ineffectiveness of some preparedness measures, but also the health inequalities prevalent in Bangladesh. The country is some way from achieving universal healthcare access, although some progress has been made in maternal health. Currently, healthcare provision is very much urban-centred and pro-rich, that is, those who can afford to pay for 'out of pocket' medical care. To effectively manage all hazards in a disaster, healthcare stakeholders must also sit on the DRR committee.

Acknowledgements

The authors are thankful for the funding provided by University of Chester to conduct this research.

References

Ahmed, I. (2019) 'The national plan for disaster management of Bangladesh: Gap between production and promulgation'. *International Journal of Disaster Risk Reduction*. 37. 101179.

Ahsan, M. N. (2016) 'Disaster preparedness at household and community levels: The case of cyclone prone coastal Bangladesh'. Unpublished Ph.D. Thesis, National Institute of Policy Studies, Japan. https://grips.repo.nii.ac.jp/?action=pages_view_main&active_action=repository_view_main_item_detail&item_id=1525&item_no=1&page_id=13&block_id=24 (last accessed on 13 April 2021).

Ahsan, M. N. and A. Khatun (2020) 'Fostering disaster preparedness through community radio in cyclone-prone coastal Bangladesh'. *International Journal of Disaster Risk Reduction*. 49. 101752.

Ahsan, M. N., A. Khatun, M. S. Islam, K. Vink, M. Ohara and B. S. H. M. Fakhruddin (2020) 'Preferences for improved early warning services among coastal communities at risk in cyclone prone southwest region of Bangladesh'. *Progress in Disaster Science*. 5. 100065.

Ahsan, M. N., K. Takeuchi, K. Vink and J. Warner (2016) 'Factors affecting the evacuation decisions of coastal households during Cyclone Aila in Bangladesh'. *Environmental Hazards*. 15(1). 16–42.

Alam, K. and M. R. Rahman (2019) 'Post-disaster recovery in the Cyclone Aila affected coastline of Bangladesh: Women's role, challenges and opportunities'. *Natural Hazards*. 96. 1067–1090.

Banglapedia (2015) 'Shyamnagar Upazila'. http://en.banglapedia.org/index.php?title=Shyamnagar_Upazila (last accessed on 11 April 2021).

Blaikie, P., T. Cannon, I. Davis and B. Wisner (1994) *At risk: Natural hazards, people's vulnerability, and disasters*. Routledge, London.

CBM (2020) 'Cyclone Amphan: Inclusive rapid needs assessment'. https://reliefweb.int/sites/reliefweb.int/files/resources/Cyclone%20Amphan%20-%20Inclusive%20rapid%20needs%20assessment%20-%20Bangladesh%20Satkhira%2C%20Patuakhali%20and%20Bagerhat%20Districts%2C%20May%2023-25%2C%202020.pdf (last accessed on 11 April 2021).

Choudhury M. U. I., M. Salim Uddin and C. Emdad Haque (2019) 'Nature brings us extreme events, some people cause us prolonged sufferings: The role of good governance in building community resilience to natural disasters in Bangladesh'. *Journal of Environmental Planning and Management*. 62(10). 1761–1781.

Chu, E., I. Anguelovski and J. Carmin (2016) 'Inclusive approaches to urban climate adaptation planning and implementation in the global South'. *Climate Policy*. 16(3). 372–392.

Communication with Communities Working Group (2020) 'COVID-19: Risk communication and community engagement update'. https://reliefweb.int/sites/reliefweb.int/files/resources/covid_19_risk_communication_and_community_engagement_update_02-08_april_2020_20200410.pdf (last accessed on 11 April 2021).

Dariagan, J. D., R. B. Atando and J. L. B. Asis (2021) 'Disaster preparedness of local governments in Panay Island, Philippines'. *Natural Hazards*. 105(2). 1922–1923.

Djalante, R., R. Shaw and A. DeWit (2020) 'Building resilience against biological hazards and pandemics: COVID-19 and its implications for the Sendai framework'. *Progress in Disaster Science*. 6. 100080.

EM-DAT (2020) 'The emergency events database'. https://www.cred.be/projects/EM-DAT.

Erbeyoğlu, G. and Ü. Bilge (2020) 'A robust disaster preparedness model for effective and fair disaster response'. *European Journal of Operational Research*. 280(2). 479–494.

Ginige, K., R. Haigh and D. Amaratunga (2010) 'Developing capacities for disaster risk reduction in the built environment: Capacity analysis in Sri Lanka'. *International Journal of Strategic Property Management*. 14(4). 287–303.

Haklay, M. (2013) 'Citizen science and volunteered geographic information: Overview and typology of participation'. In D. Sui, S. Elwood and M. Goodchild (eds) *Crowdsourcing geographic knowledge*. Springer, Dordrecht, pp. 105–122.

Hamidi, A. R., Z. Zeng and M. A. Khan (2020) 'Household vulnerability to floods and cyclones in Khyber Pakhtunkhwa, Pakistan'. *International Journal of Disaster Risk Reduction*. 46. 101496.

Haque U., M. Hashizume, K. N. Kolivras, H. J. Overgaard, B. Das and T. Yamamoto (2012) 'Reduced death from cyclones in Bangladesh: What more needs to be done'. *World Health Organisation Bulletin*. 90. 150–156.

Hasan, M. R., M. Nasreen and M. A. Chowdhury (2019) 'Gender-inclusive disaster management policy in Bangladesh: A content analysis of national and international regulatory frameworks'. *International Journal of Disaster Risk Reduction*. 41. 101324.

Hasina, S. and P. Verkooijen (2020) 'Fighting cyclones and coronavirus: How we evacuated millions during a pandemic'. *The Guardian*, 3 June. www.theguardian.com/global-development/2020/jun/03/fighting-cyclones-and-coronavirus-how-we-evacuated-millions-during-a-pandemic (last accessed on 11 April 2021).

IFRC (International Federation of Red Cross and Red Crescent Societies) (2021) 'Disaster preparedness: Working with communities to prepare for disasters and reduce their impact'. https://media.ifrc.org/ifrc/what-we-do/disaster-and-crisis-management/disaster-preparedness (last accessed on 11 April 2021).

Ishiwatari, M., T. Koike, K. Hiroki, T. Toda and T. Katsube (2020) 'Managing disasters amid COVID-19 pandemic: Approaches of response to flood disasters'. *Progress in Disaster Science*. 6. 100096.

Islam, M. T., M. Charlesworth, M. Aurangojeb, S. Hemstock, S. K. Sikder, M. S. Hassan, P. K. Dev and M. Z. Hossain (2021) 'Revisiting disaster preparedness in coastal communities since 1970s in Bangladesh with an emphasis on the case of Tropical Cyclone Amphan in May 2020'. *International Journal of Disaster Risk Reduction*. 58. 102175.

Islam, R., G. Walkerden and M. Amati (2017) 'Households' experience of local government during recovery from cyclones in coastal Bangladesh: Resilience, equity, and corruption'. *Natural Hazards*. 85. 361–378.

Izumi, T. and R. Shaw (2022) 'A multi-country comparative analysis of the impact of COVID-19 and natural hazards in India, Japan, the Philippines, and USA'. *International Journal of Disaster Risk Reduction*. 73. 102899.

Li, Z. and X. Tan (2019) 'Disaster-recovery social capital and community participation in earthquake-stricken Ya'an areas'. *Sustainability*. 11(4). 993.

Mallick, B. (2014) 'Cyclone shelters and their locational suitability: An empirical analysis from coastal Bangladesh'. *Disasters*. 38(3). 654–671.

Masud-All-Kamal, M. and S. M. Monirul Hassan (2018) 'The link between social capital and disaster recovery: Evidence from coastal communities in Bangladesh'. *Natural Hazards*. 93(3). 1547–1564.

Mazumdar, S., P. G. Mazumdar, B. Kanjilal and P. K. Singh (2014) 'Multiple shocks, Coping and welfare consequences: Natural disasters and health shocks in the Indian Sundarbans'. *PloS One*. 9(8). e105427.

Momtaz, S. and M. I. M. Shameem (2015) *Experiencing climate change in Bangladesh: Vulnerability and adaptation in coastal regions*. Academic Press, London.

Mudasser, M., M. Z. Hossain and K. R. Rahaman (2020) 'Investigating the climate-induced livelihood vulnerability index in coastal areas of Bangladesh'. *World*. 1. 149–170.

Oloruntoba, R., R. Sridharan and G. Davison (2018) 'A proposed framework of key activities and processes in the preparedness and recovery phases of disaster management'. *Disasters*. 42(3). 541–570.

Paul, B. K. (2012) 'Factors affecting evacuation behavior: The case of 2007 Cyclone Sidr, Bangladesh'. *The Professional Geographer*. 64(3). 401–414.

Paul, B. K. and S. Dutt (2010) 'Hazard warnings and responses to evacuation orders: The case of Bangladesh's Cyclone Sidr'. *Geographical Review*. 100(3). 336–355.

Peleg, K., M. Bodas, A. J. Hertelendy and T. D. Kirsch (2021) 'The COVID-19 pandemic challenge to the all-hazards approach for disaster planning'. *International Journal of Disaster Risk Reduction.* 55. 102103.

Quader, M. A., A. U. Khan and M. Kervyn (2017) 'Assessing risks from cyclones for human lives and livelihoods in the coastal region of Bangladesh'. *International Journal of Environmental Research and Public Health.* 14(8). 831.

Rahman, H. Z. and I. Matin (2020) 'Livelihoods, coping, and support during COVID-19 crisis'. Report by Power and Participation Research Centre (PPRC) and BRAC, Institute of Governance & Development (BIDG). https://bigd.bracu.ac.bd/wp-content/uploads/2020/04/Round-1_23_April_PPRC-BIGD-Final.pdf (last accessed on 12 April 2021).

reliefweb (2019) 'Bangladesh: From cyclones to multi-hazard'. https://reliefweb.int/report/bangladesh/bangladesh-cyclones-multi-hazard (last accessed on 11 April 2021).

reliefweb (2020) 'Risk communication and community engagement strategy coronavirus disease 2019 (COVID-19)'. https://reliefweb.int/sites/reliefweb.int/files/resources/covid_19_rcce_strategy_cwc_wg.pdf (last accessed on 12 April 2021).

Rodríguez-Espíndola, O., P. Albores and C. Brewster (2018) 'Disaster preparedness in humanitarian logistics: A collaborative approach for resource management in floods'. *European Journal of Operational Research.* 264(3). 978–993.

Roy, C., S. K. Sarkar, J. Åberg and R. Kovordanyi (2015) 'The current cyclone early warning system in Bangladesh: Providers' and receivers' views'. *International Journal of Disaster Risk Reduction.* 12. 285–299.

Sadeka, S., M. S. Mohamad and M. S. K. Sarkar (2020) 'Disaster experiences and preparedness of the Orang Asli families in Tasik Chini of Malaysia: A conceptual framework towards building disaster resilient community'. *Progress in Disaster Science.* 6. 100070.

Saha, C. K. (2015) 'Dynamics of disaster-induced risk in southwestern coastal Bangladesh: An analysis on tropical cyclone Aila 2009'. *Natural Hazards.* 75(1). 727–754.

Sarabia, M. M., A. Kägi, A. C. Davison, N. Banwell, C. Montes, C. Aebischer and S. Hostettler (2020) 'The challenges of impact evaluation: Attempting to measure the effectiveness of community-based disaster risk management'. *International Journal of Disaster Risk Reduction.* 49. 101732.

Shammi, M., M. Bodrud-Doza, A. R. M. T. Islam and M. M. Rahman (2020) 'COVID-19 pandemic, socioeconomic crisis and human stress in resource-limited settings: A case from Bangladesh'. *Heliyon.* 6(5). 04063.

Shongjog (2020) 'COVID-19/Coronavirus – communication tools for the general population of Bangladesh'. www.shongjog.org.bd/resources/i/?id=af8f1a37-a8e1-4edf-a38c-b40ee07d5c10 (last accessed on 11 April 2021).

Simonovic, S. P., Z. W. Kundzewicz and N. Wright (2021) 'Floods and the COVID-19 pandemic – a new double hazard problem'. *Water.* 8(2). 1509.

Skinner, C. and R. Rampersad (2014) 'A revision of communication strategies for effective disaster risk reduction: A case study of the south Durban Basin, KwaZulu-Natal, South Africa'. *Jàmbá.* 6(1). 1–10.

Soloman, S. (2014) *Mobilizing social support networks in times of disaster, trauma and its wake.* Routledge, London.

Sultana, P., P. M. Thompson and A. Wesselink (2020) 'Coping and resilience in riverine Bangladesh'. *Environmental Hazards.* 19(1). 70–89.

Syed, M. A., M. S. Shrestha and V. Khadgi (2021) 'Chapter 33 – last mile communication of multihazard early warning – a case study on Bangladesh'. In I. Pal, R. Shaw, R. Djalante and S. Shrestha (eds) *Disaster resilience and sustainability.* Elsevier, Amsterdam.

UNDRR (2015) 'Sendai framework for disaster risk reduction 2015–2030'. www.undrr.org/publication/sendai-framework-disaster-risk-reduction-2015-2030 (last accessed on 11 April 2021).

UNDRR (2020) 'Disaster risk reduction in Bangladesh status report 2020'. UNDRR, Asian Disasters Preparedness Centre. https://reliefweb.int/sites/reliefweb.int/files/resources/Disaster%20Risk%20Reduction%20in%20Bangladesh%20Status%20Report%202020.pdf (last accessed on 11 April 2021).

United News of Bangladesh (2020) 'Cyclone Amphan: Locals demand sustainable embankment'. https://unb.com.bd/category/Bangladesh/cyclone-amphan-locals-demand-sustainable-embankment/51961 (last accessed on 11 April 2021).

UN Women (2020) 'As Bangladesh battles COVID-19 and the aftermath of Super Cyclone Amphan, women's organizations lead their communities through recovery'. www.unwomen.org/en/news/stories/2020/6/feature-bangladesh-womens-organizations-in-COVID-19-and-cyclone-recovery (last accessed on 11 April 2021).

USAID (2013) 'Second time around, Bangladesh finds success in preparing for the worst'. www.usaid.gov/news-information/frontlines/risk-resilience-and-media/second-time-around-bangladesh-finds-success (last accessed on 9 April 2021).

Wedawatta, G., U. Kulatunga, D. Amaratunga and A. Parvez (2016) 'Disaster risk reduction infrastructure requirements for south-western Bangladesh'. *Built Environment Project and Asset Management*. 6(4). 379–390.

The World Bank (2019) 'Bangladesh – multipurpose disaster shelter project: Shelter from the storm'. http://documents.worldbank.org/curated/en/841391568291583959/Bangladesh-Multipurpose-Disaster-Shelter-Project-Shelter-from-the-Storm (last accessed on 11 April 2021).

Wright, N., L. Fagan, J. M. Lapitan, R. Kayano, J. Abrahams, Q. Huda and V. Murray (2020) 'Health emergency and disaster risk management: Five years into implementation of the SENDAI framework'. *International Journal of Disaster Risk Science*. 11(2). 206–217.

Yadav, D. K. and A. Barve (2019) 'Prioritization of cyclone preparedness activities in humanitarian supply chains using fuzzy analytical network process'. *Natural Hazards*. 97(2). 683–726.

11
A review of the flooding events in southern Brazil
Challenges and opportunities

Francisco Henrique de Oliveira, Frederico Rudorff, Guilherme Braghirolli, Guilherme Linheira, Maria Carolina Soares, Raidel Baez Prieto, Regina Panceri, Renan Furlan de Oliveira and Victor Luís Padilha

Introduction

Floods are one of the types of natural disasters that have generated the most negative impacts on society over the last few decades (Zhang and Wang, 2022; Arinta et al., 2020; Ibrahim and Mishra, 2021; Roslan et al., 2021). In the context of developing countries, these impacts are aggravated by the accelerated population growth and intense urbanization that occurred in the same period. This panorama becomes even more worrying when we consider the effects of climate change and the consequent increase in the occurrence of recent extreme meteorological events.

Focusing on the last 20 years (2001–2020), there were 999 natural disaster events in Europe, of which 951 were weather related, meaning they belonged to the disaster subgroups meteorological, hydrological, or climatological. Combined, these events killed over 150,000 people, affected over 11 million others and costed over US$217 billion (EM-DAT, 2021). In the context of hydrological disasters, floods are among the phenomena with the greatest destructive power, since they can affect large areas for long periods of time, causing human losses and high material damage due to the high energy associated with the flow of large volumes of water into the ground.

In Brazil, floods have been a serious and recurring issue. Official records show that the phenomenon occurred 5,309 times between 1991 and 2019, affecting approximately 18 million inhabitants, generating losses of around US$3.34 billion, in addition to causing 588 deaths.

Brazilian floods may be correlated to several factors, including climatic and historical factors. The country has a coastline of approximately 7,500 km in length, with a territory of 8.5 million km², most of which is in the tropical zone. In general terms, it is possible to affirm that the country has a hot and humid climate, with significant amounts of precipitation throughout the year across nearly the entire national territory.

In addition to climatic factors, there is the pattern of territorial occupation in Brazil, where the urbanization process has been largely marked by unorganized expansion violating urban and environmental regulations. In practical terms, the removal of natural ground cover as well as its waterproofing has been recurrent; there is also the occupation of marginal strips of water courses and other naturally flooded areas.

Evidently, the impacts caused by floods do not show the same characteristics throughout the Brazilian territory, either because of the phenomenon itself or the structure/preparation of state agents in the control and mitigation of the event. There are places where the impacts end up generating more severe damages due to the economic significance of the relevant areas. In this context, the state of Santa Catarina (Figure 11.1) currently occupies the sixth position in the country's economic indexes of gross values (GDP) and fourth in relative values (GDP per capita) (SDE, 2020). The state has had a history of floods of great magnitude, registered in the years of 1974, 1983, 1984, 1992, 1997, 2008, 2009, 2011, 2013, 2014, and 2018. In economic terms, the impacts reported between 1995 and 2019 reached the amount of approximately US$313 million (CEPED, 2022).

Faced with this reality, the state of Santa Catarina has committed financial, human, and technological resources to deal with disasters. This proactive stance was consolidated as of 2008, when the state was hit by a flood of great proportions, considered the biggest disaster occurrence in Santa Catarina, in terms of the damages and the number of people affected, to date.

The actions taken by the state after the 2008 episode made the state a national reference in risk and disaster management (Ferreira et al., 2017). Therefore, this chapter is aimed at considering aspects of the great floods that affected the state, with emphasis on the 2008 episode, as well as the current administrative organization, policies, and actions of the Civil Defence of the State of Santa Catarina and the advances in implementing flood-fighting technological resources.

Overview of flood events in Santa Catarina and the 2008 event

The state of Santa Catarina, located in the southern region of Brazil, occupies an area of approximately 96,000 km², inhabited by an estimated population of 7.3 million (IBGE, 2021). The state territory is subdivided into six administrative regions, as shown in Figure 11.2. The most expressive economic sector in Santa Catarina is commerce and services, accounting for 66% of the state GDP, followed by industry with 27.1% and agriculture with 6.9% (SDE, 2020).

In climatological terms, using the Köppen classification, Santa Catarina's climate can be classified as humid mesothermal (no dry season), with annual precipitation of around 1,500 mm, which is evenly distributed over the seasons (Pandolfo et al., 2002). However, it is essential to highlight that the state is subject to the influences of the 'El Niño' and 'La Niña' phenomena, which consist, respectively, of positive and negative fluctuations in the temperature of the Pacific Ocean that interfere with the normal rainfall regime of the state (Monteiro, 2001).

According to World Bank and Centro Universitário de Estudos e Pesquisas Sobre Desastres/ Universidade Federal de Santa Catarina (CEPED/UFSC) data, natural disasters in Santa Catarina between 1995 and 2014 resulted in material damages and losses totalling US$3.45 billion, which is equivalent to an average annual loss of US$173 million, and accounts for 0.4% of the state's GDP and the biggest loss per capita and per km² in Brazil.

Regarding flood events, between 1991 and 2019, 468 occurrences were reported in the state of Santa Catarina. The regions with the highest concentration of occurrences are the northern region, mountain region, and Itajaí Valley (Santa Catarina, 2021, p. 9). However, in addition to this mentioned period, Santa Catarina has a history of disasters resulting from flooding events that span 1974 to 2018, as shown in Figure 11.3.

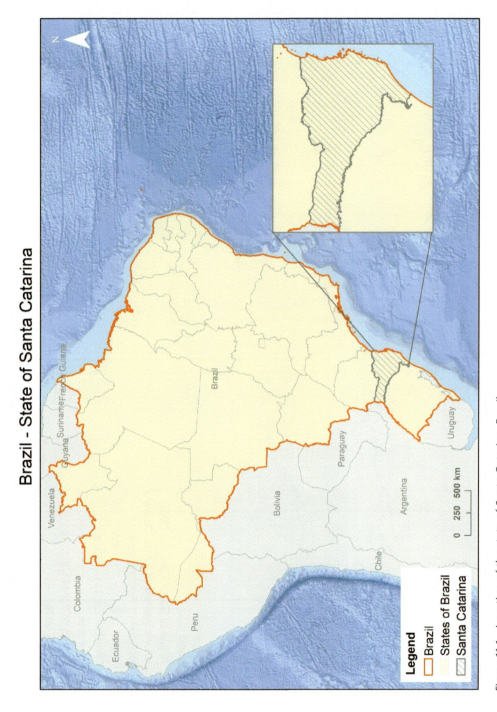

Figure 11.1 Location of the state of Santa Catarina, Brazil

Francisco Henrique de Oliveira et al.

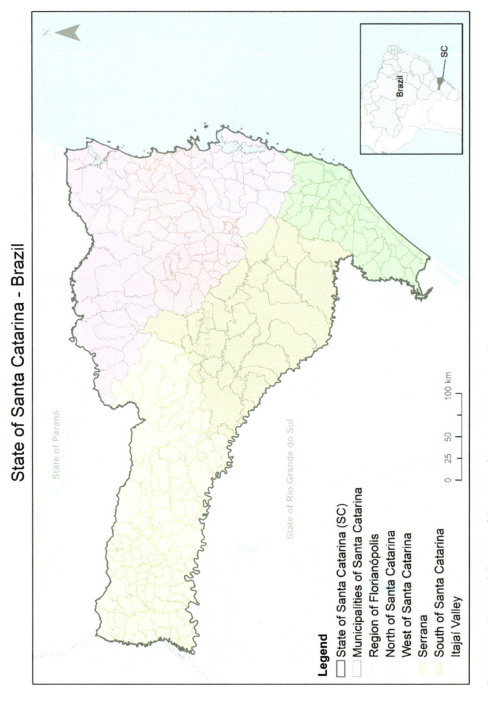

Figure 11.2 Regional division of the state of Santa Catarina, Brazil

A review of the flooding events in southern Brazil

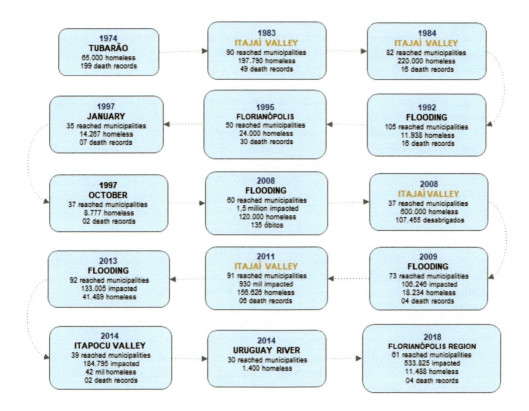

Figure 11.3 Main flood events in Santa Catarina between 1974–2018
Source: Author, based on Santa Catarina (2021)

Among the 30 disasters with the greatest impact on the state, including those in the Santa Catarina Historical Disaster Profile report (Santa Catarina, 2022), 15 are due to floods. To highlight these events historically, Figure 11.3 presents a timeline showing the main flood events that occurred in the state. The figure also shows the dimension of the impacts caused by the events based on the number of affected municipalities, number of homeless people, and deaths.

The eastern portion of the state shows higher concentration of flood events, despite the single occurrence in the western region, as shown in Figure 11.4. The concentration can be explained by the geomorphological characteristics of the eastern portion of Santa Catarina's territory. The hydrographic basins have steep slopes in the highest areas, which reach 1,000 m in altitude, and flat areas near the river mouths, located at sea level. Thus, there is a natural tendency for water to accumulate in flat areas, especially when heavy and persistent rains occur. Another important point to be considered, which involves socioeconomic factors in the eastern part of the state, is the prevalence of the most populous municipalities and, therefore, the potential for greater losses when impacted.

Based on the recurrence of reported flood episodes and considering the physical-geographical and socioeconomic characteristics of the municipalities in Santa Catarina, CEPED prepared a map showing the degree of risk and the resulting impacts of flood events for Santa Catarina. Figure 11.5 presents the spatialization of the mentioned classification.

As already mentioned, the 2008 flood event in Itajaí Valley was considered a turning point for Santa Catarina due to the magnitude of the damage caused. This 2008 environmental event demanded a new stance from the government, necessitating the need to rethink the actions and strategies of risk and disaster management adopted until then.

The series of climatological events addressed in this section comprises a sequence of progressive floods, which occurred in November and December of 2008, after an intense volume of precipitation. According to a report jointly drafted by the World Bank and Santa Catarina State Administration, the phenomenon occurred due to a high-pressure system located on the high seas, causing an abnormal volume of precipitation on the coast of Santa Catarina and inland. Regarding the impact of the phenomenon, the volume of rainfall reported in the municipality of Blumenau in November 2008 was approximately 1,000 mm compared with the historical average of 150 mm for the same period (Banco Mundial, 2012, p. 13). Therefore, the accumulation of water, the occupation of the slopes, and the formation of several urban centres culminated in mass movements and floods, resulting in human and material damages and losses. The 2008 event reached 60 municipalities of the state, of which 14 declared an emergency situation[1] and state of public calamity.[2] Both situations are declared by the local administrator (mayor of the municipality); however, the declaration of public calamity is considered when the destructive effect of the natural phenomenon (flood, for example) is more intense than in the declaration of emergency. The event negatively impacted 1.5 million citizens of Santa Catarina, causing 120,000 cases of homelessness and 135 deaths (Santa Catarina, 2022, p. 18).

Material losses amounted to approximately US$930 million, including losses in infrastructure affecting land and marine systems and the social sector. Damage and destruction to 73,000 housing units were recorded, in addition to the damage to the agriculture, livestock, fishing, forestry production, industry, and commerce sectors (Santa Catarina, 2022, p. 19).

In the Itajaí Valley region alone, shown in Figure 11.4, 37 municipalities were affected (27 municipalities declared a state of emergency and another 12 declared a state of public calamity). Among the human damage to the Itajaí Valley, more than 600,000 were affected (with 107,455 homeless), in addition to US$784 million in material losses (Santa Catarina, 2022, p. 18). The impact of the 2008 event can be better understood by Figures 11.6, 11.7, 11.8, and 11.9.

Civil Defence and its (re)actions

The Civil Defence of Santa Catarina was created on May 18, 1973, as a state coordinating agency, and its structure has evolved over the decades. In 2011 it gained the status of secretary of state, and in 2019 it became directly linked to the governor's office. An important step for the Civil Defence of Santa Catarina was the creation of the State Fund for Civil Defence and Protection (FUNPDEC – *Fundo Estadual de Proteção e Defesa Civil*), in 1990. The Fund's purpose is to capture, control, and apply financial resources to execute preventive and disaster preparedness, emergency relief and assistance, recovery and reconstruction of essential services for populations affected by disasters, and institutional strengthening and support for the State Civil Defence and Protection System (SIEPDEC – *Sistema Estadual de Proteção e Defesa Civil*).

For a long time, the Civil Defence actions were directed towards disaster management. The paradigm shifted after one of the worst disasters in history in 2008, which culminated in the creation of the Secretary of State for Civil Defence in 2011 and the implementation of the Integrated Centre for Risk and Disaster Management (CIGERD – *Centro Integrado de Gerenciamento de Riscos e Desastres*). CIGERD was inaugurated on May 18, 2018, the day when the Civil

A review of the flooding events in southern Brazil

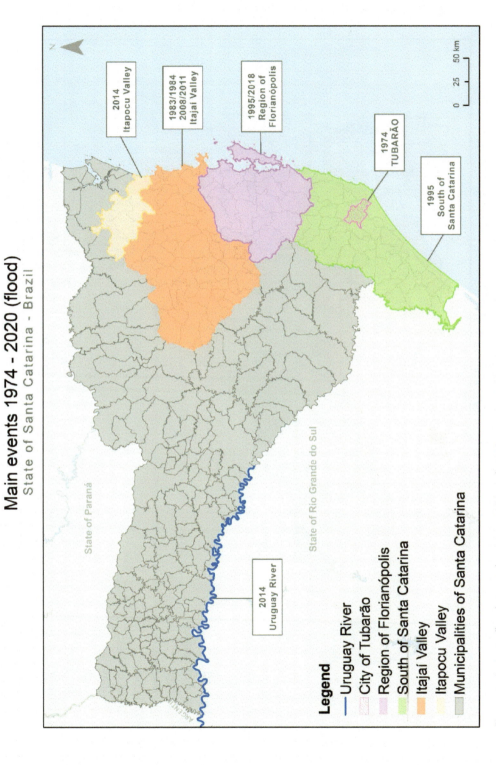

Figure 11.4 Main flood events between 1974 and 2020 in Santa Catarina

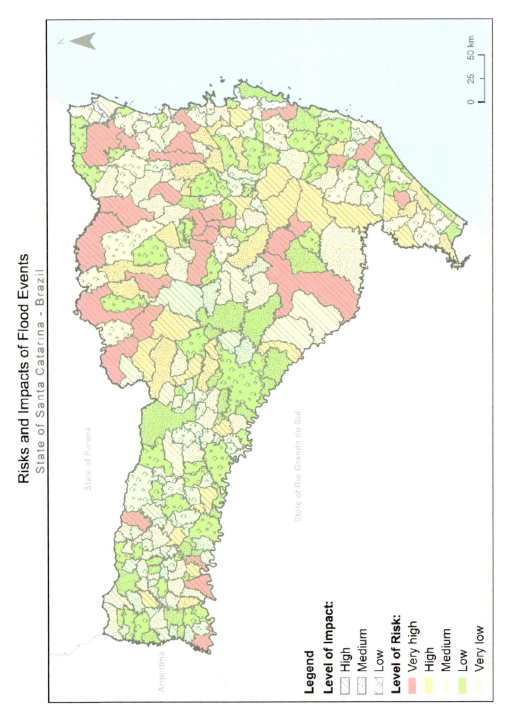

Figure 11.5 Risks and impacts of flood events in Santa Catarina

A review of the flooding events in southern Brazil

Figure 11.6 2008 floods – Itajaí Valley/SC
Source: Banco Mundial (2012)

Figure 11.7 Landslide caused by the rains of 2008 – Blumenau/SC
Source: Gilmar de Souza (2008)

Figure 11.8 Floods of 2008
Source: Jornal Expresso (2011)

Figure 11.9 Landslide caused by the rains of 2008 – Blumenau/SC
Source: Jornal O Município – Blumenau

Defence of Santa Catarina marked 45 years of history of investment in actions of prevention, mitigation, preparedness, response and reconstruction.

The Centre is part of a new proposal for the Civil Defence and protection system under development in Santa Catarina. Under it are the Monitoring and Alert Centre; the Integrated Operations Centre; the Public Information Centre; and the Education, Risk, Disaster and

Financial Management Directorates, in addition to 20 Regional Civil Defence Coordination Units that provide direct support to municipalities.

The Centre's main mission is to promote, facilitate, and support the joint efforts of the various sectors and government agencies involved in risk reduction in the territory of Santa Catarina. Its conception and execution were based on a benchmarking study carried out on different disaster management centres around the world such as in Japan, the United States, Chile, and Brazil (CEMADEN, CENAD), as well as on technological companies, including Dígitro and Google.

A reference in crisis and natural disaster management, CIGERD is self sufficient and prepared to maintain communication with the 20 regional CIGERDs as well as integrating government sectors for crisis management and faster response in risk situations. Regional centres were distributed based on geographic location, population, and recurrence of natural phenomena. These regional centres located in 20 municipalities are responsible for providing assistance in Civil Protection and Defence actions in their regions as well as in the municipalities belonging to their area of operation.

The advancement of the Santa Catarina Civil Defence preparedness after the 2008 disaster

According to the Ministry of National Integration, the natural disaster that occurred in 2008 in Santa Catarina is recognized as one of the worst disasters in the history of the state. It directly affected 1.5 million people, of which 82,770 were displaced and 38,261 were rendered homeless. Sixty municipalities declared an emergency situation and 14 declared a state of public calamity, and 135 lives were lost, with 97% of deaths resulting from landslides and about 429,000 people affected. Economic losses of about R$ 4.75 billion covered the infrastructure, social, and productive sectors. The costs of the disaster were five times greater than the total investment of Santa Catarina State Administration throughout 2008, with social and productive losses resulting from the 73,000 housing units either totally destroyed or damaged. Meanwhile, 40,000 of the affected houses belonged to low-income classes. In the productive sector, the total cost of the flood disaster was approximately R$ 1.39 billion, of which R$ 539 million was related to damages (direct impacts) to agriculture, livestock, fishing, and forestry production, while R$ 741 million of losses and damages came from the industrial sector and commerce.

The World Bank identified that the most significant impacts occurred in municipalities with the most fragile economies and the most limited capacity to respond. Regarding response capacity, the state of Santa Catarina encountered several difficulties in responding to this event due to its magnitude and the lack of public preparedness. Faced with the challenge, a mobilization/organization began in the sense of structuring of the State Civil Defence in order to change the directions and history of prevention, mitigation, and protection.

This event created a radical change in the planning and thinking of civil defence in Brazil, leading to the creation of the following groups and agencies.

1. Reaction Group (*Grupo REAÇÃO*): created in December 2008 to advise the administration on the definition of priorities related to the reconstruction of the damage caused by the rains.
2. Technical-Scientific Group: created in 2009 to assess and identify the causes, effects, and adoption of preventive measures against natural disasters in Santa Catarina.

3. Pact for Santa Catarina: started in 2010 with the administration's plan spanning 2011 to 2014, with the following priorities:

- Creating the Secretary of State for Civil Defence;
- Structural long-term investments;
- Providing the state with a structure for prompt response;
- Working with information sources that allow forecasting and issuing of alerts;
- Developing efficiency on a national reference level.

Civil Defence policies

The policies adopted by the Civil Defence of Santa Catarina address the need for a specific action plan with criteria for prioritizing interventions in disaster risk reduction considering the sectoral needs of the state. From the construction of strategic guidelines, objectives, and goals of the Santa Catarina Civil Defence and Protection Plan (PPDC – *Plano de Proteção e Defesa Civil de Santa Catarina*), the main actions to be included in the elaboration of the State Plan for Disaster Risk Reduction (PERRD – *Plano Estadual de Redução de Risco de Desastres*) and a model were presented with prioritization criteria focusing on the relevance of the intervention, scope, technical effectiveness, and financial viability.

The final document, to be published, contains the definition of monitoring instruments, evaluation and sustainability mechanisms of the PPDC, governance mechanisms, identification of strategies, and sources of financing and definition of the Framework for Risk Reduction of Disasters in Santa Catarina.

Also noteworthy is the monitoring of six progress indicators that represent the human and economic damage that impacts the population of Santa Catarina most. Thus, there is a complete analysis of the Santa Catarina scenario regarding disasters, identifying the most relevant ones, in addition to an analysis of the national and state legal context for civil defence and protection and its relationship with Santa Catarina's policies and strategies. Also, there is analysis of eight international plans from different countries and two national civil defence and (state and municipal) protection plans with the objective of supporting the construction of the structuring elements of the PPDC, based on the analysis of best practices.

Structural and non-structural measures under development

Urban planning directly affects water availability. Its effects include soil waterproofing, structural actions that cause flooding, and erosion and siltation, among others. Improper management of water in urban environments and urbanization can lead to many problems, putting human lives and properties at risk (Oneda, 2018).

In this context, instruments for controlling land use and occupation are fundamental tools for sustainable urban development and should be prioritized (Zahed Filho *et al.*, 2012). Among the possible sustainable measures for implementation in the territory is the need to develop a Master Plan for Urban Drainage. The plan is based on principles including:

- New developments cannot intensify maximum downstream flow.
- Planning and control of existing impacts must be prepared considering the basin as a whole.
- Planning horizon must be integrated into the Master Plan of the cities involved.

According to Miguez *et al.* (2012), in Brazil, the concepts of compensatory techniques have been consolidated in the urban drainage project through different measures, focusing on infiltration and storage capacity, with the objective of offsetting urban impacts on the hydrological cycle. However, with few exceptions, the design of urban drainage systems is still carried out following traditional concepts, resulting in financial, environmental, aesthetic, and health impacts and, above all, impacting the quality of life of the population (Canholi, 2015; Souza *et al.*, 2016).

As a result, the possibilities of intervention are diverse, with the purpose of mitigating the aforementioned factors. Non-structural measures can be effective, long-lasting, and even more cost-effective than structural measures. According to Canholi (2015), they can be grouped into:

- Actions to regulate land use and occupation.
- Environmental education aimed at controlling diffuse pollution, erosion, and garbage.
- Flood insurance.
- Flood warning and forecasting systems.

In this way, they are effective in the initial stages of disaster risk management but ineffective for correcting and mitigating floods in the already affected regions. Therefore, this chapter focuses on both structural and non-structural measures developed to mitigate floods in in Santa Catarina, located in the southern region of Brazil.

The first concepts introduced in urban drainage projects and plans differentiate structural and non-structural actions to solve problems of municipal flood events (Oneda, 2018). Structural measures imply the capturing, storing, and transporting of water volumes within limits established by quantification of risks and prior knowledge of flood waves, adjusted to local conditions through containment structures (Righetto, 2009).

Canholi (2015) describes that structural measures comprise engineering works that can be characterized as extensive and intensive measures. Extensive measures act on the surface of the basin, in part or in its full extension. The purpose is to combine the effects of environmental protection, improved planting, and soil conservation with flow reduction (Tucci, 2012). Extensive measures also include small storages disseminated in the basin, restoration of vegetation cover, and control of soil erosion along the drainage basin, according to Table 11.1. Its goals are to modify the relationships between precipitation and flow, and to reduce and delay the flood peaks.

Table 11.1 Extensive structural measures

Measures	Main advantage	Main disadvantage	Application
Change in vegetation cover with reforestation	Reduction of peak flood, soil erosion, and river sedimentation	Unfeasible for large areas and subject to other effects such as reduced average flow	Small rural basins
Control of soil loss through conservation measures	Reduces soil loss and silting of rivers	Unfeasible for large areas	Small rural basins

Source: Adapted from Tucci (2012)

Intensive measures are those that act on the water course directly and, according to Cordeiro *et al.* (1999), with their objective, they can be of three types:

- Flow acceleration: canalization and related works (Figure 11.2).
- Flow delay: reservoirs (detention/retention basins), restoration of natural gutters (Figure 11.3).
- Flow diversion: rectification, bypass tunnels, and diversion channels (Figure 11.4).

Structural and non-structural measures help reduce vulnerability. Structural measures aim to increase the intrinsic safety of communities through construction activities. Non-structural measures comprise a set of strategic and educational measures and are related to cultural and behavioural change.

Reducing the risk of disasters involves preparing for future adverse events; focusing on monitoring, evaluating, and understanding the risk of disasters; and sharing this information and how it is generated. In compliance with the Sendai Framework (Aitsi-Selmi *et al.*, 2015) target of substantially increasing the availability of and access to multi-hazard early warning systems and disaster risk information and assessments to people by 2030, the Civil Defence expanded its meteorological coverage with the implementation of three meteorological radars, which cover 100% of Santa Catarina's territory, and a GOES-16 satellite, which provides conditions for monitoring the entire state. In nearly real time, the Civil Defence can generate short-term forecasts and early warnings to speed up response and save more lives. A fourth radar is being implemented, which will be installed in the northern region, in the city of Joinville.

Figure 11.10 Channelling – Example of flow acceleration
Source: Cordeiro *et al.*, 1999

A review of the flooding events in southern Brazil

Figure 11.11 Canalization – Example of flow deviation or diversion
Source: Cordeiro *et al*., 1999

Conclusions

The increase in the frequency and intensity of disasters events at a global level demands governments and societies take urgent actions of confrontation, resistance, and recovery. In particular, the state of Santa Catarina (Brazil) needs to be prepared to face increasingly recurring floods. The state of Santa Catarina's geographical location favours the occurrence of disasters, as climatic fronts descending from the Amazon meet with Antarctician climatic fronts, causing natural disasters of greater harmful impact with each occurrence.

The intense rains of 2008 were a remarkable and traumatic event for the public administration of the state of Santa Catarina, especially for the Civil Defence. In the Itajaí Valley region, 37 municipalities and over 600,000 citizens were affected, resulting in approximately 107,455 homeless people and a material loss of over US$788 million (at the time).

Of note, in 2008, 21 helicopters were used for several days to help the affected population of the Itajaí Valley. On the occasion, there were blatant gaps in geographic data and in the spatial database. This led the public administration of the state of Santa Catarina to begin valuing and supporting the Civil Defence's new pre-, in-, and post-disaster projects and actions.

After the 2008 disaster, there were massive national and state investments aimed at the general reformulation of the infrastructure of the Civil Defence Centre in Florianópolis – capital of the state. New and modern equipment of all kinds were adopted, as well as investments in (geo) technologies, radar images, and others. A fundamental consequence, however, was a policy of collaboration between the agencies of the state and the potential action of the Civil Defence

Table 11.2 Intensive structural measures

Purpose of accelerating the flow			
Measures	Main advantage	Main disadvantage	Application
Dikes and polders	High degree of protection of a specific area	Significant damage if it fails. It should not be used for high slopes due to the risk of failure	Large rivers and on the plain, where the slope is small
Change the flow conductance: reduction of roughness and increase in section	Increase in flow and flow velocity and reduction in level	Effect on a stretch of river; transfers effect downstream. It can be expensive	Small and medium rivers
Bottom channel slope changes	Extends the protected area and accelerates runoff	Negative impact on downstream river with increased erosive potential	Narrow flood area
Purpose of dampening runoff			
Measures	Main advantage	Main disadvantage	Application
All reservoirs	Flood control downstream of the reservoir	Difficult location due to expropriation of areas	Small and intermediate watersheds, depending on volume
Reservoirs with multi-purpose floodgate	More efficient at the same volume	Vulnerable to human mistakes	Multipurpose projects
Reservoir for flood control	Operation with reservoir kept dry to receive the flood	Unshared cost and difficulty in controlling the reservoir area due to low-frequency flooding	Small and medium watersheds
Purpose of diverting runoff			
Measures	Main advantage	Main disadvantage	Application
Channel deviation	Downstream flow reduction	Depends on topography and effects where flow is directed	Medium and small watersheds
Flood paths: diversion of part of the volume to flood areas	Dampen and decrease the flow	Depends on topography and effects where flow is directed	All watershed sizes

Source: Adapted from Tucci (2007)

in favour of the citizens of Santa Catarina. Integrated and articulated planning with municipal administrations began to come to life. The Civil Defence of the state started to manage training centres and training of human resources aimed at the knowledge/dissemination of techniques and technologies of resilience before the occurrence of the flood phenomenon, and help and assistance to those affected began to occur in community centres, schools, and other groups of Santa Catarina society.

A review of the flooding events in southern Brazil

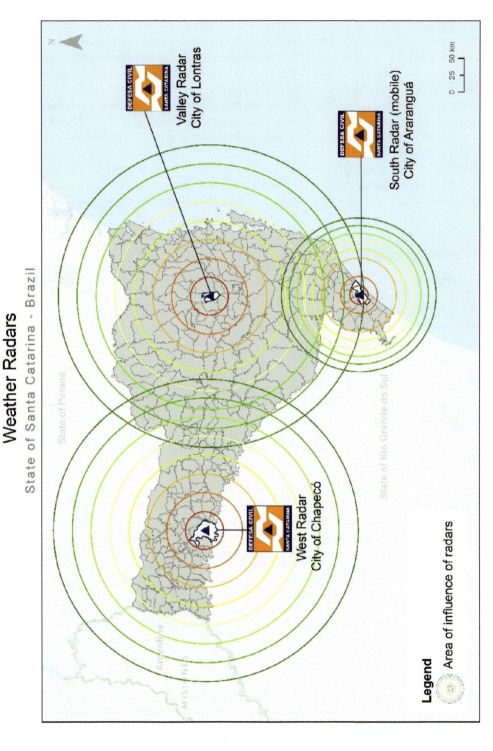

Figure 11.12   Locations of Santa Catarina Civil Defence weather radars

In addition to the urgent demand for training and preparing citizens for disasters, the importance of implementing structural and non-structural actions focused on integrated risk and disaster management was also verified. Studies point to greater attention of politics and technics to infrastructure. The issue of climate change and its relationship with the increase in the occurrence of disasters also demand immediate mitigation in the area of educational and technological interventions. Thus, it is of vital importance to insert, in an active and articulated way, the theme of risk and disaster management into the agenda of governments and society.

New alliances for the exchange of national and international experiences, technological resource profiling, and comparison became important in the learning process. Therefore, the state administration's investment and foresight in moments of crisis are considered commendable and of great wisdom.

In Brazil, disaster does not come with no warning signs; that is why we have to always be prepared to face it, uniting the efforts of the partners that make up society. Santa Catarina has already advanced a lot, but it still has room to grow in knowledge, technology, and training of resilient human resources.

Acknowledgments

This research received financial support from the Coordination of Superior Level Staff Improvement – CAPES – Brazil (PROAP/AUXPE). Words cannot express my gratitude to FAPESC - Fundação de Amparo à Pesquisa e Inovação do Estado de Santa Catarina, which has been supporting this research since 2019. Also thanks to CNPq call n° 06/2019 Agreement number: 307153/2019-3 – Research Productivity Scholarships.

Notes

1 Declaration of emergency: anomalous situation, caused by disasters, which generates damages and losses that imply the partial compromise of the public government's response capacity of the affected entity.
2 Declaration of public calamity: anomalous situation, caused by disasters, which generates damages and losses that imply a substantial compromise of the public government's response capacity of the affected entity.

References

Aitsi-Selmi, A. *et al.* (2015). The Sendai Framework for disaster risk reduction: Renewing the global commitment to people's resilience, health, and well-being. *International Journal of Disaster Risk Science*, 6, 164–176. Available: https://doi.org/10.1007/s13753-015-0050-9.

Arinta, R. R., Suyoto, S., & Emanuel, A. W. R. (2020). Effectiveness of gamification for flood emergency planning in the disaster risk reduction area. *International Journal of Engineering Pedagogy (iJEP)*, 10(4), 108–124. Available: https://doi.org/10.3991/ijep.v10i4.13145.

Banco Mundial. Governo do Estado de Santa Catarina (2012). *Avaliação de perdas e danos: Inundações bruscas em Santa Catarina – Novembro de 2008*. Brasília. Available: https://antigo.mdr.gov.br/images/stories/ArquivosDefesaCivil/ArquivosPDF/publicacoes/Inundaes-Bruscas-em-Santa-Catarina.pdf. Accessed 21st April 2022.

Canholi, Aluísio Pardo. (2015). *Drenagem urbana e controle de enchentes. 2.ed. Ampliada e atualizada*. São Paulo: Oficina de Textos, 384 pages.

CEPED. (2022). *Centro de Estudos e Pesquisas em Engenharia e Defesa Civil – UFSC*. Estudo dos impactos econômicos dos desastres no Brasil. Available: https://www.ceped.ufsc.br/relatorio-de-danos-materiais-e-prejuizos-decorrentes-de-desastres-naturais-no-brasil-1995-2014/.

Cordeiro, A., Medeiros, P. A., & Teran, A. L. (1999). *Métodos de Controle de Cheia e Erosões*. Blumenau. Available: http://ceops.furb.br/index.php/publicacoes/artigos/doc_details/5-metodos-controle-cheias1999. Accessed July 2019.

EM-DAT. (2021). *The international disaster database. Extreme weather events in Europe*. Issue n. 64, September 2021. Available: https://www.emdat.be/publications.

Ferreira, D., Albino, L., & Freitas, M. J. C. C. (2017). Mapeamento participativo para a gestão de risco de desastres: Região dos baús, Ilhota – SC. *Revista Brasileira De Cartografia*, 69(4). Available: https://seer.ufu.br/index.php/revistabrasileiracartografia/article/view/44330.

IBGE. (2021). *Instituto Brasileiro de Geografia e Estatística. Cidades e Estados*. Available: https://www.ibge.gov.br/cidades-e-estados/sc.html.

Ibrahim, T., & Mishra, A. (2021). A conceptual design of smart management system for flooding disaster. *International Journal of Environmental Research and Public Health*, 18(16), 8632. Available: https://doi.org/10.3390/ijerph1816863.

Jornal Expresso. (2011). *1 milhão de pessoas sofrem com a chuva em Santa Catarina*. Available: https://jornaloexpresso.wordpress.com/2011/09/12/1-milhao-de-pessoas-sofrem-com-a-chuva-em-santa-catarina/.

Miguez, M. G., Veról, A. P., & Carneiro, P. R. F. (2012). Sustainable drainage systems: An integrated approach, combining hydraulic engineering design, urban land control and river revitalisation aspects. In M. S. Javaid (ed.), *Drainage systems*. InTech, 21–54.

Monteiro, M. A. (2001). Caracterização climática do Estado de Santa Catarina: Uma abordagem dos principais sistemas atmosféricos que atuam durante o ano. *Geosul*, 16(31), 69–78.

Oneda, T. M. S. (2018). *Planos diretores de drenagem urbana: Uma análise comparativa entre planos de países desenvolvidos e em desenvolvimento*. 141f. Dissertação (Mestrado em Engenharia Civil) – Universidade do Estado de Santa Catarina, Joinville.

Pandolfo, C., Braga, H. J., Silva Jr, V. P. da, Mas Signam, A. M., Pereira, E. S., Thomé, V. M. R., & Valci, F. V. (2002). *Atlas climatológico do Estado de Santa Catarina*. Florianópolis: Epagri, vol. 1, 13.

Righetto, A. M. (2009). *Manejo de águas pluviais urbanas*. BES, 396.

Roslan, A. F., Fernando, T., Biscaya, S., & Sulaiman, N. (2021). Transformation towards risk-sensitive urban development: A systematic review of the issues and challenges. *Sustainability*, 13, 10631. Available: https://doi.org/10.3390/su131910631.

Santa Catarina. (2021). Governo do Estado de Santa Catarina. Fundação de Amparo à Pesquisa e Inovação do Estado de Santa Catarina. Universidade Federal de Santa Catarina. In *Perfil histórico de desastres de Santa Catarina: Identificação das principais ameaças para avaliação de risco de desastres*. Cartilha Produto 3. Florianópolis. Available: https://drive.google.com/file/d/1BTQR8QTK7wD64B0mA5aRDWrGUTYsixEz/view?usp=sharing. Acesso em 21 de abril de 2022.

Santa Catarina. Governo do Estado de Santa Catarina (2022). *Fundação de Amparo à Pesquisa e Inovação do Estado de Santa Catarina*. Universidade Federal de Santa Catarina. Avaliação de risco de desastres em santa catarina: Apresentação de resultados. Florianópolis. Accessed 21st April 2022.

SDE. (2020). *Secretaria de Estado do Desenvolvimento Econômico Sustentável. PIB 2020: Santa Catarina supera a média nacional e economia avança*. Available: https://sde.sc.gov.br/index.php/noticias/3807-pib-2020-santa-catarina-supera-a-media-nacional-e-economia-avanca.

Souza, F. A. S., Vieira, V. R., Silva, V. P. R., Melo, V. S., & Guedes, R. W. S. (2016). Estimativas dos riscos de chuvas extremas nas capitais do Nordeste do Brasil. *Revista Brasileira de Geografia Física*, 9(2), 430–439.

Souza, G. (2008). Foto. In *Prefeitura apresenta novo mapeamento das áreas com risco para deslizamento*. Available: www.nsctotal.com.br/noticias/prefeitura-apresenta-novo-mapeamento-das-areas-com-risco-para-deslizamento.

Tucci, C. E. M. (2007). *Inundações urbanas*. ABRH/RHAMA, 393.

Tucci, C. E. M. (2012). *Gestão da drenagem urbana*. CEPAL, Escritório no Brasil/IPEA, 50.

Zahed Filho, K., Martins, J. R. S., & Porto, M. F. A. (2012). *Gestão dos recursos hídricos no Ambiente Urbano*. Escola Politécnica da Universidade de São Paulo, EPUSP.

Zhang, M., & Wang, J. (2022). Global flood disaster research graph analysis based on literature mining. *Applied Sciences*, 12(6), 3066. Available: https://doi.org/10.3390/app12063066.

Section III
Risk assessments, flood mitigation and project management

12
Flood vulnerability in developing countries, international collaboration, and network visualization
A bibliometric analysis

Ahmed Karmaoui

Introduction

A large number of studies on flood vulnerability (FV) was found in the literature (Balica et al., 2012; Nasiri et al., 2016; Karmaoui et al., 2019a, 2019b). FV assessment studies play a crucial role in extreme event management in developed and developing countries using modelling and remote sensing approaches. In the context of developing countries, floods are harmful because of low levels of flood protection (Tanoue et al., 2016). The effect of flooding is felt more in these countries and is linked to natural causes, poor management practices, and urbanization, which requires the multiplication of non-structural measures of flood management (Egbinola et al., 2017). The damage induced by floods in developing countries was catastrophic. Isla Hispaniola, the island shared by Haiti and the Dominican Republic, is an example, where a flood that occurred in the River Soliette on 24 May 2004 killed over 1,000 people (Domeneghetti et al., 2015). Another example of a well-documented event that occurred in a developing country is the dramatic flood that affected Ourica watershed in the Marrakech High Atlas of Morocco on 17 August 1995 that killed 730 victims and affected 35,000 people (Karmaoui and Balica, 2019). In 2013, according to the international disaster database (EMDAT), 6,648 flood fatalities have occurred only in Nepal and India (Tanoue et al., 2016). Flood has resulted in loss of lives and income, property damage, and spread of diseases. In Southeast and Southern Asia, more than half of all urban areas were in high-frequency flood areas (Güneralp et al., 2015). The Pearl River Delta in China, Chao Phraya River in Thailand, the Ganges River in India and Pakistan, the Korean peninsula, and the Sao Paolo urban area in Brazil are examples of urban areas prone to floods, according to this same study. The high occurrence of flooding may be linked to the impact of rapid urbanization (Cai et al., 2021) and low municipal drainage systems (Yang et al., 2020). A response to urban flooding relies on a better understanding of vulnerability, as reported by Yang et al. (2020). Different mitigation strategies have been proposed in some developing countries. Dewan (2015) reported early warning awareness; post-flood rehabilitation programs;

DOI: 10.1201/9781003160823-15

rescue measures; and seminars, workshops, and training to share knowledge and adaptation measures with managers and sensitize the public.

To evaluate flood vulnerability, three dimensions can be used: exposure, susceptibility, and resilience or adaptive capacity (Karmaoui, 2019). Regarding the tools, Díez-Herrero and Garrote (2020) reported the importance of flood mapping and geographic information system (GIS) that had evolved over time. Komolafe et al. (2020) investigated flood vulnerability using integration of multi-criteria analysis the and height above nearest drainage (HAND) model, while Karmaoui and Balica (2019) explored it developing a numeric model or index in pre-Saharan region. Concerning bibliometric reviews, several papers were found in several domains, including agriculture (Luo et al., 2020); the construction industry (Alencar et al., 2020); nanomaterials (Su et al., 2021); resilience, vulnerability, robustness, and adaptive capacity of temperature in agriculture (Dardonville et al., 2020); climate change adaptation, mitigation, and resilience (Einecker and Kirby, 2020); and coastal flooding (Gao and Ruan, 2018). Studies dealing with bibliometric analysis on the evolving concepts of vulnerabilities (Giupponi and Biscaro, 2015) and with flood risk analysis and assessment, applications, and uncertainties were also recorded (Díez-Herrero and Garrote, 2020). These articles also conducted bibliometric review using analyses such as type of publication, journal, authors, co-authors, citations, sponsors, affiliations, and years.

The current chapter presents a bibliometric method to evaluate the available information accumulated on flood vulnerability. Significant focus was given to developing countries. The chapter shows how the use of Big Data extracted from the most prestigious databases may enhance the understanding and trends in research in this field. The global scientific literature on FV was quantified using the Scopus database and VOSviewer software.

In this chapter, we focused on two aspects:

- How the scientific research on FV in developing countries is structured
- What the trends are in research on this topic

Material and methods

Data

The first analysis was performed using a general key term "flood(s)" through the Scopus database. A total of 156,233 documents were selected for the period 1837–2021 (Figure 12.1). The selection of the year range is generated automatically by Scopus based on all available and indexed publications using the key terms search. A second search was specified adding to the key term "flood" the word "vulnerability", and the total number of documents was 5,920 publications for the period 1972–2021 (Figure 12.1). In order to meet the study criteria, a third search was conducted using the key terms "flood vulnerability developing countries". The bibliometric information of 897 documents was acquired and analysed using specialized software. The generated information of this last selection covered the period from 1991 to 2021. The analysis through Scopus was carried out on 8 March 2021. The different year range for these three searches is explained by the refinement search passing from general to specific searches. However, the Scopus database also allows a change in the selection of a year range to analyse.

Methodology

Using the key terms "flood vulnerability" plus "developing countries", the extracted metadata in CSV form was processed by the VOSviewer tool developed by Van Eck and Waltman (2010).

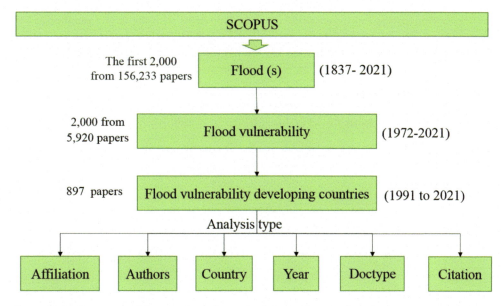

Figure 12.1 Structure of the search and acquired data

This software was conceived to visualize keyword networks, links, and trends. The generated keywords are represented by coloured circles and interlinked via coloured lines with proportional thickness. The distance between key terms depends on their closeness. The stronger the relationship between the items, the closer the distance (van Eck and Waltman, 2014).

The trend of principal aspects concerning FV was explored and presented in a trend map. The structure of research on FV in developing countries was generated using Scopus. The global collaboration and trends on flood vulnerability in developing countries were carried out using 897 documents from 1991 to 2021. The following aspects were considered: number of documents by year, number of documents by country, number of publications by affiliation, number of documents by journal, number of publications by author, doctypes, funding institutions, and publications by subject.

Network analysis, density visualization, and trends in FV in developing countries were investigated using VOSViewer. The type of analysis and counting method was "co-occurrence", and the unit of analysis was "all keywords". The threshold chosen was 5 as minimum number of a keyword. Of the 4,977 keywords, 436 met this threshold. For each of the 436 documents, the total strength of co-occurence links with other keywords was calculated. The authors with the greatest total link strength were selected. The number of documents selected was 436.

International collaboration on FV in developing countries is another aspect explored using the VOSViewer software. The type of analysis and counting method was "co-authorship", and the unit of analysis was "authors". The threshold chosen was 3 as minimum number of documents by an author. Of 2,792 authors, 71 met this threshold. For each of the 71 documents, the total strength of co-authorship links with other authors was calculated. The authors with the greatest total link strength were selected. The number of documents selected was 71.

Regarding citations, the type of analysis and counting method used was "citation", and the unit of analysis was "Documents". The threshold chosen was 10 as minimum number of citations of a document. Of 897 documents, 350 met this threshold. For each of the 350 documents, the number of citation links was calculated. The documents with the largest number of links were selected. The number of documents selected was 350. For this aspect, the publications most cited in the current study based on VOSviewer software were identified and represented.

Results and discussion

Structure of research on flood vulnerability in developing countries

The first paper recorded in the global literature based on Scopus was published in 1991 by Munasinghe et al. (1991) exploring Rio reconstruction and flood prevention in Brazil. In the last 30 years, an increasing trend was recorded, mainly after 2009 (Figure 12.2a). The United States and United Kingdom, followed by Australia, were the most influential countries in terms of research into flood vulnerability in developing countries (Figure 12.2b), while Vrije Universiteit Amsterdam, Chinese Academy of Sciences, and Vrije Universiteit Amsterdam Institute for Environmental Studies were the most productive affiliations (Figure 12.2c). *Natural Hazards*, *International Journal of Disaster Risk Reduction*, and *Sustainability Switzerland* journals published the largest number of papers on flood vulnerability in developing countries (Figure 12.2d). However, Ward, P.J. and Nicholls, R.J. were the most influential authors (Figure 12.2e) and article, book chapter, and conference paper were the main document types (Figure 12.2f). The National Science Foundation, the Seventh Framework Programme, and National Natural Science Foundation of China were the most important sponsor funds for research on FV in developing countries (Figure 12.2g). Regarding subject area of FV, environmental science, social sciences, and earth and planetary sciences recorded the highest number of publications (Figure 12.2h).

Network analysis, density visualization, and the trend of FV in developing countries

The type of analysis and counting method was "co-occurrence", and the unit of analysis was "All keywords". The threshold chosen was 5 as minimum number of a keyword. Of the 4,977 keywords, 436 met this threshold. For each of the 436 documents, the total strength of co-occurence links with other keywords was calculated. The authors with the greatest total link strength were selected, and the number of documents selected was 436.

Five clusters were recorded (Figure 12.3a); the first cluster, represented by red circles, includes key terms such as climate change, vulnerability, flood, coping strategy, exposure, and urban population. The second group of key words, represented by yellow circles, comprises key terms such as hazards, flood risk, natural hazards, disaster prevention, buildings, and coastal zones. However, the blue circles represent the cluster that groups keywords such as human and human gender, adult, geography, and seasonal variation, while green circles represent key terms such as sea level, agriculture, food supply, biodiversity, water supply, water resources, water security, and rain. Last, the purple circles (the fifth cluster) include 52 items such as disasters, storms, risk, hazard, storm surge, natural disasters, disaster risk reduction, early warning system, and urban area.

Figure 12.3b highlights density visualization through a heat map that allows representing two-dimensional data ranging from blue (very low occurrence) to red (very high occurrence). In fact, it sums up the accumulated information to reduce the complexity of the scientific field. Trend analysis (Figure 12.3c) depicts a recent focus on mapping hazard and climate change; modelling vulnerability, risk assessment, livelihood, sensitivity, and exposure; regression analysis;

Flood vulnerability, collaboration, and network visualization

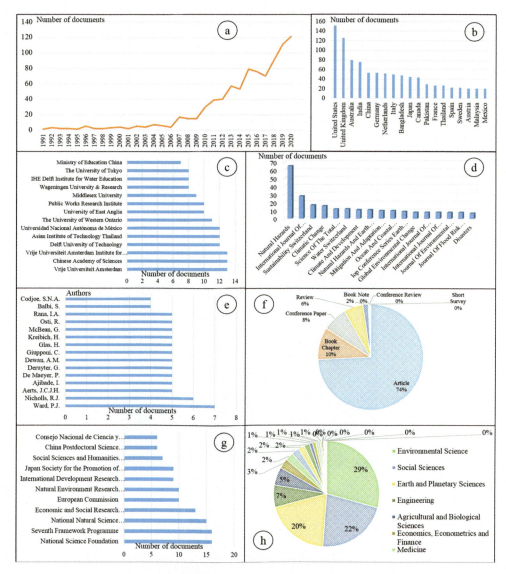

Figure 12.2 Global collaboration and trends on flood vulnerability in developing countries using 897 documents from 1991 to 2021. a, Number of documents by year; b, number of documents by country; c, number of publications by affiliation; d, number of documents by journal; e, number of publications by author; f, doc-types; g, funding institutions; and h, publications by subject

and urban floods. It seems that research is turning toward aspects of livelihood, sensitivity, and exposure to extreme flood events. Regarding the tools and techniques frequently used in flood vulnerability assessment, mapping hazards allows identification of the most vulnerable areas. Several studies have considered hazard mapping in the flood context among the most efficient techniques for prevention (Thapa et al., 2020). Modelling flood vulnerability is also among the most-used approaches, using integration of multi-criteria analysis, using, for

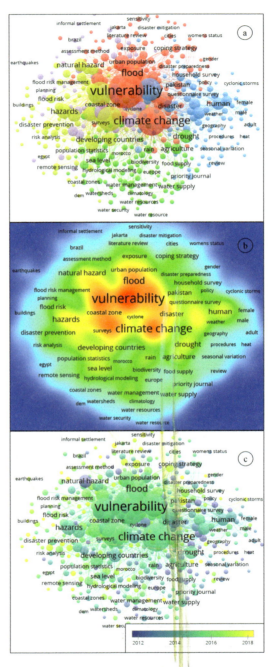

Figure 12.3 a, Network analysis; b, density visualization; and c, the trend of flood vulnerability in developing countries

example, the analytical hierarchy process (AHP) and ArcGIS to generate potential flood hazard areas (Komolafe et al., 2020). Climate change is one of the most visible aspects in the trend map linked to flood vulnerability in developing countries. In fact, in many areas worldwide, vulnerability to floods results from sea level rise due to climate change (Dinh et al., 2012).

International collaboration on FV in developing countries

The type of analysis and counting method was "co-authorship", and the unit of analysis was "authors". The threshold chosen was 3 as minimum number of documents by an author. Of 2,792 authors, 71 met this threshold. For each of the 71 documents, the total strength of co-authorship links with other authors was calculated. The authors with the greatest total link strength were selected. The number of documents selected was 71. Two clusters were recorded (Figure 12.4), the first included 5 items (authors) represented by red circles, where Ward, P.J. is the most visible with 8 documents, 479 citations, and 15 links (the highest number of links). The second cluster (green circles) comprises 3 of the most-linked authors, where Kreibich, H. published 5 documents recording 294 citations and 8 links.

Citations

For the publication quality aspect, the type of analysis and counting method used was "citation", and the unit of analysis was "documents". The threshold chosen was 10 as minimum number of citations of a document. Of 897 documents, 350 met this threshold. For each of the 350 documents, the number of citation links was calculated. The documents with the largest number of links were selected, and the number of selected documents was 350. The publications most cited in the current study based on VOSviewer software (Figure 12.5) are McLeman and Smit (2006); Few (2003); Wisner et al. (2014); Douglas et al. (2008); Haines et al. (2006), and Balica et al. (2012).

The oldest and most-cited reference with a large number of links in the field search was the work of Few (2003) entitled "Flooding, Vulnerability and Coping Strategies: Local Responses to a Global Threat" published in *Progress in Development Studies*. It reviews the current theoretical and applied findings on adaptive capacity and vulnerability in flood-prone zones and recommends research questions to increase understanding of coping strategies. The study conducted by Balica et al. (2012) is an interesting work that may help to grow our understanding of vulnerability proposing an easily accessible tool, the Coastal City Flood Vulnerability Index (CCFVI). In fact, the proposed method showed which cities are most vulnerable to coastal flooding with regard to different components such as politico-administrative, hydro-geological, and socio-economic.

Figure 12.6a depicts the approaches and tools most used in the current study based on the key term search "flood vulnerability methods". These include the GIS, analytic hierarchy process, risk perception, hydraulic models, surveys, computer simulation, digital elevation models, and index methods.

With regard to the research approaches most used in flood assessment, various types were recorded. In order to reduce the effects of floods, Cai et al. (2021) and Lyu et al. (2019) recorded four main categories: the historical data statistics method, the index system approach, GISs and remote sensing (RS) techniques, and the scenario simulation approach. RS offers critical data for monitoring flooding inundation, and GISs offer the best tools for flood risk management. (Wang and Xie, 2018). However, the analytic hierarchy process is one of the most widely used methods (Lyu et al., 2020) and is an optimal approach for integrating multi-criteria to generate

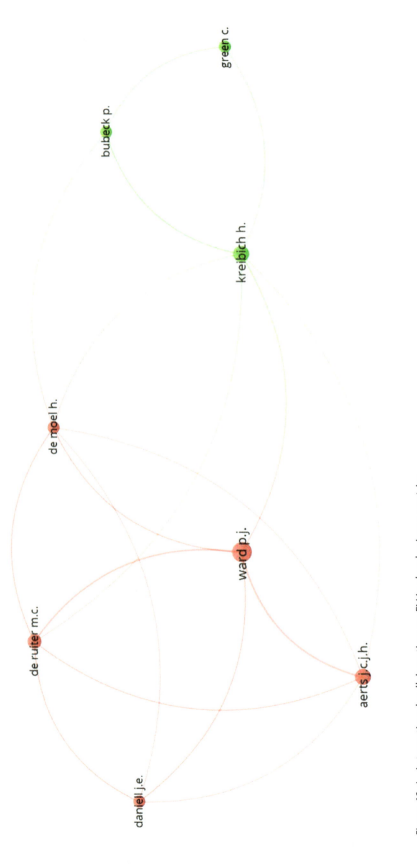

Figure 12.4 International collaboration on FV in developing countries

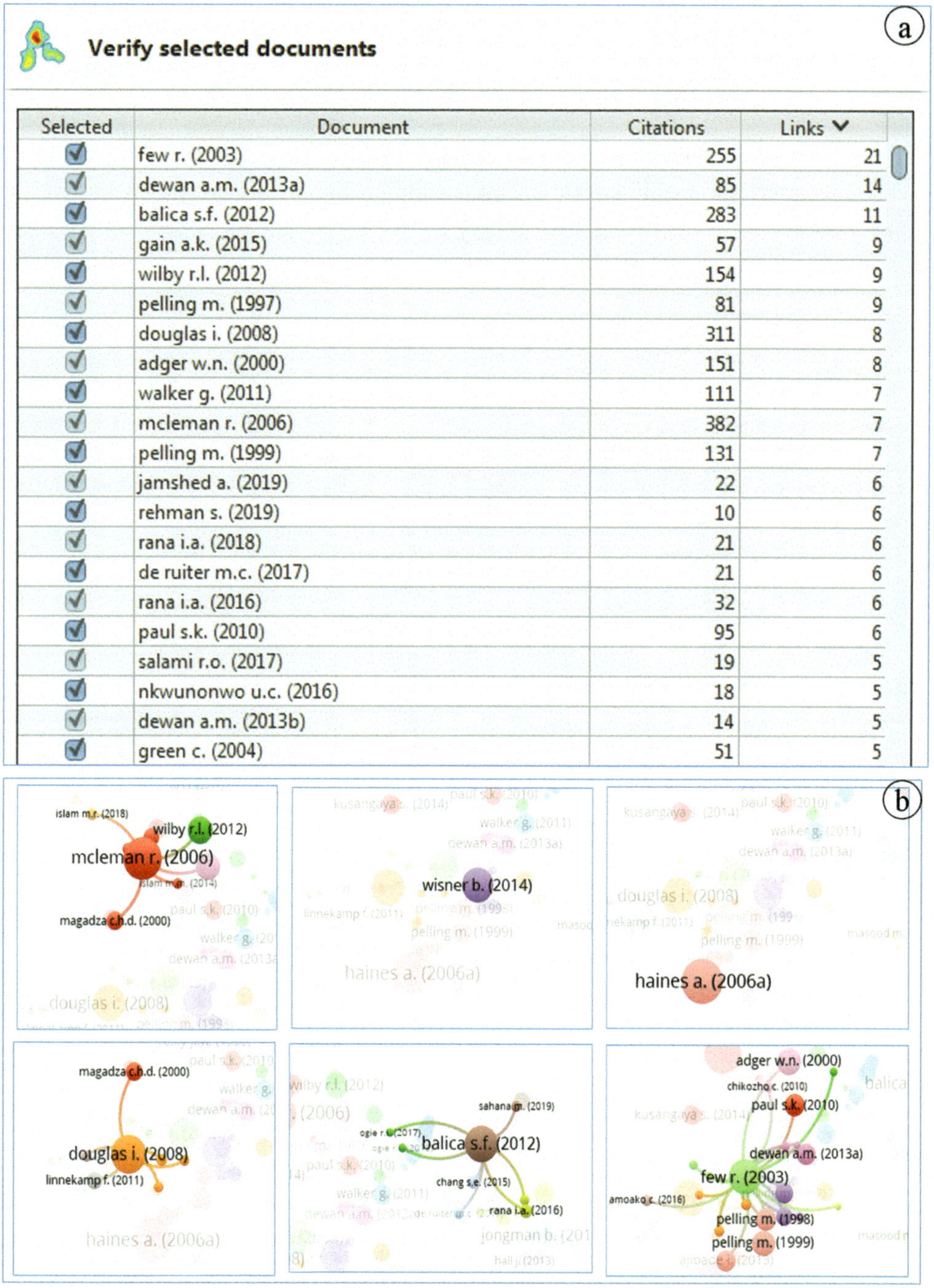

Flood vulnerability, collaboration, and network visualization

(a)

Verify selected documents

Selected	Document	Citations	Links ˅
☑	few r. (2003)	255	21
☑	dewan a.m. (2013a)	85	14
☑	balica s.f. (2012)	283	11
☑	gain a.k. (2015)	57	9
☑	wilby r.l. (2012)	154	9
☑	pelling m. (1997)	81	9
☑	douglas i. (2008)	311	8
☑	adger w.n. (2000)	151	8
☑	walker g. (2011)	111	7
☑	mcleman r. (2006)	382	7
☑	pelling m. (1999)	131	7
☑	jamshed a. (2019)	22	6
☑	rehman s. (2019)	10	6
☑	rana i.a. (2018)	21	6
☑	de ruiter m.c. (2017)	21	6
☑	rana i.a. (2016)	32	6
☑	paul s.k. (2010)	95	6
☑	salami r.o. (2017)	19	5
☑	nkwunonwo u.c. (2016)	18	5
☑	dewan a.m. (2013b)	14	5
☑	green c. (2004)	51	5

(b)

Figure 12.5 The publications most cited in the current study based on VOSviewer software: McLeman and Smit (2006), Few (2003), Wisner et al. (2014), Douglas et al. (2008), Haines et al. (2006), and Balica et al. (2012)

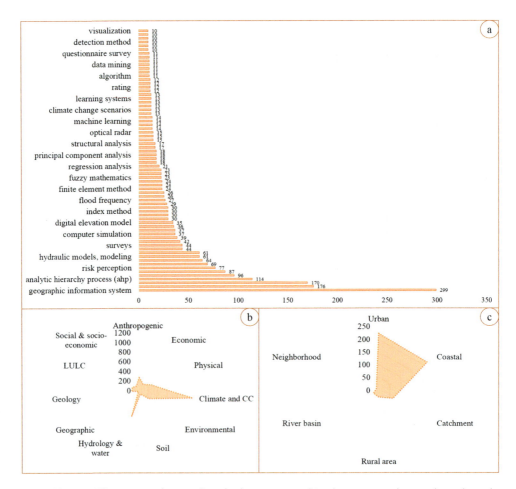

Figure 12.6 a, The approaches and tools the most used in the current chapter based on the key terms search "flood vulnerability methods" based on a search of 1,545 Scopus records from 1986 to 2021 using VOSviewer software; b, the principal components of flood vulnerability; c, the main scales of flood vulnerability

spatial vulnerable information (Hoque et al., 2019). This may subdivide flood risk into special factors (Cai et al., 2021).

Concerning the aspects or components that were the most explored in the current chapter, several categories were found. Based on a search of 1,545 Scopus records from 1986 to 2021 using VOSviewer software, the principal components of flood vulnerability (ranked in order of importance) are anthropogenic, economic, physical, climate and climate change, environmental, soil, hydrology and water, geographic, geology, land use and land cover, and social and socio-economic (Figure 12.6b). However, the main scales of flood vulnerability (Figure 12.6c) are as follows: urban, coastal, catchment, rural area, river basin, and neighbourhood.

The reports need conclusion. In fact, the flood vulnerability assessments were conducted at different spatial scales, such as urban (Balica et al., 2009), sub-catchment (Karmaoui and Balica, 2021), basin (Balica et al., 2009), mountain (Karmaoui et al., 2021; Meraj et al., 2015), and coastal scales (Balica et al., 2012). The high occurrence of flooding may be linked to the influence of climate change and rapid urbanization (Cai et al., 2021) and limitations of municipal drainage systems (Yang et al., 2020). To cope with the increasing damage of urban flooding, a better understanding of its vulnerability is needed (Yang et al., 2020). A growing number of studies dealing with flood risk assessment in plain cities and coastal cities were conducted in recent years (Sarmah et al., 2020). With regard to aspects of flood vulnerability, several components were used. Mavhura et al. (2017) investigated flood vulnerability using social-economic and institutional components, while Sebald (2010) used physical aspects, and Connor and Hiroki (2005) introduced a climatic component. However, Villordon and Gourbesville (2016) explored politico-administrative and socio-behavioural aspects, and Antwi et al. (2015) added ecological, engineering, and political aspects.

Benefits and potential challenges of the approach used

The approach documents and visualizes scientific information over the long term. The current chapter showed a long history of research on floods that was started in 1837, and the concept of flood vulnerability was first studied in 1972, according to the Scopus database (Figure 12.1). However, research on flood vulnerability in developing countries was started in 1991. In addition to the main key terms used in the current search (flood, vulnerability, developing countries), climate change, human, disaster management, and risk assessment have higher visibility and are the most linked to the search key terms. In fact, climate change was also selected as a research trend aspect associated with the flood field. The human or social dimension is also an important component that may be taken into account in all flood assessment studies. This chapter explores the most significant aspects of flood vulnerability in developing countries. It also highlights the main components of a flood vulnerability assessment, which allow understanding of the complexity of flood impacts. It may help also to gain understanding into the structure and trends of the research in the specific field. The approach is used in context to summarize information accumulated in all scientific domains indexed in prestigious databases and also for simple texts and reports. With regard to methods for assessing floods, we expect an important trend in the number of studies including mapping hazards and climate change, modelling vulnerability, risk assessment, livelihood, sensitivity and exposure, regression analysis, and urban floods in the future. Many weaknesses are associated with this approach, such as the language: we considered only English publications. In addition, our chapter was based solely on the Scopus database without considering other prestigious databases such as Web of Sciences, for example. The grey literature, the official reports of governments and international organizations, was also omitted.

Conclusion

Floods are responsible for considerable human and material losses, mainly in developing countries. Flood assessment, particularly flood vulnerability assessment, is an important element in extreme event management using modelling and remote sensing approaches. The current study provides a bibliometric method to evaluate the available information gathered in this research area. It is based on data retrieved from the Scopus database and analysed using VOSViewer software. The research in the flood field is growing due to the devastating consequences on the human and economic sectors. The chapter shows how the use of Big Data extracted from the most prestigious databases may enhance understanding and trends in research in this field. The results showed five groups of key terms about FV in developing countries, where ecological, social, and economic domains of research are correlated. Trend analysis highlights a recent focus on techniques such as mapping hazards and climate change, modelling vulnerability, risk assessment, livelihood, sensitivity and exposure, regression analysis, and urban flood scale. This trend is able to recommend the integration of these dimensions and approaches in flood vulnerability assessment.

References

Alencar, L., Alencar, M., Lima, L., Trindade, E., & Silva, L. (2020). Sustainability in the construction industry: A systematic review of the literature. *Journal of Cleaner Production*, 125730. https://doi.org/10.1016/j.jclepro.2020.125730

Antwi, E. K., Boakye-Danquah, J., Owusu, A. B., Loh, S. K., Mensah, R., Boafo, Y. A., & Apronti, P. T. (2015). Community vulnerability assessment index for flood prone savannah agro-ecological zone: A case study of Wa West District, Ghana. *Weather and Climate Extremes*, 10, 56–69. https://doi.org/10.1016/j.wace.2015.10.008

Balica, S. F., Douben, N., & Wright, N. G. (2009). Flood vulnerability indices at varying spatial scales. *Water Science and Technology Journal*, 60(10), 2571–2580. https://doi.org/10.2166/wst.2009.183

Balica, S. F., Wright, N. G., & Van der Meulen, F. (2012). A flood vulnerability index for coastal cities and its use in assessing climate change impacts. *Natural Hazards*, 64(1), 73–105. https://doi.org/10.1007/s11069-012-0234-1

Cai, S., Fan, J., & Yang, W. (2021). Flooding risk assessment and analysis based on GIS and the TFN-AHP method: A case study of Chongqing, China. *Atmosphere*, 12(5), 623. https://doi.org/10.3390/atmos12050623

Connor, R. F., & Hiroki, K. (2005). Development of a method for assessing flood vulnerability. *Water Science and Technology*, 51(5), 61–67.

Dardonville, M., Bockstaller, C., & Therond, O. (2020). Review of quantitative evaluations of the resilience, vulnerability, robustness and adaptive capacity of temperate agricultural systems. *Journal of Cleaner Production*. https://doi.org/10.1016/j.jclepro.2020.125456

Dewan, T. H. (2015). Societal impacts and vulnerability to floods in Bangladesh and Nepal. *Weather and Climate Extremes*, 7, 36–42. https://doi.org/10.1016/j.wace.2014.11.001

Díez-Herrero, A., & Garrote, J. (2020). Flood risk analysis and assessment, applications and uncertainties: A bibliometric review. *Water*, 12(7), 2050. https://doi.org/10.3390/w12072050

Dinh, Q., Balica, S., Popescu, I., & Jonoski, A. (2012). Climate change impact on flood hazard, vulnerability and risk of the Long Xuyen Quadrangle in the Mekong Delta. *International Journal of River Basin Management*, 10(1), 103–120. https://doi.org/10.1080/15715124.2012.663383

Domeneghetti, A., Gandolfi, S., Castellarin, A., Brandimarte, L., Di Baldassarre, G., Barbarella, M., & Brath, A. (2015). Flood risk mitigation in developing countries: Deriving accurate topographic data for remote areas under severe time and economic constraints. *Journal of Flood Risk Management*, 8(4), 301–314. https://doi.org/10.1111/jfr3.12095. https://onlinelibrary.wiley.com/doi/epdf/10.1111/jfr3.12095

Douglas, I., Alam, K., Maghenda, M., Mcdonnell, Y., McLean, L., & Campbell, J. (2008). Unjust waters: Climate change, flooding and the urban poor in Africa. *Environment and Urbanization*, 20(1), 187–205. https://doi.org/10.1177/0956247808089156

Egbinola, C. N., Olaniran, H. D., & Amanambu, A. C. (2017). Flood management in cities of developing countries: The example of Ibadan, Nigeria. *Journal of Flood Risk Management*, 10(4), 546–554. https://doi.org/10.1111/jfr3.12157. https://onlinelibrary.wiley.com/doi/epdf/10.1111/jfr3.12157

Einecker, R., & Kirby, A. (2020). Climate change: A bibliometric study of adaptation, mitigation and resilience. *Sustainability*, 12(17), 6935. https://doi.org/10.3390/su12176935

Few, R. (2003). Flooding, vulnerability and coping strategies: Local responses to a global threat. *Progress in Development Studies*, 3(1), 43–58. https://doi.org/10.1191/1464993403ps049ra

Gao, C., & Ruan, T. (2018). Bibliometric analysis of global research progress on coastal flooding 1995–2016. *Chinese Geographical Science*, 28(6), 998–1008. https://doi.org/10.1007/s11769-018-0996-9

Giupponi, C., & Biscaro, C. (2015). Vulnerabilities – bibliometric analysis and literature review of evolving concepts. *Environmental Research Letters*, 10(12), 123002. https://iopscience.iop.org/article/10.1088/1748-9326/10/12/123002/meta

Güneralp, B., Guneralp, I., & Liu, Y. (2015). Changing global patterns of urban exposure to flood and drought hazards. *Global Environmental Change*, *31*, 217–225. https://doi.org/10.1016/j.gloenvcha.2015.01.002

Haines, A., Kovats, R. S., Campbell-Lendrum, D., & Corvalán, C. (2006). Climate change and human health: Impacts, vulnerability and public health. *Public Health*, *120*(7), 585–596. https://doi.org/10.1016/j.puhe.2006.01.002

Hoque, M. A. A., Tasfia, S., Ahmed, N., & Pradhan, B. (2019). Assessing spatial flood vulnerability at Kalapara Upazila in Bangladesh using an analytic hierarchy process. *Sensors*, *19*(6), 1302. https://doi.org/10.3390/s19061302

Karmaoui, A. (2019). *Decision support methods for assessing flood risk and vulnerability* (pp. 1–325). IGI Global. https://doi.org/10.4018/978-1-5225-9771-1. www.igi-global.com/book/decision-support-methods-assessing-flood/223461#table-of-contents

Karmaoui, A., & Balica, S. (2021). A new flood vulnerability index adapted for the pre-Saharan region. *International Journal of River Basin Management*. https://doi.org/10.1080/15715124.2019.1583668

Karmaoui, A., Minucci, G., & Ben Salem, A. (2019a). Composite indicators as decision support method for flood analysis: Flood vulnerability index category. In *Decision support methods for assessing flood risk and vulnerability*. IGI Global. https://doi.org/10.4018/978-1-5225-9771-1.ch002

Karmaoui, A., Moumane, A., & Akchbab, J. (2019b). Environmental hazards assessment at pre-Saharan local scale: Case study from the Draa Valley, Morocco. In *Decision support methods for assessing flood risk and vulnerability*. IGI Global. https://doi.org/10.4018/978-1-5225-9771-1.ch012

Karmaoui, A., Zerouali, S., Ayt Ougougdal, H., & Shah Ashfaq, A. (2021). A new mountain flood vulnerability index (MFVI) for the assessment of flood vulnerability. *Sustainable Water Resources Management*, *7*(6), 1–13.

Komolafe, A. A., Awe, B. S., Olorunfemi, I. E., & Oguntunde, P. G. (2020). Modelling flood-prone area and vulnerability using integration of multi-criteria analysis and HAND model in the Ogun River Basin, Nigeria. *Hydrological Sciences Journal*, *65*(10), 1766–1783. https://doi.org/10.1080/02626667.2020.1764960

Luo, J., Han, H., Jia, F., & Don, H. (2020). Agricultural co-operatives in the western world: A bibliometric analysis. *Journal of Cleaner Production*, *273*, 122945. https://doi.org/10.1016/j.jclepro.2020.122945

Lyu, H. M., Shen, S. L., Zhou, A., & Yang, J. (2019). Perspectives for flood risk assessment and management for mega-city metro system. *Tunnelling and Underground Space Technology*, *84*, 31–44. https://doi.org/10.1016/j.tust.2018.10.019

Lyu, H. M., Zhou, W. H., Shen, S. L., & Zhou, A. N. (2020). Inundation risk assessment of metro system using AHP and TFN-AHP in Shenzhen. *Sustainable Cities and Society*, *56*, 102103. https://doi.org/10.1016/j.scs.2020.102103

Mavhura, E., Manyena, B., & Collins, A. E. (2017). An approach for measuring social vulnerability in context: The case of flood hazards in Muzarabani district, Zimbabwe. *Geoforum*, *86*, 103–117. https://doi.org/10.1016/j.geoforum.2017.09.008

McLeman, R., & Smit, B. (2006). Migration as an adaptation to climate change. *Climatic Change*, *76*(1), 31–53. https://doi.org/10.1007/s10584-005-9000-7

Meraj, G., Romshoo, S. A., Yousuf, A. R., Altaf, S., & Altaf, F. (2015). Assessing the influence of watershed characteristics on the flood vulnerability of Jhelum basin in Kashmir Himalaya. *Natural Hazards*, *77*(1), 153–175. http://dx.doi.org/10.1007/s11069-015-1775-x

Munasinghe, M., Menezes, B., & Preece, M. (1991). Rio reconstruction and flood prevention in Brazil. *Land Use Policy*, *8*(4), 282–287. https://doi.org/10.1016/0264-8377(91)90018-E

Nasiri, H., Yusof, M. J. M., & Ali, T. A. M. (2016). An overview to flood vulnerability assessment methods. *Sustainable Water Resources Management*, *2*(3), 331–336. https://doi.org/10.1007/s40899-016-0051-x

Sarmah, T., Das, S., Narendr, A., & Aithal, B. H. (2020). Assessing human vulnerability to urban flood hazard using the analytic hierarchy process and geographic information system. *International Journal of Disaster Risk Reduction*, *50*, 101659. https://doi.org/10.1016/j.ijdrr.2020.101659

Sebald, C. (2010). Towards an integrated flood vulnerability index: A flood vulnerability assessment. Master of Science (MSc). *International Institute for Geo-Information Science and Earth Observation*, p. 87.

Thesis number: 2010-25, DOI: 10.13140/RG.2.1.4933.8000. Available on: https://www.researchgate.net/publication/298789832_Towards_an_integrated_flood_vulnerability_index_-_A_flood_vulnerability_assessment

Su, W., Zhang, H., Xing, Y., Li, X., Wang, J., & Cai, C. (2021). A bibliometric analysis and review of supercritical fluids for the synthesis of nanomaterials. *Nanomaterials*, *11*(2), 336. https://doi.org/10.3390/nano11020336

Tanoue, M., Hirabayashi, Y., & Ikeuchi, H. (2016). Global-scale river flood vulnerability in the last 50 years. *Scientific Reports*, *6*(1), 1–9. https://doi.org/10.1038/srep36021

Thapa, S., Shrestha, A., Lamichhane, S., Adhikari, R., & Gautam, D. (2020). Catchment-scale flood hazard mapping and flood vulnerability analysis of residential buildings: The case of Khando River in eastern Nepal. *Journal of Hydrology: Regional Studies*, *30*, 100704. https://doi.org/10.1016/j.ejrh.2020.100704

Van Eck, N. J., & Waltman, L. (2014). Visualizing bibliometric networks. In *Measuring scholarly impact* (pp. 285–320). Springer. https://doi.org/10.1007/978-3-319-10377-8_13

Van Eck, N. J., & Waltman, L. (2010). Software survey: VOSviewer, a computer program for bibliometric mapping. *Scientometrics*, *84*(2), 523–538. https://doi.org/10.1007/s11192-009-0146-3

Villordon, M. B. B. L., & Gourbesville, P. (2016). Community-based flood vulnerability index for urban flooding: Understanding social vulnerabilities and risks. In Gourbesville, P., Cunge, J., Caignaert, G. (eds), *Advances in hydroinformatics* (pp. 75–96). Springer, Singapore. https://doi.org/10.1007/978-981-287-615-7_6

Wang, X. W., & Xie, H. J. (2018). A review on applications of remote sensing and geographic information systems (GIS) in water resources and flood risk management. *Water*, *10*, 608. https://doi.org/10.3390/w10050608

Wisner, B., Blaikie, P., Cannon, T., & Davis, I. (2014). *At risk: Natural hazards, people's vulnerability and disasters*. Routledge. www.routledge.com/At-Risk-Natural-Hazards-Peoples-Vulnerability-and-Disasters/Blaikie-Cannon-Davis-Wisner/p/book/9780415252164

Yang, Q., Zhang, S., Dai, Q., & Yao, R. (2020). Assessment of community vulnerability to different types of urban floods: A case for Lishui city, China. *Sustainability*, *12*(19), 7865. https://doi.org/10.3390/su12197865

13

Flood risk assessment in developing countries

Dealing with data quality and availability

*Srijon Datta, Shahpara Nawaz, Md. Nazmul Hossen,
Mir Enamul Karim, Nure Tasnim Juthy,
Md. Lokman Hossain and Md. Humayain Kabir*

Introduction

Flood risk

The definition of risk is used from various perspectives. The most common technical definition of risk is the combination of the probability of an event and its negative consequences (UNISDR, 2009). Society for Risk Analysis provided some qualitative definitions where risk is related to exposure or consequences of the activity and associated uncertainties (Aven et al., 2018). As one of the natural disasters, flood is also associated with some risks, which concern different disciplines: hydrogeology, sociology and economics, and everyone has a viewpoint to interpret it. In hydrology, it is related to the probability of annual maximum discharge (Solin and Skubincan, 2013). From the sociological perspective, the risk is the probability of people and social hotspots (hospitals, schools, houses) being affected, and in the economic dimension, it can be the possibility to cause damage to assets (Meyer et al., 2009). The multidimensional nature of the risk is expressed by the definition of the UN (1992), where risk is the anticipated loss (of lives, injuries, property damage and economic activity disruption) caused by a particular hazard in a specific area and time. Here, flood risk generally contains two fundamental components (Solin and Skubincan, 2013), a product of the likelihood of an event happening (= hazard) and the consequences if it occurred (= vulnerability) (Kron, 2005).

Flood risk assessment

The United Nations International Strategy for Disaster Reduction (UNISDR) defined risk assessment as a qualitative or quantitative method for determining the nature and extent of risk by analysing potential hazards and evaluating existing conditions of vulnerability that together could negatively affect exposed people, property, services, livelihoods and the environment on which they rely (UNISDR, 2009). Flood risk assessment (FRA) is a set of techniques concerning the quantitative analysis (e.g. estimate of economic damage and loss of life) and evaluation of flood risk (Díez-Herrero and Garrote, 2020b).

Importance of understanding flood risk

Every country has natural disasters, and flood is one of the most rapid and abrupt among them and can cause extreme widespread destruction with varying magnitude around the world. The impacts of floods can be physical damage to structures, livelihood disruptions, and an outbreak of diseases, and they can sometimes be beneficial for soil, providing nutrients. About 19% (1.47 billion) of the world's population lives in high-risk flood zones, where most of them (about 1.36 billion) are located in South and East Asia and 89% live in low- and middle-income countries. In terms of exposure, an estimated 5.3 trillion USD of economic activity is directly located in significant flood risk areas by a report of the World Bank (Rentschler and Salhab, 2020).

Importance of FRA in disaster management

Communities typically emerge and settle around rivers and flood-prone coastal areas due to the availability of water for domestic, agricultural, transportation, employment and commercial purposes. For this reason, risk assessment of the areas is needed, as it can suggest management options, where the goal is to maximize the benefits rather than trying to control floods. Flood risk assessment is crucial for any development site to find the major flood risks so that future damage can be reduced. From a thorough risk assessment report, it is easy to select the framework of disaster management plans and suggest proper mitigation measures. Further, necessary steps would be undertaken to implement the specific mitigation measures to alleviate the impact of flooding. Moreover, FRA will support people to understand the negative impacts of flood events as well as helping them to be prepared to cope with extreme situations.

Flood risk in developing countries

In lower-income countries, floods often cause unmitigated loss and hardship, as infrastructure systems of drainage and flood protection are less developed (Rentschler and Salhab, 2020). According to the Aqueduct Global Flood Risk Country Ranking of the World Resources Institute (WRI), the top 15 countries are considered least developed or developing, and about 80% of the total population is affected every year (Winsemius and Ward, 2015). According to the Emergency Events Database (EM-DAT) data on total affected people, estimated damage, deaths and occurrences of flood disasters during 1900–2021 (to November), most of the countries on the top ten list are developing countries (UN, 2020; EM-DAT, 2021).

From the WRI rankings, India (first), Bangladesh (second), Indonesia (sixth) and Nigeria (tenth) are classified as developing economies by the United Nations (Winsemius and Ward, 2015; UN, 2020). The World Bank listed Bangladesh (51.9 m), India (225.3 m), Indonesia (76.0 m) and Nigeria (38.8 m) among the top ten countries with the most people (in million, m) exposed to the significant flood risk. Here, Bangladesh is also listed in the top countries, with a large share (31.6%) of the total national population exposed to flood risk. Meanwhile, India (898.6 million USD) and Indonesia (498.7 million USD) are listed in the top ten countries of economic value at risk. Moreover, considering poverty (1.90 dollars/day), with 35.4 m, 15.2 m and 3.9 m people in India, Nigeria and Bangladesh, respectively, were in the top countries with poor people exposed to significant flood risk (Rentschler and Salhab, 2020). So, as developing countries, we will be focusing on these four countries throughout our chapter.

Objectives of the chapter

The objectives of this chapter are to (i) identify recent FRA-related works in four developing countries (i.e., Bangladesh, India, Indonesia and Nigeria) for up-to-date methods with used models, data sets and databases, (ii) examine the status of user data concerning quality and availability. Here all the information related to objectives is extracted from scientific articles after a systematic review.

Overview of existing research

Overview of flood events and their impacts in developing countries

From the global data source EM-DAT, we have an overview of flood events that happened in developing countries like Bangladesh, India, Indonesia and Nigeria. In the land area of Bangladesh, India, Indonesia and Nigeria, respectively, a total of 97, 313, 244 and 56 flood events occurred 1900–October 2021 (EM-DAT, 2021; Tellman et al., 2021; Cloud to Street, 2022). This puts Bangladesh (ninth), India (first) and Indonesia (third) in the top list of global flood occurrences. India is the third among the top ten countries of the world that suffered the most damage and second among the countries with the highest death toll, where Bangladesh is third.

Though the flood frequency has increased, the mortality rate by year has decreased in Bangladesh. In 1999, five flood events occurred in Bangladesh which cost 196 lives. More than three flood events happened in the years 1983, 1994, 1995 and 1999. In 1974, 28,700 people lost their lives in one flood event which created famine and food shortage, with about 579.2 million USD total damages. However, in the 1998 flood, 4,300 million USD damage was done, with 1,050 dead people, and in 1960, 10,000 people lost their lives (Chowdhury, 2000; EM-DAT, 2021).

India suffered from 313 flood events with more than 86,844 million USD economic damages and loss of 76,875 people during 1926–2021 (EM-DAT, 2021). During 1953–2011, an average of 32.43 million people were affected and about 16,500 people were killed annually (Bahinipati and Patnaik, 2020). In the last 20 years, India faced 183 events, where more than 8 events occurred in 10 individual years. In 2005 and 2006, there were a total of 34 floods (17 floods in each year) with total death tolls of 2,129 and 1,194. In 2013, India lost 6,453 people in five flood events with a total of 1,362 million USD damages. In 2014, the highest amount of damages (16,465 million USD) was caused by seven events, with the cost of 622 deaths. Overall, flood frequency and death rate have increased by the year (EM-DAT, 2021).

In Indonesia, the overall frequency of floods is increasing from 1953–2020 (Isa et al., 2018; EM-DAT, 2021). Indonesia lost a total of 7,333 persons with 8,191.35 million USD damages from a total of 244 flood events. In the last 20 years, 17 individual years faced more than 6 flood events, and 25 events occurred in 2020, with 199 human losses. Indonesia suffered from more than 100 million USD of damages in the following years: 1996, 2000, 2002, 2006–2007, 2013–2016 and 2019 (EM-DAT, 2021), where a total of 645 people died in eight flood events in 2006, with 107.3 million USD damages. In 2013, ten events brought 3,006 million USD damages, with only 89 deaths (EM-DAT, 2021).

From 1985, Nigeria lost 2,031 people with total damage of 1,021.42 million USD. The highest number of flood events (six) occurred in 2000, with a loss of only four people and total damage of 6.71 million USD. Nigeria had six flood events in the years 1994, 2012, 2010, 2015, 2018 and 2020 has more than 25 million USD in losses. The highest number of people, 363, were found dead in only one flood event with 500 million USD in damage, and more than 7 million people were affected in 2012, the highest among all the floods (Tami and Moses, 2015; EM-DAT, 2021).

All floods are not equally dangerous in terms of the death toll and economic damage. So more economic damage does not mean more death. It mainly depends on the flood type, location of occurrence and if it is associated with other factors like heavy rainfall or a flood management plan.

Evolution or trend of FRA studies in developing countries

After a bibliometric study of flood risk analysis and assessment based on Web of Science databases, Díez-Herrero and Garrote (2020a) stated that the approaches in FRA are more scientific in terms of articles or journal papers (84.5% of global FRA studies) than technological or industrial in terms of patents (0.1% in the whole database). They also found notable growth in FRA studies in the last 25 years.

Though developing countries have been affected by floods for a long time, according to the Scopus database, the first initiative in the FRA study was started in 2000 in Bangladesh. But after that, ten years onward, merely one publication was indexed from Bangladesh (Figure 13.1). The contribution of Indian and Indonesian researchers started after 2015. In the years 2001, 2002, 2004, 2008, 2009, and 2014, though, there is no publication in this research area. Starting from 2018, it increased, and in 2021, it peaked at 13 studies. The highest number of studies was found for Bangladesh (total 29: 22 articles and 7 conference papers), followed by India (total 23: 20 articles, 2 book chapters and 1 conference paper), Indonesia (total 8: 7 conference papers and 1 article) and Nigeria (total 6: 5 articles, 1 review paper).

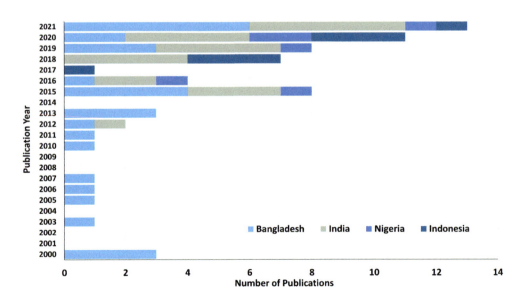

Figure 13.1 Evolution of FRA publications in developing countries (Bangladesh, India, Nigeria and Indonesia) from 2000–2021 in the Scopus database

Field of applications and research areas of FRA in developing countries

The application of FRAs in the world has evolved in recent years, with more than ten main fields such as planning (land, urban), resilience, communication, emergency, natural water retention, nature-based, perception and warning (Díez-Herrero and Garrote, 2020b).

FRAs in Bangladesh are mainly concerned with coastal and riverine floods and flash floods. FRA studies here revolve around preparedness (Haque et al., 2021), location-based damage mitigation (Alam et al., 2021), future climate change–induced flood scenarios (Roy et al., 2021a), development of useful assessment approaches in the data-scarce region (Mondal et al., 2020), possible risk in dike breaching by cyclone (Islam et al., 2019) and characteristics of wetland floods to formulate management strategies (Suman and Bhattacharya, 2015).

In India, FRA studies are mainly applied for prioritizing areas for ecological restoration and conservation (Rehman et al., 2021), flood mitigation strategy and preparing a policy framework to reduce future flood hazards (Bipinchandra et al., 2020; Pathak et al., 2020; Roy et al., 2021b). Besides, these studies also promote participatory planning for reducing flood events and also demand further research for designing a scheme of sustainable food management (Islam and Ghosh, 2021). The majority of the studies on FRAs focused on flood hazard, vulnerability and flood risk mapping or zoning (Shivaprasad Sharma et al., 2017; Saha and Agrawal, 2020; Manna et al., 2019; Ghosh and Kar, 2018; Chakraborty and Mukhopadhyay, 2019). Moreover, FRA is also used in initiating flood index insurance plans (Matheswaran et al., 2019) and flood risk of glacial lake outburst (Allen et al., 2016).

Flood risks with land-use changes on an island (Nafishoh et al., 2018), developing a decision support system for urban flooding (Febrianto et al., 2016; Fariza et al., 2017, 2020) and flood risk relation to land use policy and flood management (Handayani et al., 2020) were the major focus for FRA studies in Indonesia.

In Nigeria, FRA studies are mainly applied for understanding the effect of barrier islands, urban development and planning (Obiefuna et al., 2021), flood mitigation and decision making over multiple factors of floods (Zulkhairi et al., 2019). Most of the flood risk–related works in Nigeria are focused on assessing flood impacts through hazard mapping and modelling, exposure and vulnerability assessment (Komolafe et al., 2015).

Methodology

Scope and rationale for the chapter

In this chapter, the recent development of FRA methods by developing countries is discussed to explore the most commonly used databases. From a current study, it is found that most of the reviews of FRA studies are focused on methodological approaches, spatial scale analysis, uncertainties of application of the study, mapping, global scale risk assessment and economic damage (Díez-Herrero and Garrote, 2020a). In one study, it is indicated that local-scale research on flood and drought in data-scarce regions like Pakistan, India, Inner Mongolia, Nigeria, Arkansas, Sweden, Kenya, Central Asia and East Africa mostly uses global or planetary scale data sets (Lindersson et al., 2020). Though there are studies about global-scale data sets of precipitation (Sun et al., 2018) and free data sets allowing spatial (and temporal) analyses of floods (Lindersson et al., 2020), studies related to the quality and availability of local data sets regarding FRA are not available so far. So it has been attempted to gather information about data sets used in FRAs by developing countries such as what kind of data are available and accessible and their quality in terms of temporal and spatial scale.

Systematic literature review

"Flood" and "risk" were used in the logical query of the Scopus database to get basic information (title, publication year and abstract) of flood risk–related works in the selected four developing countries. Then, this information was again filtered by going through the title and abstract to include studies regarding flood risk (hazard, vulnerability, susceptibility and inundation) mapping and indexes and also imposing publication year range 2015–2021 to have the most up-to-date research in FRA. The final selection of papers (a total of 29) was done after going through the full studies (if accessible) to avoid further irrelevancy. To check the availability and quality, we also visited the local database websites.

Research questions:

- Identification of types of existing methods with used data sets and databases
- Status of used data concerning availability/accessibility and quality

Results and discussion

Methods, models or techniques used in FRA in developing countries

Bangladesh

Researchers in Bangladesh incorporated statistical models and remote sensing technologies to assess flood risk in different parts of the country. Roy et al. (2021a) aimed specifically to predict inundation on the floodplain of Areal Khan river as a result of different climatic projections of RCP 8.5 using the Soil & Water Assessment Tool (SWAT) and HEC-RAS model. Furthermore, in these Haor regions, a study was found to perform GIS-based analyses and analytic hierarchy process (AHP) for estimating flooding impacts and predicting flood-prone areas (Haque et al., 2021). Alam et al. (2021) studied flash flood susceptibility assessment for distinct hilly watersheds of Cox's Bazar, Bandarban and Chattogram using GIS tools, focusing on morphometric parameters to predict flash flood susceptibility and coastal flood risk. Another study aimed to assess risk of riverine flood for households by developing and establishing the Riverine Flood Disaster Risk Index (RFDRI) with the help of a household questionnaire survey, key informant interviews (KII), principal component analysis (PCA) and logistic regression for Nilphamari, Rangpur and Gaibandha Districts of Bangladesh (Mondal et al., 2020). Islam et al. (2019) developed a methodology to identify critical dike breaching locations in Kuakata of Patuakhali District with a predicted flood risk map and probabilistic flood map for such a catastrophe in the area using HEC-RAS. Suman and Bhattacharya (2015) developed MIKE-Flood models and 1D–2D models to simulate the flooding extent in Haors and rivers. Kwak et al. (2015) estimated the spatial and temporal change of flood risk in light of affected population and damaged rice crops for the Ganges Brahmaputra Meghna (GBM) basin using the GIS-based flood inundation depth (FID) model to simulate a flood-inundated area and a distributed hydrological block-wise TOPMODEL (BTOP) model for river discharge (Table 13.1).

India

The analytical hierarchy process model was applied in many studies. In the northern Rarh plains of India, a study was done to conceptualize community-based risk assessment and its drivers

Table 13.1 Models and data used for FRA in developing countries with their main output

Major output	Models	Used data	Study area	Reference
Bangladesh				
Prediction of floods and floodplain area with climate change scenarios	HEC-RAS, SWAT	DEM, river cross-section, discharge and daily water level data, projected climate and population data	Areal Khan River and its floodplains	(Roy et al., 2021a)
Flash flood susceptibility	GIS and remote sensing	DEM, precipitation patterns and rainfall	Cox's Bazar, Bandarban and Chittagong Districts	(Alam et al., 2021)
Flash flood zoning	AHP and GIS and remote sensing	DEM, satellite images, soil data, river data, rainfall, flood level data	North-eastern Haor lands of Bangladesh	(Haque et al., 2021)
Flood risk of households	household questionnaire survey, KII, PCA and logistic regression	Population data, socio-demographic data	Nilphamari, Rangpur and Gaibandha Districts	(Mondal et al., 2020)
Risk map probabilistic flood map	HEC-RAS	River network, alignment, cross-section, water level, discharge, DEM, land use, geometric properties of dikes – crest level, alignment, design crest level of the dike for future development; storm surge height, flood depths of previous events; RCP8.5 with SLR of 0.63 m, satellite images	Kuakata, Kolapara Upazilla and Patukhali Districts	(Islam et al., 2019)
Temporal and spatial dynamics of flood risk representing affected people and damaged rice crop	GIS-based flood inundation depth (FID) model, BTOP model	DEM, topographic data, land cover, population data, water level and stage-damage curves	GBM basin	(Kwak et al., 2015)
Flood index and severity	MIKE	DEM, discharge, water level data	North-eastern region (Haor)	(Suman and Bhattacharya, 2015)
India				
Intensity of ecological vulnerability and risk	AHP and multiple regression	Satellite images, DEM, rainfall, temperature, land use/land cover, soil data	Bhagirathi sub-basin	(Rehman et al., 2021)
Flood hazard extent	AHP, Inverse Herfindahl–Hirschman index (HHI) and Gibbs–Martin (GM) index	Disaster data – frequency, depth, duration; population – density, female-male ratio and dependency ratio	Northern Rarh plains, western Murshidabad	(Islam and Ghosh, 2021)

(Continued)

Table 13.1 (Continued)

Major output	Models	Used data	Study area	Reference
Flood susceptibility, vulnerability and risk	AHP, GIS and remote sensing	DEM; rainfall; geology and soil map; population – density, household density; distance from the river to road, flood shelter and hospital	Jalpaiguri foothill region, Himalaya	(Roy et al., 2021b)
Flood inundation map	STI-FM, HEC-RAS, GIS and remote sensing	DEM, satellite images	Imphal River Basin, Manipur, Himalayas	(Bipinchandra et al., 2020)
1. Flood extent map 2. Flood hazard and risk maps	InVEST model, HEC-HMS, HEC-GeoRAS, GIS and remote sensing	DEM, satellite images, precipitation and temperature data, soil and land use map	Mumbai	(Pathak et al., 2020)
1. Identifying flood-affected LULC classes 2. Flood risk mapping	AHP and GIS and remote sensing	District and tehsil map, DEM	Prayagraj District	(Saha and Agrawal, 2020)
Risk map	AHP, GIS and remote sensing	District map including the CD blocks and villages/towns, demographic and socioeconomic data: literacy, working population; land use/land cover classification, DEM	Coochbehar District of West Bengal	(Chakraborty and Mukhopadhyay, 2019)
Risk map	Formulating of Thiessen polygons and autoregressive distributed lag model	River water level, geographical attributes of the river basin, rainfall pattern, inundation depth, dominant socio-economic activity, real-life data about the parameters influencing flood vulnerability	Eastern Region	(Manna et al., 2019)
Risk zoning	GIS and remote sensing	Rainfall; inundation extent; population data – gender, occupation; rice yield and production; flood data – affected people, houses, blocks, villages, agricultural and non-agricultural area, crop loss	Bihar	(Matheswaran et al., 2019)
Risk map	AHP, GIS and remote sensing	DEM, distance from river and road, flood shelter; population data – growth, household density, literacy rate; rainfall data; landform category – land use	Malda District of West Bengal	(Ghosh and Kar, 2018)
Flood risk assessment	Risk triangle model	Population data – density, housing density, age classification, sex, working population, literacy, service centre distribution (health, emergency education and finance)	Srinagar City	(Alam et al., 2018)
Flood hazard zone mapping	Analysis of the spatio-multi-temporal historical data sets	Satellite images, historical flood data	Assam	(Shivaprasad Sharma et al., 2017)

Focus	Method	Data	Location	Reference
Natural and anthropogenic causes of flood	HEC-RAS	Geo-reference map, satellite images, DEM, land use assessment, slope and gradient, population	Ambala City	(Saini et al., 2016)
1. GLOF risk assessment 2. Future prediction GLOF hazard	GIS and remote sensing	Satellite images, DEM, population, female population, disabled population, literacy rate, unemployment, education facilities, banking services, home renters, medical facilities	Himachal Pradesh	(Allen et al., 2016)
Indonesia				
1. Estimation of risk and economic loss 2. Future projection	Quantitative risk assessment and GIS	Population and future projection, land cover, road network, Google Earth images, food supply ecosystem services index (IJEPBP), ecoregion map, food production at regency/city levels, average cost of housing construction	Java Island	(Nafishoh et al., 2018)
Risk map	FMCDM and GIS	District base map, disaster data – duration, inundation height, affected population, rainfall data, drainage data	Sidoarjo, East Java Province	(Fariza et al., 2020)
1. Effect of urbanization 2. Influence of land use policy and flood management	Spatial (GIS) and non-spatial analysis	Satellite images; delineated watershed; rural and urban land class; historical disaster data – year, location, time of occurrence, duration, inundation height and damage/loss status	Four river basins of the northern coast of Central Java	(Handayani et al., 2020)
Risk map	AHP-natural break	Population data – sex, age, handicapped, permanent residents' houses, GDP and poverty level; forest class – natural, protected, production, swamp and shrubland; public facilities; productive land; disaster handling capacity of districts; disaster data – date, time, impact, point of occurrence and height of the flood	East Java	(Fariza et al., 2017)
Risk map	AHP-natural break, GIS and remote sensing	District base map, administrative divisions, rainfall, drainage, population, history of floods	Surabaya city, East Java	(Febrianto et al., 2016)
Nigeria				
Effect of barrier islands, urban development and planning	CAESAR, GIS and remote sensing	Land cover and topography; satellite images; DEM; sea-level rise data – historical, projection; storm surge data – historical, projection; height	Lekki Peninsula, southeastern side of metropolitan Lagos, Lagos coastline	(Obiefuna et al., 2021)
1. Effect of flood causative factors 2. Flood risk index	Mi-AFRA, GIS and remote sensing	DEM, satellite images, precipitation data, historical flood data	Niger State, north-central part Nigeria	(Zulkhairi et al., 2019)
1. Hazard map 2. Flood-prone index	Hydrological modelling, GIS and remote sensing	DEM, satellite images, soil data, rainfall data, topographic map, administrative boundary and settlements	ALGA of Ebonyi State, southeastern Nigeria	(Aja et al., 2020)

depending on hazard and vulnerability where the AHP was integrated with the Herfindahl–Hirschman (HHI) and Gibbs-Martin (GM) indexes for weighting parameters (demographic, social, infrastructural and economic vulnerability) (Islam and Ghosh, 2021). A study was conducted in the Bhagirathi sub-basin to identify ecological vulnerability and risk due to flood by using AHP and GIS-based site-specific parameters (Rehman et al., 2021). Here, the AHP was used to weight and influence the parameters (biological richness; indexes – vegetation, soil, salinity, water, and disturbance; slope; rainfall and temperature) on ecological vulnerability. Also, the AHP model on the GIS platform was applied in generating flood susceptibility, vulnerability and risk maps to sort out a mitigation plan in the Jalpaiguri foothill region, Himalaya (Roy et al., 2021b). Saha and Agrawal (2020) intended to identify flood-affected zones in the Prayagraj District of the country with flood-affected land use land cover classes and assess flood risk using AHP. A study in the Coochbehar district of West Bengal focused on risk mapping and analysing hazard intensities and vulnerabilities spatially by using AHP in the GIS environment to recognize policy priorities for flood management (Chakraborty and Mukhopadhyay, 2019). Ghosh and Kar (2018) aimed to evaluate the probability and magnitude of flood hazard occurrences and the degree of vulnerability for risk zoning in Malda District using AHP within a GIS environment (Ghosh and Kar, 2018) (Table 13.1).

At the sub-watershed of the Imphal river basin, Manipur, flood hazard risk was assessed with a spatiotemporal image fusion model (STI-FM) and hydrologic engineering centre river analysis system (HEC-RAS) to construct flood inundation maps (Bipinchandra et al., 2020). Also, the hydrological engineering centre-hydrologic modelling system (HEC-HMS), HEC-RAS with the integrated valuation of ecosystem services and tradeoffs (InVEST) model was used to construct hazard and risk maps from social, environmental, economic and infrastructure criteria (Pathak et al., 2020). Additionally, a geospatial approach was implemented for evaluating potential flood risk and its impact on land use and local communities in the urban environment of Ambala City where HEC-HMS, HEC-GeoRAS, Arc Hydro and MIKE models were adopted for this purpose (Saini et al., 2016) (Table 13.1).

Probabilistic modelling of flood hazards was used in the eastern region of India, where the Thiessen polygon was used for area segmentation, and an autoregressive distributed lag model was used to forecast rainfall. Then, together with river and ground variables, modelled runoff and consequent inundation depth, flood forecasting and FRA were done (Manna et al., 2019). Bihar's districts were divided into four flood risk zones using cluster and hot spot analysis in a GIS environment, and then an area with high flood risk potential and impact was chosen for the trial of a flood index insurance system (Matheswaran et al., 2019). In a study conducted for Srinagar city in Jammu and Kashmir, a risk triangle model was used to evaluate parameters reflecting hazard (intensity), exposure (spatial) and vulnerability (sensitivity) using satellite images, a digital elevation model (DEM) and socioeconomic and demographic data (Alam et al., 2018). Furthermore, another study done at the Kopili River Basin in Assam used spatiotemporal historical data sets with satellite images to generate hazard zone maps (Shivaprasad Sharma et al., 2017). The study of Allen et al. (2016) set forth automating risk assessment of glacial lake outburst floods (GLOFs) integrating a climate risk paradigm accounting for both physical and socioeconomic determinants and evaluating the current situation and future forecast of hazards using GIS and remote sensing (Allen et al., 2016) (Table 13.1).

Indonesia

The quantitative risk of flood-affected people (vulnerability and hazard) and economic loss of Java Island were assessed, and the elements at risk (population, house building and agricultural land) were projected until 2030 based on a land change scenario by spatial modelling through GIS (Nafishoh et al., 2018). A fuzzy multi-criteria decision-making (FMCDM) model was developed to apprehend levels of urban flood risk in Sidoarjo in East Java Province with the help of GIS (Fariza et al., 2020). The effect of urbanization and flood events, the influence of land use policy and flood management in flood risks concerning four river basins of the northern coast of Central Java were also assessed using a combination of spatial and non-spatial analysis, where flood-related data were interpreted with land-use change analysis by Handayani et al. (2020). In East Java, to produce risk maps for spatial decision support systems, risk-level analysis was done by using AHP and natural break classification on a GIS platform or database server (Febrianto et al., 2016; Fariza et al., 2017) (Table 13.1).

Nigeria

At Lekki Peninsula on the Lagos coastline, a study regarding dynamic changes of land cover and assessment of flood risks in terms of urban development and application of barrier islands was done to conceptualize coastal management (Obiefuna et al., 2021). Here, cellular automation evolutionary slope and river (CAESAR), a 2D hydrodynamic simulation model with AutoCAD raster design, ArcGIS (for geo-referencing, digitization, mosaic and edit) and ENVI 5.0 (for unsupervised and supervised) software were used to assess landscape dynamics and evaluate risks of storm surge flooding (Obiefuna et al., 2021). At the Abakaliki local government area (ALGA) of Ebonyi State in south-eastern Nigeria, a flood hazard map was developed to provide information support for early warning, risk preparedness and response using a modified hydrologic model (flow-based) in the GIS environment (Aja et al., 2020). Another study in Niger State in the north-central part of Nigeria, a multicriteria index and analytical hierarchy process on flood risk assessment (Mi-AFRA) was developed using AHP to assess flood risks considering some factors (topographical – slope, elevation, hydrological – precipitation, flow direction, flow accumulation, topographic wetness and vegetation) for a flood risk index (FRI) using spatial analysis (Zulkhairi et al., 2019) (Table 13.1).

Data used for FRA studies in developing countries

Most FRA studies in developing countries involve GIS and remote sensing activities. The AHP and different hydrological models are mostly found to be integrated with the general methodological procedure. In these kinds of methodologies, spatial, meteorological (time series), river morphological, geographical, socio-economic, disaster-related and climate model data are usually required (Figure 13.2). According to the objectives and field of application, these data can be of different spatial resolutions and a long period.

Status of available data on FRA studies in developing countries

In some developing countries, there are authorities or agencies that are responsible for maintaining databases related to disasters, for example, ministries of the census, agriculture, environment, forestry, land and survey; national or regional agencies, institutes of meteorology, climatology, geophysics, soil survey, land use planning, water modelling and development, agricultural research, space research organizations, public work services, statistics and disaster management with local sub-divisions (Table 13.2). These kinds of databases and data collected from them are listed here. Data availability–related discussions are lower in Nigerian studies than in other countries. Sometimes flood event data were also collected by searching internet-based

Common Methods of FRA Studies Integrating AHP	Common Objectives of FRA Studies
Used data for this kind of method • **Disaster Data** - date, time, frequency, depth, inundation height, duration, impact, point of occurrence, historical data and disaster handling capacity of the area • **Spatial Data** - satellite images, DEM, Land use/land cover, forest area classification, productive land, area map including small blocks, administrative division, • **Geography Data** – soil data, geology map, drainage network, distance of road, shelters and hospital from river and place of interest • **Demographic Data** - population, density, female-male ratio, dependency ratio, household density, Literacy, working population • **Socio-Economic Data** - public facilities, GDP and poverty level • **Meteorological Data** – time series of rainfall, temperature data	• Flood zoning or mapping – risk, susceptibility, vulnerability, inundation. hazard, extent, probabilistic based on index • Temporal and Spatial land use changes of flood affected area • Economic loss estimation • Future flood projection – inundation, risk area

Common Methods of FRA Studies Integrating AHP (cont.)	Unique Methods
	QRA Required data: Population and future projection, land cover, road network, satellite image, ecoregion map, food production at regency/city levels, Food Supply Ecosystem Services Index, and average cost of the housing construction
Common Methods of FRA Studies Incorporating HEC-RAS, HEC-HMS, SWAT, MIKE Models	**FMCDM** Required data: District basemap, disaster data – duration, inundation height, affected population; rainfall data; drainage data
Used data for this kind of methods • **Spatial Data** - satellite images, DEM, Soil and land use • **Morphological Data** - drainage network, river cross section, discharge, water level data, alignment, geometric properties of dike - crest level, alignment, design crest level of the dike for future development; • **Demographic Data** – current and projected population data • **Climate Data** – climate model data • **Meteorological Data** – time series of rainfall, temperature data • **Disaster Data** - storm surge height, inundation height of previous floods	**CAESAR** Required data: Land cover and topography; satellite imageries; DEM, Sea Level Rise data – historical, projection; storm surges data – historical, projection, height
	Mi-AFRA Required data: DEM, satellite images, precipitation data, historical flood data
	FID Require data: DEM, topographic data, land cover, population data, water level & stage-damage curves

Figure 13.2 Models or techniques used by different FRA studies (with required data) in developing countries

news articles, mass media websites and other institutions to validate the data released by the government (Handayani et al., 2020). The data gathered and analysed from the studied research were treated following the respective studies' objectives. Nonetheless, to have a thorough qualitative assessment of data quality, flooding event monitoring websites must be visited. In terms of quality, resolution (temporal and spatial) and consistency of data are important. Though there is limited spatial data on Bangladesh, Indonesia and Nigeria, India has well-established websites and satellites for distributing spatial and topographical data by geo-platform of the Indian Institute of Remote Sensing (IIRS), which are also equivalents of ASTER and SRTM DEM in resolution. However, from the studies, spatial resolution less than 30 m for DEM, 0.25 degrees for precipitation and a 1:50,000 or 25,000 ratio for the map of a specific area were not mentioned to be available in different sources. There are variations in temporal resolution (daily, monthly and yearly) and the extent of data.

It is found that most of the studies used DEM data and satellite images from international databases (Table 13.3). Moreover, land use, soil, precipitation and climate model data were also used from available global databases. The data found in different studies could be ordered via online request form of the respective authority's website or by direct contact. Data used in different studies are sometimes free for research usage, or sometimes researchers have to buy them. Lindersson

Table 13.2 Available local sources of databases and data used from the sources

Sources	Data used from specific sources in different studies
Bangladesh	
Bangladesh Water Development Board (BWDB)	River network, cross-section, discharge and water level data, river data, water level and stage-damage curves
Bangladesh Agricultural Research Institute (BARI)	Soil data, river data
Bangladesh Meteorological Department (BMD)	Rainfall data
Bangladesh Bureau of Statistics (BBS)	Population and housing census
Institute of Water Modelling (IWM)	River network; alignment; cross-section; water level; discharge; DEM; land use; geometric properties of dike – crest level, alignment, design crest level of the dike for future development; storm surge height
Ministry of Land (MoL)	Land use
India	
Indian Institute of Remote Sensing (IIRS)	Vegetation disturbance index, biological richness, vegetation (2005–2006), slope, 50 m spatial resolution
Indian Meteorological Department (IMD)	Time series of rainfall (0.25 degree gridded data) and temperature, inundation extent
Irrigation and WaterWays Directorate	Demographic data – population density, female–male ratio, and dependency ratio, annual reports, river water level, geographical attributes of the river basin,
Kandi final report 2012	Historical flood data – frequency, depth, duration
District Census Handbook	District map including the CD blocks and villages/towns, demographic and socioeconomic data
Bihar Statistical Handbook 2012	Demographics
Geological Survey of India	Geomorphological data of rivers
Survey of India (SoI)	District and tehsil map, topographical map
Census of India	Population density, household density, literacy rate, average annual rainfall
Official website of Jalpaiguri	Flood shelter details
Indian National Bureau of Soil Survey and Land-Use Planning	Soil data/map
Indian Space Research Organization – Cartosat	DEM
District Human Development Reports	Socioeconomic activity data
Flood reports	Number of flood-affected people, houses, blocks, villages, the agricultural and non-agricultural affected area, crop loss, house damage, public property damage, economic loss, crop production
District Disaster Management Plan 2016	Absolute height, slope, landform category, distance from the river, rainfall pattern and distance from river confluence, population growth, household density, land use, distance from road distance from flood shelter, literacy

(*Continued*)

Table 13.2 (Continued)

Sources	Data used from specific sources in different studies
Indonesia	
Central Bureau of Statistics (BPS – Badan Pusat Statistik or Statistics Indonesia)	Population and economic census, digital statistical maps, house distribution, average cost of the housing construction, land class, GDP
Regional Disaster Management Agency (BPBD – Badan Penanggulangan Bencana Daerah)	District base map, flood inundation height, duration of flood inundation, affected population data
National Disaster Management Agency (BNPB – Badan Nasional Penanggulangan Bencana)	inaRISK – spatial data related to disaster risk; disaster geoportal; all disaster information data – date, time, impact, point of occurrence and height of flood and capacity level of each district in handling natural disasters
Reports from local (city/regency) government, Disaster Management Board (DMB) of Central Java Province	Location (name of villages), duration, depth, and damage/loss status
Meteorology, Climatology and Geophysics Agency (BMKG – Badan Meteorologi, Klimatologi, Dan Geofisika)	Rainfall data, history of floods
Public work service Sidoarjo (Dinas PU Sidoarjo)	Amount of drainage data, public facilities
Budiyono et al., 2016	Vulnerability curve or stage-damage curve
Agriculture service	Productive land
Ministry of Environment and Forestry	Natural forest, protected forest, production forest, swamp and shrubland delineated watershed
Nigeria	
Office of the Surveyor General of the Federation (OSGOF)	Land cover and topography map (1:25,000, 1984–1985)
Nigerian Meteorological Agency (NiMet), Abakaliki area synoptic station	Rainfall data
Ministry of Lands and Survey, Abakaliki	Topographic map, administrative boundary and settlements

et al. (2020) stated that the global databases have a significant amount of geospatial data which captures mostly hydrological hazards and more exposure than vulnerability aspects. But some challenges like time and source variant data sets can make these kinds of studies difficult.

Flood-related global databases

We found that most developing countries have flood-related observed data sources for a relatively short period. Considering this limitation, we have listed some global and regional data sources from which we can get long-time series observed, synthetic and reanalysis data. The most common global flood data providers include the NASA data portal, Dartmouth Flood Observatory and Copernicus Climate Data Store. Apart from this, there are some regional data sources, particularly for Asian countries (Table 13.4). The subject areas covered from these sources include climatology, hydrology, water governance, water use and SDGs tracking. The temporal scale of these sources covers daily, monthly and yearly statistics of past, present and future climate and hydrology. However, data sources are not homogenous in terms of temporal and spatial scale over the regions.

Flood risk assessment in developing countries

Table 13.3 International database used by developing countries

Data	Sources
DEM	SRTM – Consultative Group for International Agricultural Research (CGIAR-CSI)
	ASTER/ASTER GDEM – Japanese Ministry of Economy, Trade and Industry and the United States National Aeronautics and Space Administration (NASA)
	HydroSHEDS
	USGS earth explorer
Satellite images	USGS earth explorer – Landsat, Sentinel
	Google earth
	Modis
Land cover data	USGS earth explorer
	Global land cover data set by national mapping organizations
Rainfall data	Climate-data.org
	Tropical Rainfall Measuring Mission (TRMM)
	Global Satellite Mapping of Precipitation (GSMaP)
Climate model data	RCP 8.5 scenario – CMIP5 daily precipitation and maximum/minimum temperature data from EC-EARTH3 – European Centre of Medium-Range Weather Forecast (ECMWF)
Population data – distributed, projected	LandScan
	Socioeconomic Data and Applications Center (SEDAC)
Soil data	Harmonized World Soil Database (HWSD)
SLR data	IPCC assessment report and data
Previous flood depth data	Japan Society of Civil Engineers (JSCE)
Open street map	Road network

Table 13.4 Commonly used global data sources (in alphabetical order) for flood risk assessment

Data portals and flood-related websites	Weblinks
APHRoDITE water resources (Asian data set)	http://aphrodite.st.hirosaki-u.ac.jp/products.html
Asian Disaster Preparedness Center (ADPC)	https://servir.adpc.net/publications/flood-extent-mapping
Coupled Model Intercomparison Project Phase 5 (CMIP5)	https://pcmdi.llnl.gov/mips/cmip5/
Coupled Model Intercomparison Project Phase 6 (CMIP6)	https://pcmdi.llnl.gov/CMIP6/
CGIAR Consortium for Spatial Information (CGIAR-CSI)	http://srtm.csi.cgiar.org/
Climate Hazards Center InfraRed Precipitation with Station data (CHIRPS) data set for Asia	www.chc.ucsb.edu/data/chirps
Climate Risk Atlas (ARClim)	https://arclim.mma.gob.cl/
Copernicus Climate Data Store	https://cds.climate.copernicus.eu/
Center for Climate and Resilience Research (CR)[2] Climate Explorer	http://explorador.cr2.cl/
GDEx	http://gdex.cr.usgs.gov/gdex/
Dartmouth Flood Observatory (DFO)	http://floodobservatory.colorado.edu/
Data Pool and DAAC2Disk	https://lpdaac.usgs.gov/data_access/data_pool

(Continued)

211

Table 13.4 (Continued)

Data portals and flood-related websites	Weblinks
Emergency Events Database (EM-DAT)	www.emdat.be/
ERA5, ERA-Land	https://cds.climate.copernicus.eu/
Extreme Rainfall Detection System (ERDS)	http://erds.ithacaweb.org/
Giovanni Version 4	http://giovanni.gsfc.nasa.gov/giovanni/
Global Disaster Alert and Coordination System (GDACS)	www.gdacs.org
Global Flood Monitoring System (GFMS)	http://flood.umd.edu/
Global Flood Detection System – Version 2	www.gdacs.org/flooddetection/
Global Precipitation Measurement (GPM)	http://pmm.nasa.gov/GPM
Information Technology for Humanitarian Assistance, Cooperation, and Action (ITHACA)	www.ithacaweb.org/
IPCC Data Distribution Centre (DDC)	www.ipcc-data.org/index.html
MODIS Near Real Time Global Flood Mapping	http://oas.gsfc.nasa.gov/floodmap/
NASA DEVELOP program	http://develop.larc.nasa.gov/
NASA SERVIR Global Program	www.nasa.gov/mission_pages/servir/index.html
NOAA Physical Sciences Laboratory (PSL)	https://psl.noaa.gov/
Our World in Data	https://ourworldindata.org/
Precipitation Measurement Missions Data and Access	http://pmm.nasa.gov/data-access/downloads/gpm
Red Cross Disaster Mapping	http://maps.redcross.org/website/Links/ARC_Disaster_Links_Hazards.html
Reverb	http://reverb.echo.nasa.gov/reverb
River Watch Version 2 Satellite River Discharge Measurements	http://floodobservatory.colorado.edu/DischargeAccess.html
Sea level change	https://sealevel.nasa.gov/
SERVIR CREST Hydrology Model Viewer	http://ags.servirlabs.net/crestviewer/
SERVIR Product catalogue	https://servirglobal.net/Product-Catalog
Socioeconomic Data and Applications Center (SEDAC)	http://sedac.ciesin.columbia.edu
The Global Flood Database	https://global-flood-database.cloudtostreet.ai/
Tropical Rainfall Measuring Mission (TRMM)	http://trmm.gsfc.nasa.gov/
Tropical Rainfall Measuring Mission (TRMM) Current Heavy Rain, Flood and Landslide Estimates Tool	http://trmm.gsfc.nasa.gov/publications_dir/potential_flood_hydro.html
World Bank Data Catalogue	https://datacatalog.worldbank.org/
WorldClim	www.worldclim.org/data/index.html
World Resources Institute (WRI) Aqueduct 3.0	www.wri.org/resources/data-sets/aqueduct-global-maps-30-data
Worldview	https://earthdata.nasa.gov/labs/worldview/

Source: Author's compilation from different sources (official website of NASA, ECMWF, etc.).

Conclusions and summary

It is evident that in developing countries like Bangladesh and India, the number of studies on FRA continues to increase, and efforts are being made to maintain the database properly. However, there are a few limitations, such as a lack of completeness and consistency in the database maintenance process. Gaps exist in the compilation of all pertinent data concerning a single flood event and the monitoring of collecting continuous data daily during a flood event. Indonesia is also in the way of building a better disaster flood database (like inaRISK). But Nigeria is far behind in FRA research, though it recently established and started to provide flood-related data by respective organizations. In terms of FRA research in developing countries, most of the studies are based on GIS and remote sensing platforms and data. Most of the data used in the FRA studies were from local databases, except DEM, satellite images and climate model data. Some of the studies mentioned limitations, challenges and suggestions regarding the data and databases.

Limitations of existing FRA studies in developing countries

- Data scarcity like vulnerability maps, sea-level rise (SLR) data on local and regional extent and finer-resolution data creates limited application of numerical modelling (Obiefuna et al., 2021).
- Though coarser resolution data (90 and 30 m) have limitations if used in hydrodynamic modelling, better data are not freely available (Obiefuna et al., 2021).
- Information in the base map is not enough to represent the flood risk situation (e.g. based on one parameter like inundation height, flood velocity and flood risk level) and also compare with the new model with many variables (Fariza et al., 2020).
- Predicted flood hazard scenarios need to be validated using the historical records of flash floods in Bangladesh. The unavailability of data hinders such procedures. Finer resolution of DEM data is also a prerequisite for the field application of the generated flash flood susceptibility data (Mondal et al., 2020).
- For generating FRMs and PFMs due to breaching of dikes during a cyclone, Islam et al. (2019) found bathymetric data for the sea and storm surge levels of previous cyclones to be unavailable. Lack of data was also noted for existing and the earlier data on breaching of the dikes. The logical assumptions for these data were anticipated as the possible source of error of the study.
- Most of the basins are ungauged and lack information on their past hydrological behaviour (Alam et al., 2021).
- Limitation of data and the lack of a very high-resolution satellite image for a large study area might show a slight difference in the final output (Haque et al., 2021).
- Irregular discharge measurement (like twice a month), which does not represent emergencies.

Challenges faced for high-quality data access/availability

- Without compulsory rules, local governments sometimes skip reporting about the events (Handayani et al., 2020).
- Without funds, researchers cannot have the liberty to use paid models and software, which have more features than free online numerical models (Obiefuna et al., 2021).

Suggestions for overcoming the challenges

- The limitation of using coarser resolution spatial data can be reduced by finer data (LIDAR) (Obiefuna et al., 2021), though they are not always free.
- Flood-prone area identification and mapping can achieve a high accuracy level by using high-resolution spatial data sets, accurate local data and machine learning techniques (Haque et al., 2021).
- Irregular discharge measurement can be solved by rating curve development for specific rivers.

References

Aja, D., Elias, E. and Obiahu, O. H. (2020) 'Flood risk zone mapping using rational model in a highly weathered Nitisols of Abakaliki Local Government Area, South-eastern Nigeria', *Geology, Ecology, and Landscapes*, 4(2), pp. 131–139. doi: https://doi.org/10.1080/24749508.2019.1600912

Alam, A., Ahmed, B. and Sammonds, P. (2021) 'Flash flood susceptibility assessment using the parameters of drainage basin morphometry in SE Bangladesh', *Quaternary International*, 575–576, pp. 295–307. doi: https://doi.org/10.1016/j.quaint.2020.04.047

Alam, A., Bhat, M. S., Farooq, H., Ahmad, B., Ahmad, S. and Sheikh, A. H. (2018) 'Flood risk assessment of Srinagar city in Jammu and Kashmir, India', *International Journal of Disaster Resilience in the Built Environment*, 9(2), pp. 114–129. doi: https://doi.org/10.1108/ijdrbe-02-2017-0012

Allen, S. K., Linsbauer, A., Randhawa, S. S., Huggel, C., Rana, P. and Kumari, A. (2016) 'Glacial lake outburst flood risk in Himachal Pradesh, India: An integrative and anticipatory approach considering current and future threats', *Natural Hazards*, 84(3), pp. 1741–1763. doi: https://doi.org/10.1007/s11069-016-2511-x

Aven, T., Ben-Haim, Y., Boje Andersen, H., Cox, T., Droguett, E. L., Greenberg, M., Guikema, S., Kröger, W., Renn, O., Thompson, K. M. and Zio, E. (2018). *Society for risk analysis glossary*. McLean, VA: Committee on Foundations of Risk Analysis. Available at www.sra.org/wp-content/uploads/2020/04/SRA-Glossary-FINAL.pdf

Bahinipati, C. S. and Patnaik, U. (2020) 'Does development reduce damage risk from climate extremes? Empirical evidence for floods in India', *Water Policy*, 22(5), pp. 748–767.

Bipinchandra, M., Romeji, N. and Loukrakpam, C. (2020) 'Flood hazard risk assessment and mapping of a sub-watershed of Imphal River Basin, Manipur, India: A multi-resolution approach', *Applications of Geomatics in Civil Engineering*, pp. 155–163, Springer.

Budiyono Y., Aerts, J. C. J. H., Tollenaar, D. and Ward, P. J. (2016) 'River flood risk in Jakarta under scenarios of future change', *Natural Hazards and Earth System Sciences*, 16, pp. 757–774. doi: https://doi.org/10.5194/nhess-16-757-2016

Chakraborty, S. and Mukhopadhyay, S. (2019) 'Assessing flood risk using analytical hierarchy process (AHP) and geographical information system (GIS): Application in Coochbehar district of West Bengal, India', *Natural Hazards*, 99(1), pp. 247–274.

Chowdhury, M. R. (2000) 'An assessment of flood forecasting in Bangladesh: The experience of the 1998 flood', *Natural Hazards*, 22(2), pp. 139–163.

Cloud to Street. (2022) 'The global flood database: A project of cloud to street, flood observatory and Google Earth outreach'. Available at: https://global-flood-database.cloudtostreet.ai/?fbclid=IwAR1_LfTuJ7_x66E53ccF__zqkDlfp4OeYw7rTjxpY-l_7lfchNaeq64TV_A#interactive-map (Accessed: 17 January).

Díez-Herrero, A. and Garrote, J. (2020a) 'Flood risk analysis and assessment, applications and uncertainties: A bibliometric review', *Water*, 12(7), pp. 1–24. doi: https://doi.org/10.3390/w12072050

Díez-Herrero, A. and Garrote, J. (2020b) 'Flood risk assessments: Applications and uncertainties', *Water*, 12(8), pp. 1–11. doi: https://doi.org/10.3390/w12082096

EM-DAT (2021) 'Total affected from flood disasters, total estimated damages (in US$) from flood disasters, total deaths from flood disasters and global occurrences from flood disasters in 1900 to 2021'. Available at: www.emdat.be (D. Guha-Sapir) (Accessed: 2 November).

Fariza, A., Basofi, A., Prasetyaningrum, I. and Ika Pratiwi, V. (2020) 'Urban flood risk assessment in Sidoarjo, Indonesia, using fuzzy multi-criteria decision making', *Journal of Physics: Conference Series*, 1444, p. 012027. doi: https://doi.org/10.1088/1742-6596/1444/1/012027

Fariza, A., Rusydi, I., Hasim, J. A. N. and Basofi, A. (2017) 'Spatial flood risk mapping in East Java, Indonesia, using analytic hierarchy process – natural breaks classification'. *2nd International Conferences on Information Technology, Information Systems and Electrical Engineering (ICITISEE)*, pp. 406–411, 1–2 November. doi: https://doi.org/10.1109/ICITISEE.2017.8285539

Febrianto, H., Fariza, A. and Hasim, J. A. N. (2016) 'Urban flood risk mapping using analytic hierarchy process and natural break classification (Case study: Surabaya, East Java, Indonesia)'. *International*

Conference on Knowledge Creation and Intelligent Computing (KCIC), pp. 148–154, 15–17 November. doi: https://doi.org/10.1109/KCIC.2016.7883639

Ghosh, A. and Kar, S. K. (2018) 'Application of analytical hierarchy process (AHP) for flood risk assessment: A case study in Malda district of West Bengal, India', *Natural Hazards*, 94(1), pp. 349–368. doi: https://doi.org/10.1007/s11069-018-3392-y

Handayani, W., Chigbu, U. E., Rudiarto, I. and Putri, I. H. S. (2020) 'Urbanization and increasing flood risk in the northern coast of Central Java – Indonesia: An assessment towards better land use policy and flood management', *Land*, 9(10). doi: https://doi.org/10.3390/land9100343

Haque, M. N., Siddika, S., Sresto, M. A., Saroar, M. M. and Shabab, K. R. (2021) 'Geo-spatial analysis for flash flood susceptibility mapping in the North-East Haor (Wetland) region in Bangladesh', *Earth Systems and Environment*, 5(2), pp. 365–384. doi: https://doi.org/10.1007/s41748-021-00221-w

Isa, M., Sugiyanto, F. X. and Susilowati, I. (2018) 'Community resilience to floods in the coastal zone for disaster risk reduction', *Jamba (Potchefstroom, South Africa)*, 10(1), p. 356. PubMed. doi: https://doi.org/10.4102/jamba.v10i1.356 (Accessed 2018).

Islam, A. and Ghosh, S. (2021) 'Community-based riverine flood risk assessment and evaluating its drivers: Evidence from Rarh Plains of India', *Applied Spatial Analysis Policy*, pp. 1–47. doi: https://doi.org/10.1007/s12061-021-09384-5

Islam, M. F., Bhattacharya, B. and Popescu, I. (2019) 'Flood risk assessment due to cyclone-induced dike breaching in coastal areas of Bangladesh', *Natural Hazards and Earth System Sciences*, 19(2), pp. 353–368. doi: https://doi.org/10.5194/nhess-19-353-2019

Komolafe, A. A., Adegboyega, S. A. and Akinluyi, F. O. (2015) 'A review of flood risk analysis in Nigeria', *American Journal of Environmental Sciences*, 11(3), pp. 157–166. doi: https://doi.org/10.3844/ajessp.2015.157.166

Kron, W. (2005) 'Flood risk = hazard • values • vulnerability', *Water International*, 30(1), pp. 58–68. doi: https://doi.org/10.1080/02508060508691837

Kwak, Y., Gusyev, M., Arifuzzaman, B., Khairul, I., Iwami, Y. and Takeuchi, K. (2015) 'Effectiveness of water infrastructure for river flood management: Part 2 – flood risk assessment and its changes in Bangladesh', *Proceedings of the International Association of Hydrological Sciences*, 370, pp. 83–87. doi: https://doi.org/10.5194/piahs-370-83-2015

Lindersson, S., Brandimarte, L., Mård, J. and Di Baldassarre, G. (2020) 'A review of freely accessible global datasets for the study of floods, droughts and their interactions with human societies', *WIREs Water*, 7(3), p. e1424. doi: https://doi.org/10.1002/wat2.1424

Manna, P., Anis, M. Z., Das, P. and Banerjee, S. (2019) 'Probabilistic modeling of flood hazard and its risk assessment for Eastern region of India', *Risk Analysis*, 39(7), pp. 1615–1633. doi: https://doi.org/10.1111/risa.13333

Matheswaran, K., Alahacoon, N., Pandey, R. and Amarnath, G. (2019) 'Flood risk assessment in South Asia to prioritize flood index insurance applications in Bihar, India', *Geomatics, Natural Hazards Risk*, 10(1), pp. 26–48.

Meyer, V., Scheuer, S. and Haase, D. (2009) 'A multicriteria approach for flood risk mapping exemplified at the Mulde river, Germany', *Natural Hazards*, 48(1), pp. 17–39.

Mondal, M. S. H., Murayama, T. and Nishikizawa, S. (2020) 'Assessing the flood risk of riverine households: A case study from the right bank of the Teesta River, Bangladesh', *International Journal of Disaster Risk Reduction*, 51, p. 101758. doi: https://doi.org/10.1016/j.ijdrr.2020.101758

Nafishoh, Q., Meilano, I. and Riqqi, A. (2018) 'Quantitative flood risk projection in Java Island, Indonesia'. *IEEE Asia-Pacific Conference on Geoscience, Electronics and Remote Sensing Technology (AGERS)*, pp. 1–6, 18–19 September. doi: https://doi.org/10.1109/AGERS.2018.8554093

Obiefuna, J., Adeaga, O., Omojola, A., Atagbaza, A. and Okolie, C. (2021) 'Flood risks to urban development on a coastal barrier landscape of Lekki Peninsula in Lagos, Nigeria', *Scientific African*, 12, p. e00787. doi: https://doi.org/10.1016/j.sciaf.2021.e00787

Pathak, S., Liu, M., Jato-Espino, D. and Zevenbergen, C. (2020) 'Social, economic and environmental assessment of urban sub-catchment flood risks using a multi-criteria approach: A case study in Mumbai City, India', *Journal of Hydrology*, 591, p. 125216. doi: https://doi.org/10.1016/j.jhydrol.2020.125216

Rehman, S., Hasan, M. S. U., Rai, A. K., Avtar, R. and Sajjad, H. (2021) 'Assessing flood-induced ecological vulnerability and risk using GIS-based in situ measurements in Bhagirathi sub-basin, India', *Arabian Journal of Geosciences*, 14(15), pp. 1–17. doi: https://doi.org/10.1007/s12517-021-07780-2

Rentschler, J. and Salhab, M. (2020) *People in harm's way: Flood exposure and poverty in 189 Countries*. Policy Research Working Paper No. 9447. Washington, DC: World Bank, p. 28.

Roy, B., Khan, M. S. M., Islam, A. K. M. S., Mohammed, K. and Khan, M. J. U. (2021a) 'Climate-induced flood inundation for the Arial Khan River of Bangladesh using open-source SWAT and HEC-RAS model for RCP8.5-SSP5 scenario', *SN Applied Sciences*, 3(6), pp. 1–13. doi: https://doi.org/10.1007/s42452-021-04460-4

Roy, S., Bose, A. and Chowdhury, I. R. (2021b) 'Flood risk assessment using geospatial data and multi-criteria decision approach: A study from historically active flood-prone region of Himalayan foothill, India', *Arabian Journal of Geosciences*, 14(11), pp. 1–25. doi: https://doi.org/10.1007/s12517-021-07324-8

Saha, A. K. and Agrawal, S. (2020) 'Mapping and assessment of flood risk in Prayagraj District, India: A GIS and remote sensing study', *Nanotechnology for Environmental Engineering*, 5, pp. 1–18. doi: https://doi.org/10.1007/s41204-020-00073-1

Saini, S. S., Kaushik, S. P. and Jangra, R. (2016) 'Flood-risk assessment in urban environment by geospatial approach: A case study of Ambala City, India', *Applied Geomatics*, 8(3), pp. 163–190. doi: https://doi.org/10.1007/s12518-016-0174-7

Shivaprasad Sharma, S. V., Roy, P. S., Chakravarthi, V., Srinivasarao, G., & Bhanumurthy, V. (2017). Extraction of detailed level flood hazard zones using multi-temporal historical satellite data-sets – A case study of Kopili River Basin, Assam, India, *Geomatics, Natural Hazards and Risk*, 8(2), pp. 792–802. doi: 10.1080/19475705.2016.1265014

Solin, L. and Skubincan, P. (2013) 'Flood risk assessment and management: Review of concepts, definitions and methods', *Geographical Journal*, 65(1), pp. 23–44.

Suman, A. and Bhattacharya, B. (2015) 'Flood characterization of the Haor region of Bangladesh using flood index', *Hydrology Research*, 46(5), pp. 824–835. doi: https://doi.org/10.2166/nh.2014.065

Sun, Q., Miao, C., Duan, Q., Ashouri, H., Sorooshian, S. and Hsu, K. L. (2018) 'A review of global precipitation data sets: Data sources, estimation, and intercomparisons', *Reviews of Geophysics*, 56(1), pp. 79–107. doi: https://doi.org/10.1002/2017RG000574

Tami, A. G. and Moses, O. (2015) 'Flood vulnerability assessment of Niger Delta states relative to 2012 flood disaster in Nigeria', *American Journal of Environmental Protection*, 3(3), pp. 76–83.

Tellman, B., Sullivan, J. A., Kuhn, C., Kettner, A. J., Doyle, C. S., Brakenridge, G. R., Erickson, T. A. and Slayback, D. A. (2021) 'Satellite imaging reveals increased proportion of population exposed to floods', *Nature*, 596(7870), pp. 80–86.

UN (1992) *Internationally agreed glossary of basic terms related to disaster management*. Geneva: United Nations Department of Humanitarian Affairs (UN DHA).

UN (2020). *Country classifications, statistical annex, world economic situation prospects, United Nations*. New York: United Nations.

UNISDR (2009). *UNISDR terminology on disaster risk reduction*. Geneva, Switzerland: United Nations International Strategy for Disaster Reduction (UNISDR).

Winsemius, H. and Ward, P. (2015) 'Aqueduct global flood risk country rankings', *World Resources Institute*. Available at: www.wri.org/data/aqueduct-global-flood-risk-country-rankings

Zulkhairi, M., Azman, T. and Ndanusa, A. (2019) 'Multicriteria index and analytical hierarchy process on flood risk assessment: Application in Niger State, Nigeria', *International Journal of Innovative Technology and Exploring Engineering (IJITEE)*, 8(8S), pp. 674–678.

14

Flood risk management projects

Financing and project implementation, the case of Ibadan Urban Flood Management Project

Adedayo Ayodele Ayorinde, Abiodun Adefioye, Olakunle Oladipo and Oluseyi O. Fabiyi

Introduction

Background information

One of the challenges of global climatic perturbation is flooding in coastal and urban areas. Flood disasters often result in untold economic losses and loss of lives. Most developing countries lack the capacity to prevent and mitigate the effects of flood disasters in urban areas. Apart from low resources to embark on flood prevention and flood mitigation projects, the willpower to achieve flood-resilient urban areas is lacking in most active urban gatekeepers and the governments of most developing countries. Low-income earners often locate in vulnerable zones where land is cheap.

Containing urban flooding requires integrated approaches, which are broadly classified as structural and non-structural (Howe & White, 2004; Neuvel & Van Den Brink, 2009). The structural approach requires construction of embankments and channelization, while the non-structural approach includes the value orientation of people and subsidies for vulnerable urban dwellers to relocate to well-drained areas.

Ibadan city is drained by three major rivers, River Ogunpa, River Ona and River Ogbere, each with numerous tributaries. Besides River Ogunpa, which has been partly channelized by the federal government of Nigeria with some minor rivers discharging water into it, nearly all the other rivers and their tributaries contributed significantly to the last major flood disaster in Ibadan in 2011. The major areas affected are Odo-Ona, Odo-Ona Elewe, Orogun, Agbowo, Apata, Ajibode, University of Ibadan, Ogbere-Babanla, Ogbere Moradeyo, Onipepeye and Eleyele Dam/Water Works. The total length of streams and rivers found in the 11 local government areas making up Ibadan metropolis is 3,168.64 km (Government of Oyo State Flood Report, 2011). See Figure 14.1.

Ibadan city has witnessed several significant flood disasters since 1933. The series of recorded flood disasters in Ibadan city were in 1951, 1955, 1960, 1963, 1969, 1973, 1978, 1980, 1982, 1984 1986, 1987, 1997 and the climax in 2011. Agbola *et al.* (2012) noted that though the rainfall of

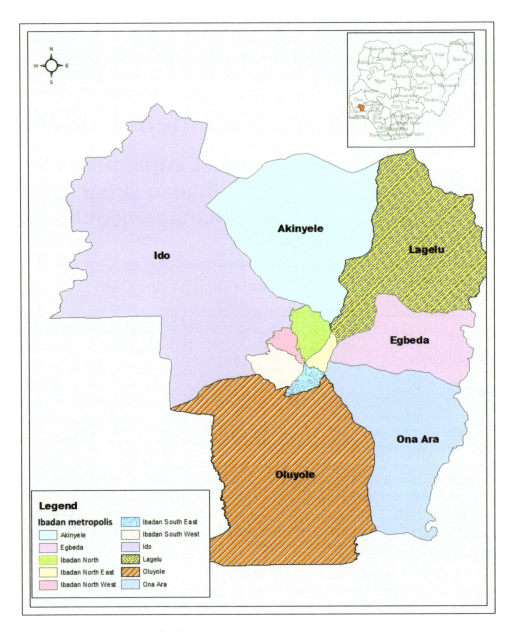

Figure 14.1 Major rivers in Ibadan

August 26, 2011, was not the highest in the recorded history of Ibadan city, the monetary value of damage to property that resulted from the event was by far the highest. Ibadan city has been plagued by incessant flood disasters since circa 1860 and has lost huge human and material resources to this menace. Therefore, a number of piecemeal approaches have been applied by successive governments to address the hazards with limited success. One of the efforts of previous administrations of government to control the flood disasters was the channelization of Ogunpa stream – one of the river channels that have been affected by perennial ten-year return period. Some of these projects were not only poorly conceived but were also poorly implemented. Even the relatively more comprehensive interventions in the aftermath of the 1980 Ogunpa flood disaster were short lived and soon fizzled out. The flood disaster in 2011 in Ibadan triggered a number of activities that culminated in the government of Oyo state of Nigeria approaching the World Bank for a credit facility to implement a comprehensive flood risk management project in Ibadan city.

The efforts of most succeeding governments to curtail flooding in Ibadan have been merely palliative and not far reaching enough. However, after the flood disaster of 2011, the Oyo state government approached the World Bank for a credit facility to proffer lasting solutions to the recurrent flooding in Ibadan city. The credit facility was to the tune of 200 million USD and was to take structural and non-structural measures for the taming of the scourge of flood disasters in Ibadan city. The project commenced on 23 February 2015 and was to last for eight years, during which the project was expected to be completed and delivered. The remaining sections of the chapter describe different segments of the implementation strategies adopted for the project and the success story of the project so far.

Operations of Ibadan Urban Flood Management Project

Units under Ibadan Urban Flood Management Project

The Ibadan Urban Flood Management Project (IUFMP) comprises the following units:

- Technical Unit: The technical unit is the engineering hub of the project. It includes specialists and consultants in the engineering profession, including civil, hydraulic, mechanical and infrastructural engineering professionals. The unit is responsible for the day-to-day supervision of the structural component of the project.
- Environmental and Social Safeguards Unit: Environmental and social safeguards experts are to ensure compliance with health, safety and environmental considerations of the project. They ensure that all hired contractors and staff engaged for the project conform to global best practices in service delivery with respect to social and environmental safety.
- Monitoring and Evaluation Unit: The M&E unit ensures that the project implementation strictly adheres to the work plan and specifications.
- Finance Unit: The finance unit is responsible for disbursement and accounting for all expenses in the project, including inflows and outflows.
- Procurement Unit: The procurement unit is responsible for all contracts, especially those that relate to purchases, recruitment of consultants, contractors and all purchasing orders.
- Communication Unit: The communication unit interfaces between the project and the public, bringing the activities of the project to the public and informing them about the areas by which they are directly or indirectly affected before and after the project is implemented. The unit is also responsible for implementing behavioural change communication as a complement to the structural interventions being put in place under the project.

These units are responsible for the day-to-day running of the project, which includes supervision and the award of contracts to achieve the following key component areas.

Key project components

- **Component-1:** Flood Risk Identification, Prevention and Preparedness Measures: The objective of this component is to assess flood risk in the city of Ibadan, plan risk reduction

measures and finance preventive structural and non-structural measures to enhance flood preparedness;
- **Component-2:** Flood Risk Reduction: The objective of this component is to ensure flood risk mitigation through structural measures by financing public infrastructure investments for flood mitigation and drainage improvements; and
- **Component-3:** Project Administration and Management Support: The objective of this component is to finance incremental operational costs related to the implementation of the project for goods, equipment, staff, travel and the project management unit's consultant services.

The overall project objective and implementation process

The principal objective of the project are to improve the capacity of Oyo state to effectively manage flood risk in the city of Ibadan. Therefore, the project implementation unit (PIU) is charged with the responsibility of project management and implementation, procurement and financial management, environmental and social safeguards, due diligence, monitoring and evaluation of the project, as well as reporting on progress and implementation issues to the project steering committee. The PIU works closely with relevant ministries, departments and agencies (MDAs) at both federal and state government levels and seeks their contributions to both project inputs and outputs.

The institutional arrangement for decision-making and project implementation as documented in the project appraisal document (PAD) for IUFMP revolves around the PIU, the project technical committee (PTC) and the project steering committee (PSC), with the latter being the project's highest decision-making body headed by His Excellency, the executive governor of Oyo state. These committees work hand in hand with the technical assistance team of the World Bank Group headed by the task team leader (TTL).

The project area

Ibadan city is located in southwestern Nigeria. It is the capital of Oyo state and is reputed to be the largest indigenous city in tropical Africa south of the Sahara. Ibadan is centred about latitude 7° 25' north and longitude 3° 5' east approximately 145 kilometres north of Lagos. Ibadan is directly connected to many towns in Nigeria and its rural hinterland by a system of roads, railways and air routes.

Ibadan was established by Lagelu after the destruction of the first settlement near Awotan in the neighbourhood of Apete in the Ido local government area. The presence of hills makes the site of the city easily defensible, while its location close to the boundary between forest and grassland makes it a melting pot for people and products of the forests as well as those of the grassland areas. However, Ibadan was resettled about 1820 as a camp by soldiers of Ife, Ijebu and Oyo.

Ibadan is an ancient city and has been a regional capital as far back as the end of the 19th century. The Ibadan metropolitan area consists of 11 local government areas which are made up of the core traditional areas and new areas. The new areas are medium-density areas and peripheral development or urban sprawl.

Since its founding in the 1800s, Ibadan has been an important rallying point for people in the southwest of Nigeria. Due to its historical and administrative position in Nigeria, Ibadan has continuously witnessed an influx of people which has contributed to its rapid growth in population, physical expansion and land coverage. In terms of demographic growth, Ibadan experienced a geometrical increase between 1851 and 1921. By 1856, the population was estimated at

60,000 (Hinderer, 1856), which rose to over 200,000 in 1890 (Millson, 1891), 238,094 in 1921 and 386,359 in 1931. The 1991 census in Nigeria put the population at 1,222,570 with a density of 475.11 persons per square kilometre. Its population is estimated to be about 2,550,593 according to 2006 estimates by the National Population Commission. Its projected population by 2010 using a 3.2% growth rate is about 2,893,137. (Government of Oyo State Flood Report, 2011). The estimated population for 2021 is 4,196,367 (Oyo State Bureau of Statistics), with a growth rate of 3.35%

In terms of physical expansion and land coverage, this pre-colonial urban centre has expanded very fast, sprawling daily into the hinterland. It was noted that developed land in Ibadan increased from only 10 km^2 in 1830 to 12 km^2 in 1931, 30 km^2 in 1963, 112 km^2 in 1973, 136 km^2 in 1981 and 214 km^2 in 1988. An aerial photograph in 1964 showed that the city had spread beyond the drainage basins of Ogunpa and Kudeti and to the catchment area of Ogbere stream in the eastern side. In recent years, the city spread has extended to Odo-Ona Kekere village in the south to Iroko/Motunde villages in the north, Asejire in the east and Bakatari in the west (Government of Oyo State Flood Report, 2011). Flooding is not new to the city. Historical records have it that the city has been experiencing an increasing number of floods events during the last 50 years. One of the major causes of flooding in the city is land use, which is primarily residential, coupled with the poor planning and an unregulated approach to building development in the city. Most houses have been built on drainage without the minimum river setback benchmark, so there are houses built near rivers and obstructing the free flow of flood water.

A poor attitude to waste disposal and management has made flooding a common occurrence in Ibadan. Indiscriminate dumping of waste blocks culverts and bridges and prevents free flow of storm water. This results in backflows, which overtop bridges or make their way into houses close to the river banks.

Another major cause of flooding in Ibadan is the geology. The terrain is an undulating pattern, made up of intersecting valleys and hills. Flood waters from the hillside freely flow into the valleys, dotted by mainly residential buildings, thus causing major flash floods.

Financing structure and disbursement of IUFMP

IUFMP financing structure

As contained in the project appraisal document, the total estimated project cost of the Ibadan Urban Flood Management Project is US$220 million, made up of the World Bank-facilitated US$200 million International Development Association (IDA) credit facility and US$20 million counterpart fund from the borrower/recipient, which is the Oyo state government. The scheduled disbursement of the credit facility was calibrated in two ways: across the three components of the project as well as across the eight years of the project's lifespan. Tables 14.1 and 14.2 present the disbursement schedules for both:

Funding sources

There are two major sources of fund for the project, the IDA and counterpart funds.

The IDA fund

The IDA fund is the World Bank credit facility of US$200 million, which is drawn down into a designated account on a report-based system; that is, based on a six-month disbursement

Table 14.1 Fund allocation for key components of project

Component name	Cost (US$ millions)
Component 1: Flood Risk Identification, Prevention and Preparedness Measure	43.0
Component 2: Flood Risk Reduction	149.0
Component 3: Project Administration and Management	8.0
Total	200.0

Table 14.2 Expected disbursements on yearly basis (in US$ millions)

Fiscal year	2015	2016	2017	2018	2019	2020	2021	2022	2023
Annual	9.00	11.30	26.20	27.00	32.20	34.10	32.10	28.10	0.00
Cumulative	9.00	20.30	46.50	73.50	105.70	139.80	171.90	200.00	200.00

plan/budget to be requested by the project from the World Bank. This is in contrast to the transaction-based disbursement/budget system that most World Bank-assisted projects in Nigeria are placed on, which translates to requesting money draw-downs only on an as-needed basis. The IDA fund is assigned to funding Components 1 and 2 as well as some parts of Component 3 of the project.

The report-based system is such that requires the project to present a six-month fund disbursement plan in line with existing and upcoming contracts and commitments.

The counterpart fund

The counterpart fund is the Oyo state government's financial commitment to the project. It is deducted directly by the Federal Ministry of Finance (on behalf of the federal government of Nigeria) from the monthly allocation to Oyo state as prepared by the Federation Accounts Allocation Committee (FAAC). This is used for some elements of Component 3 of the IUFMP, specifically procurement of office supplies and furniture, information and communication technology (ICT) equipment, transport vehicles for the project implementation unit, office running costs, hiring of external financial and technical audits and preparation and implementation of resettlement action plans (RAPs), among others.

Contract management approach adopted in IUFMP

The role of contract management techniques in project financing is to estimate a disbursement forecast based on anticipated or planned performance of ongoing works. As draw-down of funds in the account continues in accordance with contract performance, a minimum value is set that automatically alerts the donor to replenish the designated account with another tranche of funds. This is achieved through financial software.

In terms of best practice, the contract management software is wired into the financial software, which enables the pattern of spending/disbursement to be viewed at a glance from the dashboard. Monthly reports are therefore automatically generated from both the contract and financial management software.

The responsibility for establishing and maintaining acceptable financial management (FM) arrangements is conferred on the existing Oyo State Project Financial Management Unit

(PFMU), which is a multi-donor and multi-project financial management platform, jointly set up by the World Bank and the Oyo state government. This common FM platform features robust systems and controls, and it has won the World Bank Team's performance awards a couple of times.

Implementation strategies of IUFMP

The IUFMP has been carefully designed to run in a systematic and consistent manner in line with best practices in project development and implementation. The project has a results framework upon which success will be gauged.

Result framework for the project

This results framework contains 11 indicators, 5 of which are quantitative and the remaining 6 qualitative. The quantitative indicators are direct project beneficiaries, female percentage, land area protected, flood-prone sites made flood resilient and population protected by restoring the safety of Eleyele Dam, while the qualitative indicators are effective use of a flood control asset management plan, adoption of a flood control asset management plan, improved institutional coordination on flood risk management in Ibadan, improved capacity for flood forecasting and warning and a flood risk management capital investment plan adopted for targeted sites. All these are done within the context of the three project components: flood risk identification, prevention and preparedness measures; flood risk reduction; and project administration and management support.

The framework design under this project allows for a phased approach combining the immediate implementation of priority investments within the city while the medium- to long-term investment program is being formulated.

Implementing structural and non-structural components of the project

The formulation of effective and sustainable urban flood risk management for Ibadan is a long, complex and costly process. Similarly, promoting an integrated and effective urban flood risk management program combining both structural and non-structural measures requires a good understanding of available alternatives based on the future growth of the city and acceptable risk of the communities. Such measures must be based on transparent cost-benefit analysis to facilitate the prioritization of financing on the most urgent and effective of these measures.

It is worthy of note that IUFMP interventions cut across two major strata, as mentioned earlier: structural and non-structural. Structural interventions include improvement of priority sites, especially those destroyed by the 2011 floods. A number of sites are listed in this category. Selection criteria for priority sites include: opportunity to reduce local flood risk, need to reconstruct flood-damaged infrastructure and need to re-establish or improve community connectivity.

Prioritization of project sites

The selection criteria for targeted sites are: investment types, consistency with strategic masterplans, socio-economic impact, community participation and compliance with project implementation manual. The scope of work centres on rehabilitation of culverts, drains and approach roads. Four of these sites received first attention. In addition, another set of 13 sites were targeted and interventions were carried out, with direct project beneficiaries totalling about 118,016 people.

The Eleyele Dam, situated on the Ona River, which was devastated by the 2011 floods and accounted for the loss of lives and properties downstream from the dam, was targeted for rehabilitation. The dam had never received turnaround maintenance in its over 70 years of existence. The damaged spillway, a potential threat to about 5,065 people downstream, posed a major integrity issue and could lead to a dam break. The dam supplies treated water for almost half of the city's inhabitants and is a source of water for factories and institutions within and around the metropolis. As of May 2019, the rehabilitation of the facility had been completed and a maintenance arrangement which includes instituting a dam management body has been put in place.

Going further on the structural angle is also channelization of major rivers in the city and rehabilitation of associated and stand-alone structures such as bridges and culverts. A number of such rivers are Agodi, Kudeti, Ogbere and Orogun. This is referred to as the First Pool of Works, while the Second Pool of Works, which constitutes the last lap of the project implementation, is the construction of 17 dikes in the upper section of the Ona River in Akinyele local government area, Oyo state, Nigeria: channelization of over 13 kilometres of the same river and rehabilitation of associated hydraulic structures.

The non-structural interventions include preparation of various studies aimed at addressing long-term flood risk management in the city. The first of these was the preparation of the Ibadan City Masterplan. The document sets out a pattern of development in the city for the next 25 years addressing housing issues, urban renewal, establishment of growth centres and a new circular road of over 100 kilometres to decongest the main heart of the city and also boost growth in the outer parts of Ibadan. The project provides policy guidelines to control building regulation and prevent building on flood plains, well-stocked spatial data sets on different aspects of the city and a robust pathway to economic development for the inhabitants of Ibadan. The document has been completed, and the necessary legislative backing to see to its implementation is being worked on, thereby making Ibadan a planning area.

The integrated flood risk management and drainage masterplan for Ibadan was carefully developed. It is a compendium of all crossing structures, their state, the amount needed to rehabilitate or construct new ones, locations, recommended hydraulic structures and other data. It provides a document guiding drainage intervention in the city. The phases of intervention are broken down into a five-year phase each, detailing costs needed for them. The first phase of the intervention is being carried out by IUFMP, while financing for subsequent phases will be sourced by the government.

Another non-structural intervention is the Ibadan Solid Waste Management Masterplan. It has been established that the number-one cause of flooding in the city is the indiscriminate dumping of refuse into streams and canals, which moves down to clog hydraulic structures and block the flood pathway, thereby overtopping and finding its way into surrounding buildings and farmlands. To this end, the project is at an advanced stage of seeing that the study is conducted and the necessary legislation that will back it up is provided for.

Capacity building and sustainability strategy of the project

A key sub-component of IUFMP is project administration. This activity focuses mainly on training and capacity development of the staff of the PIU and relevant MDAs. Particular attention is paid to training on fiduciary details and safeguards. Capacity building also involves study tours, specific studies and workshops and partnerships between national educational centres and universities. The project also supports the implementation unit in office running costs (PIU staff salaries, electricity, water, internet, telephone, fuel, etc.), thus enabling better planning and implementing of activities on project sites.

Procurement process in IUFMP

It is important to note that procurement is a key activity in the project. Every year, a procurement plan is prepared and approved by the state government, after which it is transmitted to the World Bank for clearance/no objection. Any activity not captured in the procurement plan will not be implemented in that procurement year. Procurement cuts across works, consultancies, goods and services. There are various procurement packages for different activities based on the financial thresholds. These include international competitive bidding, national competitive bidding and shopping. Some steps in the national competitive bidding are listed in the following:

- The procurement specialist prepares specification, bills of quantities (BoQs) and drawings in liaison with the user department and submits to the project coordinator for approval.
- Submits to procurement consultant for review.
- Approval from state government.
- Advertisement.
- Public bid opening.
- Project coordinator constitutes evaluation committee.
- Committee carries out evaluation.
- Bid evaluation report submitted for approval.
- Award notification sent to bidder.
- Signing of contract.
- Implementation of works.

Environmental and social safeguard activities

During construction, some local stakeholders are likely to be negatively affected, but these impacts are expected to be manageable. Most of the activities are not expected to result in major losses or acquisition of land or restriction of access to livelihoods. The project seeks to mitigate impact through a comprehensive stakeholder, communication and participatory engagement strategy combined with a strong capacity-building program at the local and city levels. Project-affected persons (PAPs) are identified based on classification of impacts, such as loss of buildings and disturbances to economic activities. The project carries out enumeration, valuation and verification before compensation is paid. The aim is for the project to leave them better than it met them. As such, a fair compensation package, based on the World Bank template, is adopted for this. Environmental impacts are mitigated through consistent inspection of contractors' site yards and work area. Pollution of air samples and water quality, as well as spillage of chemicals from plants and equipment, are forestalled, and in cases where these happen, the contractor is made to make amends. Dust is also controlled through regular wetting of roads and work areas to mitigate health hazards.

In approaching the social and environmental components of the project, it is mandatory that before any major civil work is embarked upon, the project prepares what is called safeguard instruments. There are studies which indicate the risk and impact on people and the environment in the course of the implementation. Some of these instruments include the resettlement action plan. The category of impact on people and the environment dictates the types of instruments to be prepared. These instruments factor in the compensation plan for relieving project-affected persons to maintain or restore their livelihoods as project implementation commences.

As the project progresses, lessons learnt and success stories are communicated to stakeholders and project beneficiaries. Best practices are also documented for replication in other similar projects. The project boasts a robust communication strategy which cascades information to different audiences both within and outside the project areas. This comes in the form of newsletters, publications, jingles, social media activities and press briefings. This activity has helped to paint the project in good light.

Challenges, success stories and recommendations of IUFMP

In this section, we examine the problems encountered so far in the implementation of IUFMP as well as the project's prospects and potential for replication in other developing countries with similar economic outlooks as Nigeria.

IUFMP's identified challenges

IUFMP is classified as an A Category project in the World Bank portfolio, given the high risk index of its interventions, especially the now almost-80-year-old Eleyele Dam which has been rehabilitated as part of the project's works. Also, it is a multi-faceted and multi-sectoral project, whose interventions touch on structural and non-structural measures, spanning dam rehabilitation, building bridges and culverts, channelization of rivers, opening up communities hitherto ravaged or disconnected by floods, setting up a first-of-its-kind flood forecast and early warning system, capacity building for technocrats and preparation of strategic masterplans, among others.

A requirement for delivering such a complex and intricate project as IUFMP is human resources in the right quantity and quality. The first set of personnel drafted to the project implementation unit were seconded from the relevant ministries, departments and agencies of the mainstream civil/public service of the Oyo state government. While this provided the project with a team of readily available workforce, it also came with its challenges, chief of which was the problem of inadequate capacity to meet with the demands of a donor-funded project.

Though part of the project design was for the PIU to be reinforced by a project management consultancy (PMC) for the first four years of the project's lifespan, this was hamstrung by the inability of both parties (the Oyo state government and the shortlisted PMC firm) to reach a consensus on the fee regime. To address this, the project's regulatory stakeholders later resolved to engage individual consultants for key functions to bridge the gap.

Yet another challenge was the fact that the bulk of IUFMP interventions are taking place within the heavily built-up Ibadan metropolis, thus hindering the pace of works on account of the imperative of having due consideration for impacts on lives and livelihoods. As a principle, the first consideration of places of intervention is to explore avoidance of impacts where possible. Where this is non-feasible, due diligence is keenly followed in the management of the impacts, part of which includes emplacement of a resettlement action plan or its abbreviated version, depending on the extent of the impacts.

Also, the project implementers are often exposed to the palpable worries of the community stakeholders on the possibility of interventions in their localities being caught up in the political undercurrents of the state, especially in the event of change of government, policy somersaults and so on. It is often an arduous task convincing them that the project, by its design, enjoys some measure of guarantee that shields it from undue political interference.

Over time, it has become obvious that the eight-year timeline specified for the IUFMP implementation is grossly inadequate to cover the outlined interventions. This is even more so given that a substantial chunk of the first three to four years was spent on project preparation and take-off. Worse still, the outbreak of the COVID-19 pandemic around the sixth and seventh years of the implementation, which could be rightly described as the project's prime, has stalled not a few interventions within the metropolis.

Finally, one cannot but mention the unchanging attitude of people to proper waste disposal as a challenge to the implementation of IUFMP, more so when such acts are noticeable at places where new interventions have already been carried out. With the massive awareness being created on the adverse effects of indiscriminate dumping of municipal waste on water channels and flood plains vis-à-vis the past unsavoury experiences of the community beneficiaries on flooding, it would have been expected that the newly constructed flood control infrastructure would not be clogged with waste. The fact that this is not the case in some instances calls for concern about the long-term sustainability of the highly capital-intensive interventions.

IUFMP's achievements

Notwithstanding the challenges highlighted in the preceding section, IUFMP no doubt has a lot of prospects and holds much promise for flood risk management in other parts of Nigeria and indeed other countries with similar ecological and economic outlooks as Nigeria. A test case of the foregoing statement is the affirmation that IUFMP is a first-of-its-kind FRM project in Africa in the portfolios of the World Bank. It therefore offers a good study for replication in other affinity localities suffering the same inundation problem as Ibadan.

On the part of the Oyo state government, a robust sustainability plan, or asset management plan, is being designed to provide a workable framework for maintaining the gains of the project. Part of the considerations is the likelihood of the project metamorphosing into a top-notch agency or a directorate which will be sufficiently empowered to carry on the mandate.

It is also worthy of note that, over the course of the project trajectory, a working relationship, sealed with a memorandum of understanding (MoU), has been cultivated with the University of Ibadan, aimed at encouraging relevant academics and scholars to make research-based inputs into the project implementation. Provisions have been made for this to still be sustained even after the life of the project.

Recommendations towards improving implementation mechanisms for IUFMP and similar projects

- Improvement on response time between the project implementation unit and the World Bank Team on correspondence exchanged through e-mail towards removing all forms of delay in project implementation.
- Regular meetings (weekly) to discuss freely ongoing project activities/pending issues and taking decisions/steps on a way forward between the PIU and the World Bank Team through virtual meetings.
- Regular meetings of the project steering committee of IUFMP chaired by His Excellency, the governor of Oyo state, to ensure timely approval/response to various project activities by appropriate government functionaries, policy makers and technocrats from the line MDAs involved in the implementation of the project.
- Regular site monitoring and meetings to ensure that the ongoing channelization works in the First and Second Pool of Works under the long-term investments emanating from the

drainage masterplan for Ibadan being undertaken by the contractors and the supervising consultant are timely done to ensure the successful completion of the works before the project closure date of June 2022.

Conclusion

The Ibadan Urban Flood Management Project is one of the major urban flood control infrastructural projects carried out in West Africa based on the far-reaching measures and the volume of infrastructural development associated with the project. The cost of implementing the project is also very significant, at a total of 220 million US dollars. The project adopted an integrated approach of combined structural measures and non-structural-social measures to tackle the menace of flood problems in the project area. The project is in the final stage of implementation and has significant lessons learnt to showcase to the academic communities.

The chapter presented operational activities of the Ibadan Urban Flood Management Project and the achievements as well as significant impacts the project made to the residents of Ibadan city and Oyo state, Nigeria.

References

Agbola, B. S., O. Ajayi, O. J. Taiwo and B. W. Wahab (2012). The August 2011 Flood in Ibadan, Nigeria: Anthropogenic Causes and Consequences. *International Journal of Disaster Risk Science*, 3 (4): 207–217.

Eleyele Dam Rehabilitation and Intake Works Completion Report. https://documents1.worldbank.org/curated/pt/566181485759738747/pdf/SFG2854-EA-P130840-Box402881B-PUBLIC-Disclosed-1-26-2017.pdf.

Government of Oyo State Report on the Assessment of the 26 August 2011 (2011). Flood Disaster in Ibadan Metropolis. *Nigerian Tribune*, October. Source newspaper.

Hinderer, D. (1856). Dictionary of African Christian Biography. https://dacb.org/stories/nigeria/hinderer-david/.

Howe, J. and I. White (2004). Like a Fish Out of Water: The Relationship Between Planning and Flood Risk Management in the UK. *Planning Practice and Research*, 19 (4): 415–425.

IUFMP Integrated Flood Risk Management and Drainage Master Plan Report. https://ewsdata.rightsindevelopment.org/files/documents/40/WB-P130840_eOttA5M.pdf.

IUFMP 4 and 13 Priority Sites Design and Completion Reports. https://feedbackoysg.com/the-ibadan-urban-flood-management-project-your-questions-answered/.

Millson, A. (1891). The Yoruba Country, West Africa. *Proceedings of the Royal Geographical Society and Monthly Record of Geography*, 13: 577–587.

Nigeria Meteorological Services Agency 2021 Seasonal Climate Prediction (SCP). https://public.wmo.int/en/resources/meteoworld/nigeria%E2%80%99s-2021-seasonal-climate-prediction#:~:text=The%20Minister%20of%20Aviation%2C%20Senator,normal%20rainfall%20totals%20are%20envisaged.

Nigerian Hydrological Service Agency (NIHSA), 2016–2019 Annual Flood Outlook. https://nihsa.gov.ng/wp-content/uploads/2021/05/2021-AFO.pdf.

Neuvel, J. M. M. and A. Van Den Brink (2009). Flood risk management in Dutch local spatial planning practices. *Journal of Environmental Planning and Management*, 52 (7): 865–880.

Oyo State Bureau of Statistics Estimated Population, 2018–2021. https://en.wikipedia.org/wiki/Oyo_State.

15
Current knowledge, uncertainties, and aspirations of flood risk management policy for developing countries

Ugonna C. Nkwunonwo

Introduction

Widespread flooding prompts the need to help communities cope with uncertainties, risks, and disasters. This is of value to flood risk and disaster management within the contexts of global climate change and increasing demographic pressure in recent times. Climate change is forcing increased probability of flooding globally, while the increasing exposure of socio-economic variables scales up flood risk and disaster expectations both now and in the future (Haer *et al.*, 2020). Presently, the global footprint of flood disasters raises the need to improve scientific knowledge about the typologies and attributions of flood hazards and the socio-economic consequences on societies (Coronese *et al.*, 2019). So it is vital to improve public awareness of flood risk, promote community participation in flood management policy, and build community resilience – aimed to prevent loss of human lives, reduce destruction of vital resources, and forestall major disruption of economic activities. The United Nations' agenda for disaster risk reduction recognises these dynamics in formulating policies and strategies to ensure a safer and more sustainable future (Aitsi-Selmi *et al.*, 2015; UNDRR, 2020). This agenda and the optimistic 'cathartic' idea of 'living with floods rather than fighting them' inspire more efficient flood risk management strategies, which have advanced over the years (Dieperink *et al.*, 2016; Wolff, 2021).

Flood risk management (FRM) is both an ideological and realistic response to the effects of flooding on people and the environment. Its sustainability lies in its ability to reduce both the probability of flooding and its consequences (Pender & Faulkner, 2010). Therefore, in principle, FRM is a shift from the traditional flood control mindset (often involving engineering structures) to an integrated, risk-based framework. Extant research acknowledges FRM as a far more strategic, proactive, and sustainable procedure than the traditional approach, since it aims to mitigate loses because of flooding based on a variety of resources and techniques (Pender & Faulkner, 2010; Mai *et al.*, 2020). All practical techniques for tackling flooding are science based (Schanze, 2006; Morrison *et al.*, 2018). However, FRM creates a framework whereby both science and non-science strategies and stakeholders can interact and formulate policies to deal with flooding within a variety of spatial and temporal scales (Cvetkovic & Martinović, 2020).

Despite several years of developments in disaster management science, current FRM does not reflect sufficient preparations in many developing countries to meet the actual flooding crises that confront them today, let alone address the future uncertainties (Boulange et al., 2021; Dedekorkut-Howes et al., 2021; Tabari, 2021). This is primarily a major stimulus for research in flood risk and disaster management and a key component of the Sendai framework for action (2015–2030). A significant number of human populations and communities lack the capacity to cope with flood hazards or recover from losses (Nur & Shrestha, 2017). Institutional efforts in flood risk governance are below the expected thresholds to initiate proactive responses to increasing flood threats, whilst they lack the capacity to stimulate multi-agency participation in flood disaster management (Schwanenberg et al., 2018). How to actualise FRM within the context of sustainable community livelihoods, associated with ecosystem services, and land uses in flood-prone areas is still an ongoing research problem (Juarez-Lucas & Kibler, 2016; Juarez-Lucas et al., 2016; Strazzera et al., 2021; Huang et al., 2021). Lack of access to quality topographic, hydrologic, and flood damage data, with insufficient transferable and reproducible flood modelling tools, got attention in Nkwunonwo et al. (2020) as major constraints to FRM in developing countries.

Globally, the rapid development in geospatial science and infrastructure is at the forefront of various state-of-the-art methods of FRM and science. Free and open-source geospatial data sets and software are now commonplace and increasingly creating research and technical possibilities, with alternatives to the data issues that plague FRM in developing countries (Albano et al., 2017; Ekeu-wei & Blackburn, 2018; Meadows & Wilson, 2021). However, the value of these scientific innovations is indeterminate, whilst their utility is disproportionately sluggish in developing countries. Their availability is asymmetric, whilst local policy and politics did not align well with sustainable development initiatives which sustain FRM policy in developed societies (Rehman et al., 2019).

Against the foregoing background, this chapter examines and critiques FRM policies in developing countries and highlights strengths, weaknesses, and opportunities. Drawing from previously published work and flood management reports, the present chapter assimilates lessons from global best practices to spotlight opportunities that can contribute to realistic flood risk management in the developing countries. Using Nigeria, which is in the heart of West Africa, as a reference for flooding and its management, this chapter reviews the significance of formulating policies based on indigenous approaches and solutions endogenous to flood risk and disaster in the developing countries (Makondo & Thomas, 2018). Finally, it recommends developing bespoke models and approaches that will take advantage of free and open-source geospatial infrastructure, training local technical capacity, and building the local infrastructure to reinforce sustainable development objectives in developing countries.

Methods and data

Limitations in FRM in developing countries arguably reflect a major constraint with current science, political discussions, and stakeholder engagement to address the increasing threats of flooding on people, the environment, and developments. Solutions to these limitations is part of an active disaster and risk management research, especially in view of the need to keep at equilibrium the delicate FRM balance between science, politics, and stakeholder whilst aiming to realise the aspirations of FRM in the developing countries. To understand what the situation is in these areas, this chapter undertakes a review of literature. It frames its objectives around five

Flood risk management policy for developing countries

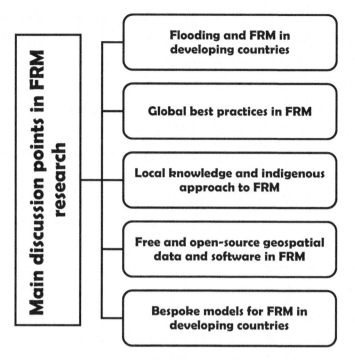

Figure 15.1 The key debates in flood risk management within developing countries

key themes (shown in Figure 15.1): (1) flooding, flood risk, and FRM in developing countries; (2) some global best practices in FRM fail in developing countries – why some of them succeed and why some of them fail; (3) local knowledge and indigenous approaches to FRM in developing countries; (4) free and open-source geospatial data and software in FRM; and (5) bespoke models for FRM in developing countries.

Flood disaster data from global disaster reports and the Centre for Research on the Epidemiology of Disasters (CRED) EM-DAT international disaster database provided much-needed evidence on the pervasiveness, attributions, and consequences of flooding in developing countries. Technical reports from the Sendai framework (2015–2030), UNDRR, KULTURisk, and EU flood risk directive highlighted most of the state-of-the-art flood management approaches and global best practices in FRM. This chapter discuses key issues extracted from a thorough perusal of these documents. It critically reviews the facts and then provides a rationale on which it based an argument for the most practicable approach of FRM in developing countries.

Discussion

An overview of flooding and flood risk management efforts in developing countries

Flooding means that water is occupying places that are normally described as dry land, threatening people, the physical environment, and developments. This description holds for developing countries, although it often requires the need to understand the basic science that undergirds flooding. Data from the EMDAT database show that within the last five years, flooding in developing countries has affected over 1 million people, disrupted significant livelihoods and economic activities, and caused economic loss worth hundreds of billions of US dollars. Basically, the complex interaction of key hydrological components – rainfall, infiltration, evapotranspiration, runoff, overland flow – underpins flooding. Research has developed on a theory of flood formation because of the interaction of hydrological components (see, for example, Chow et al., 1988; Naughton et al., 2012). Of these components, changes in rainfall patterns in recent times because of climate change are increasingly a major source of concern for flooding in developing countries. An understanding of the typologies of flooding in developing countries enables formulation of risk and disaster management efforts. Based on hydrological framing, fluvial, coastal, flash, and pluvial are the four main types of flooding that occur in developing countries (Figure 15.2).

Fluvial flooding is the oldest type of flooding in history. It affects developing countries that are surrounded by major rivers, for example, Bangladesh, Japan, Indonesia, places in northern and southern Africa, and many parts of the Caribbean countries where people live very close to major rivers. In Nigeria, the River Niger and River Benue are major sources of fluvial flooding in Lokoja and parts of the north central geopolitical zone. Coastal flooding affects many places in coastal areas. There is a major concern in developing countries regarding the benefits of coastal areas and how to apply FRM strategies within the context of ecosystem services (Juarez-Lucas et al., 2016). Flash flooding in developing countries is often because of unannounced rainfall, heavy rainfall that tilts the balance of the local hydrological system. Such rainfall can raise the amount of water in rivers so that they overtop their defences. It can also overwhelm local drainage systems and attenuate overland flow and surface runoff. Pluvial flooding is a special case of flash flooding that affects mainly urban areas. It is now the most widespread flooding in developing countries and comes from heavy rainfall, which is low-frequency–long-duration rainfall or short-duration–high-frequency rainfall. Overwhelmed urban drainage systems and impervious surfaces in urban areas, which make water infiltration into the soil almost negligible,

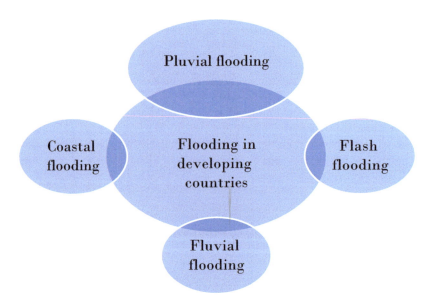

Figure 15.2 Typologies of flooding in developing countries

characterise pluvial flooding. A major concern with this type of flooding is the high rate at which urban areas are growing in developing countries and the level of demographic pressure.

Developing countries collectively have some common features which explain widespread flooding, risk, and disaster management in them. Figure 15.3 highlights the relationship between developing countries, flooding, and FRM. First, many places in developing countries are highly vulnerable to climate change, so more flooding and other extreme weather events are likely to occur, and with rapid urbanisation and population growth, they will increase the human and socio-economic consequences (Campbell-Lendrum & Corvalán, 2007). The presence of slum areas suggests many people are living and finding their livelihoods in flood-prone areas. Second, unstable leadership, weak political willpower, and institutional failures undermine the enforcement of environmental laws and other activities towards flood risk prevention and preparedness (Nkwunonwo et al., 2016; Al Mamun & Al Pavel, 2014). Third, underdevelopment and a high rate of poverty expose many people to a wide range of activities leading to flooding in urban areas, such as roadside car washing, indiscriminate dumping of household waste, and violation of housing and building regulations (Nkwunonwo, 2016). Poor developmental infrastructure and inadequate access to public social amenities including hospitals, schools, financial institutions, recreational facilities, and open spaces means that a community's resilience and capacity to cope with flooding are limited (Leal Filho et al., 2019). Finally, a lack of access to clean drinking water and sanitation facilities, traffic congestion, and a high number of road accidents mean that during flooding, the chances of disease outbreak and many deaths are high and response activities are difficult to undertake.

Nigeria arguably typifies a developing country regarding flooding and FRM. Existing institutions and legal frameworks are in many ways offshoots of colonial powers. The country, like many developing countries, is still struggling to meet the United Nations Sustainable Development Goals in terms of human and economic development. This translates into the way the country addresses threats of flooding. Nigeria has experienced annual flooding for many years, mostly the flash and pluvial types (Nkwunonwo, 2016; Egbinola et al., 2017). The most devastating flood for the country was the 2015 episode, which affected over 30 states. Lagos state, which is the hub of economic development in the country, has experienced most of the floods, while more frequent floods occur in Niger, Adamawa, Oyo, Kano, and Jigawa states, possibly because of the influence of the rivers Niger, Benue, Ogun, and Hadeja. Figure 15.4 is an original map produced using the data from the Nigerian Environmental Agency on the hotspot areas of flooding in Nigeria. Reducing vulnerability and building the recovery capacity of flood victims are critical issues. Rapid population growth, urbanisation, poor urban planning, and increased frequency and intensity of rainfall because of climate change worsen the probability of flooding and the intensity of its consequences. Lack of flood data and other remote causes which are yet to be identified constrain FRM efforts in Nigeria. More robust and scientific approaches to flood risk reduction, such as flood modelling and vulnerability assessment, are lacking.

Oladokun and Proverbs (2016) reviewed and characterised Nigeria's FRM practices and linked its present scope to lack of a unified action, lack of institutional planning and coordination, poor infrastructure, poor drainage system, high urban poverty, low-level literacy, and cultural barriers. These issues set a higher standard for action towards an integrated process for community development. The authors promoted the idea of a multidisciplinary platform for developing effective strategic planning and operational frameworks for FRM, gaining the interest of more local entrepreneurs in the development of FRM solutions and services, as well as incorporating FRM ideas and practices into the country's education system. Such recommendations can go a long way in meeting the urgent needs of Nigeria's FRM, but they also

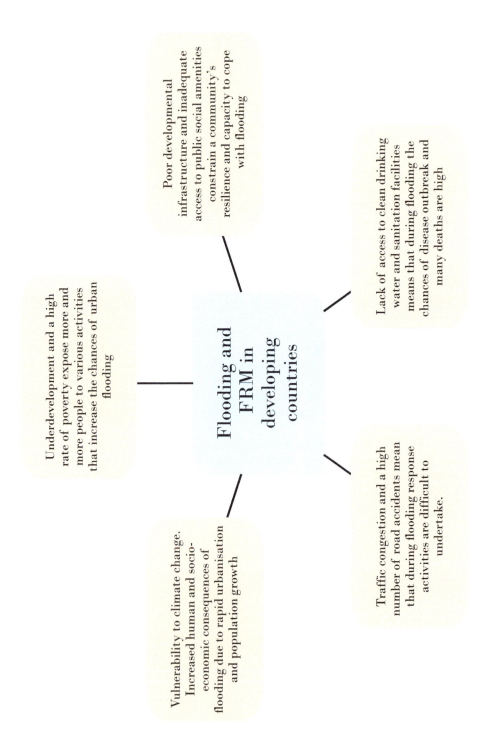

Figure 15.3 The relationship between developing countries and perception of flooding and FRM

Flood risk management policy for developing countries

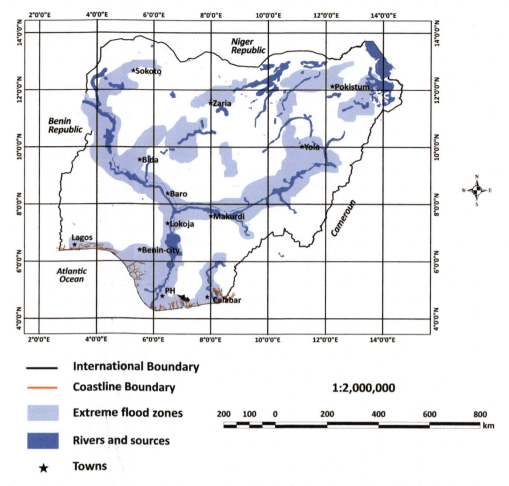

Figure 15.4 Spatial distribution of areas most affected by extreme floods in Nigeria

Note: This map was created using ArcGIS software by Esri. ArcGIS and ArcMap are the intellectual property of Esri and are used herein under license. Copyright © Esri. All rights reserved. For more information about Esri software, please visit www.esri.com.

highlight the significance of the present study in leveraging local opportunities – such as the major sources of lifeline for many Nigerians, cultural and indigenous practices, and growth of a home-based economy in Nigerian cities – to promote a sustainable FRM for the country.

Examples of global best practices in FRM in developing countries

Extreme consequences of recent flooding events have shown the importance of improvement and innovation in FRM. In line with global policy on disaster management, for example, the UNDRR and Sendai framework, innovations in flood risk and disaster management should achieve the following objectives: (1) improve the scientific and technical knowledge of flooding and its risk management; (2) improve flood risk communication strategies to enable risk financing, political discussions, and a multi-agency engagement in FRM; (3) support the means to build a corporate coping capacity and resilience to flooding; and (4) promote the use of state-of-the-art prevention measures, such as early warning systems, flood hotspot mapping and planning, insurance policies, and green infrastructure. This chapter examines three of the best practices developed, applied, and validated using specific case studies in the developing countries. This is to highlight technical and knowledge issues that need to be addressed.

Integrated flood risk management (IFRM) is one of the globally accepted strategies to tackle flooding using both structural and non-structural measures. IFRM creates room to both galvanise the benefits of flooding whilst aiming to mitigate its consequences. The Netherlands' flood and water management system, with its delta works designed to protect from even the rarest flooding events (for example 1:5000-year flooding), is driven by the principles of IFRM. Dutch flood management and other useful evidence of IFRM in developed countries are examples of best practices in flood and disaster management that are often elusive in developing countries. However, the three cases this chapter reviews highlight significant innovation in the theory and practice of FRM.

Ecosystem services and FRM

Juarez-Lucas et al. (2019) proposed a means to implement IFRM to manage flood risk in a floodplain designated for agricultural use in Candaba, the Philippines. This study combined hydrologic flood hazard estimation with both damage and benefits functions to create more comprehensive information on which to base flood risk reduction decisions. The study found that rice-crop damages associated with flood-prone land offered livelihood benefits that far exceed risks across the range of investigated flood hazards. The study highlighted solutions that reflect significant opportunities for IFRM in many developing countries, often with conflicts between flood hazards and ecosystem services. Therefore, a solution to this conflict within the context of IFRM lies in admitting that preserving the coping capacity and adaptation of communities in flood-prone areas should not be a secondary issue in FRM decisions.

Political economy and FRM

For many developing countries, FRM policy embraces urban planning to make space for rapid urbanisation, and local climate action aims to minimise most anthropogenic activities that foster climate change. In view of this policy, there are institutional framework and political and economic factors which constrain IFRM within developing countries. Flood risk and disaster management in developing countries are still sensitive to the general distribution of wealth, tenancy, and property ownership, which determine who takes responsibility for risk, government functions, and power interactions across various levels of leadership in the country. These issues were crucial in deciding on how to tackle flooding within the context of China's FRM, which Moore (2018) examined. The study showed that resolving every conflict of interest within a country is essential to implementing IFRM considering existing political and economic conditions.

Unmanned aerial vehicles (drones) and FRM

Drones are increasingly aiding cost-effective acquisition of detailed data that can determine flood source, extent, impact, and remediation measures. This is a major development in geospatial technology. Addo et al. (2018) used a drone to monitor flooding and erosion activities in a flood-prone fishing community in the Volta Delta in Ghana. The study provided some valid metrics and empirical evidence which revealed how coastal flooding and erosion are major threats to human lives and property in Ghana's Volta Delta. Irrespective of the increasing availability and diversity of drone technology in recent times, it is still at an embryonic stage in many developing countries and needs an ample standardised approach, particularly in deployment, policy development, governance, and public engagement, to fulfil its potential within FRM (Casado & Leinster, 2020). However, application to FRM is a vital opportunity to optimise the prospects of IFRM in developing countries.

Local knowledge and indigenous approaches to FRM – opportunities and challenges

Indigenous knowledge (IK) and techniques have helped local communities all over the world survive the threats of disasters for many generations (Iloka, 2016). There is ample documented evidence of how communities tackled historic and prehistoric flooding using local techniques that predate current FRM policies and protocols (Parsons *et al.*, 2019). This should not come as a surprise, since research on the application of IK has developed into a new research paradigm in anthropology (Sillitoe, 1998). Communities have undocumented knowledge of meteorology and local engineering approaches to manage flood disasters, which are based on observation, traditional practices, and belief systems which often revolve around water management (Fabiyi & Oloukoi, 2013). This IK enables them to predict flooding in real time and on a seasonal and long-term basis. There have been many debates on indigenous and traditional approaches to FRM within the contexts of climate change, although the Intergovernmental Panel on Climate Change (IPCC) only acknowledged the idea in its fifth assessment report (AR5) (Petzold *et al.*, 2020). So, scientists must probe into the prospects of indigenous approaches in addressing flood risk, especially where western praxis is rife with cumbersome and intractable burdens, which border mostly on accuracy and the use of top-end techniques that are costly and often unsustainable.

Evidently, many places within developing countries have benefitted and are still benefitting from IK and practices in tackling flooding. Neeta (2016) showed how the Lozi people of Zambia utilised IK and practices as mechanism for flood disaster risk reduction. Hooli (2016) discussed how communities in northern Namibia used IK to develop coping strategies from abrupt socio-ecological changes, multiple stressors which compounded residents' vulnerability to flooding. Fabiyi and Oloukoi (2013) discuss IK of flood risk, local adaptation strategies, and its application to flooding in Nigeria's coastal communities. Within the same case studies, Obi *et al.* (2021) applied principal component analysis on several socio-cultural variables to analyse the significance of IK in flood disaster risk reduction. The studies underlined the effectiveness of IK and showed a variety of indigenous flood control and management measures, which have significant effects on the success of FRM in Nigeria's coastal areas. Acharya (2012) introduced a gender perspective to indigenous flood management techniques in the Kailali District in Nepal. The study revealed most of the prevalent sources of gendered vulnerability to flooding and identified indigenous coping strategies.

Similar to cultural theory, IK can underpin both theoretical and practical frameworks for building flood-related resilience, as it leans on learning from experiences of natural hazards. Its major prospect finds expression in what Ebhuoma *et al.* (2019) described as 'we know our terrain'. Flooding and local terrain are endogenous to communities' IK within developing countries, as they have used and transmitted coping and adaptation knowledge from one generation to the next. With rises in flood incidences and their consequences in developing countries, IK and other indigenous adaptation strategies create opportunities to advance global FRM techniques. One major criticism is that variations and inability to predict the effects of flooding may mean that IK will not be reliable in predicting future weather events. However, IK may combine many western FRM concepts to create a hybrid IFRM framework that could help poor countries develop standard flood disaster management policies.

Free and open-source data and software for FRM in developing countries

A free and open-source geospatial (FOSG) infrastructure provides alternative data and software to aid a better understanding and characterisation of flooding and its threats. It is a ubiquitous, big step towards improving the science of flood risk and disaster management and capacity to understand and communicate the spatial and temporal dimensions of flood hazards in different parts of the world. A review of global flood management policies, for example, the European Union flood directive, will highlight the importance of FOSG data and software and various efforts to advance their availability, access, and interoperability (Albano et al., 2015). The Copernicus programme adopted an open data policy to provide full-scale access to most of the sentinel satellite data. The KULTURisk initiative documents major FOSG software being used for flood risk analysis both in Europe and elsewhere in the world. LISFLOOD, developed by Bates et al. (2010), is a prominent example. Other examples are FloodRisk by Albano et al. (2017, 2018) and TRITON, developed in Morales-Hernández et al. (2021). These tools set the stage for the study and mathematical analysis of flood risk and facilitate compliance with global best practices and policies in FRM in producing preliminary flood risk assessment and flood risk maps that delineate the current and future flood risk within each context.

FOSG data and software are fit-for-purpose tools which offer flexibility, collaboration, and cost-effective options for FRM policy in more developed societies, including the United States, Canada, Australia, and elsewhere in Europe (Aggarwal, 2016; Singh & Garg, 2016; Samela et al., 2018). Although knowledge has increased over the years with the evidence of extant research in flood analysis based on FOSG data and software, FRM techniques in developing countries have yet to optimise these possibilities. Most of the recent studies which relate to FOSG infrastructure in developing countries offer critical insight into the prospects and knowledge of FOSG and its limited application, which is despite the paucity and cost implications of legacy geospatial data for flood analysis in the region. Fisher et al. (2018) used a set of free satellite imagery (Landsat 8 and 5), digital elevation data, and SAGA open-source spatial analysis geographic information system (GIS) software to study natural resource management in eastern Indonesia. Brovelli et al. (2019) showed how open data and open software, with community geospatial science, promote the actualisation and monitoring of the United Nations Sustainable Development Goals (SDGs) in Tanzania. Guler and Yomralioglu (2021) proposed a workflow that combines quantum geographic information system (QGIS) and multi-criteria decision modelling methods to select sites for bicycle stations and lane locations.

Despite the prospects of these studies, knowledge of what motivates or limits the use of FOSG in developing countries is lacking. FOSG application is still under-represented in developing countries. Clearly, there are context-specific issues raised in the literature which do not provide a broader view of the actual situation regarding access and utility of FOSG. Teeuw et al. (2013) and Leidig and Teeuw (2015) argued that the reasons this situation prevails could not be far from limited access to the Internet of Things (IoT), which is needed for access, storage, processing, and sharing of FOSG. Besides, a general lack of awareness of the availability of free geospatial data and software, a lack of the relevant technical capacity to adapt such infrastructure, and a lack of political power to invest in improving a local geoinformatics base are complicit.

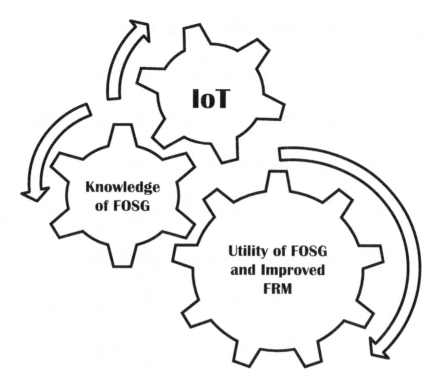

Figure 15.5 Illustration that uses a gear system to develop the logic that the IoT and poor knowledge of FOSG have significant impacts on the utility of FOSG and overall FRM in developing countries

Like a gear system, illustrated in Figure 15.5, it seems logical to hypothesise that the IoT and knowledge of FOSG mesh to create a framework for effective FRM. This means that in the future, among other policy issues, improving awareness of FOSG and making the IoT more widespread and accessible to flood risk scientists and the public will influence how local FRM frameworks apply FOSG for flood analysis.

Some of the questions raised by Choi et al. (2016) regarding what drives users in developing countries to FOSG are still pertinent to the current discussion. Quality, standard, and volume of FOSG available today create a problem of choice for end users. Still, what are the criteria for choosing a particular FOSG? What relative merits of FOSG should developing countries prioritise in making data and software decisions? Unlike the proprietary and legacy geospatial resources, FOSG has no obligation to offer support to the user community. FRM and disaster science research have acknowledged these challenges *vis-à-vis* the roles that FOSG plays in reinventing FRM. However, only a few studies have explored how to overcome the prevailing dead end and achieve meaningful FRM within poor and small island countries. Henrico et al. (2021) examined several hypotheses regarding the acceptance and use of QGIS in South Africa. Findings from the study showed that habit played the greatest role in behavioural intention to use QGIS. Adoption of QGIS in South Africa is not primarily influenced by benefits attributed to open-source software, such as cost benefits, customisability, improved reliability, quality, and security. As doubtful as it may, the study answers a major question relating to the use of FOSG data and software in developing countries, although the finding may not be true for other developing countries. Thus, it introduces a new research direction to find a generic answer.

Prospects of bespoke flood models for flood risk management

Research is rapidly responding to the obvious impacts of climate change on the uncertainties and the spatio-temporal realities of flood risk (Carter *et al.*, 2018). However, in view of understanding flood risk and disasters, there is still an increasing concern with the complex interaction between hydrology (rainfall, runoff, surface flow, drainage, and river channels), catchment (urban and rural) land surfaces and the density of human population, topography, soil types, and ecosystem services (Chow *et al.*, 1988). In line with global best practices, effective FRM, which promotes dialogue and corporation among all relevant stakeholders, combines the best science with technological capacities (Nkwunonwo, 2018; Norbury *et al.*, 2018; Cislaghi *et al.*, 2020). Examples are innovations that apply advances in remote sensing earth observation, flood monitoring, early warning systems, numerical computing, ensemble analysis, and uncertainties (Giustarini *et al.*, 2015; Noh *et al.*, 2018).

Flood modelling is at the heart of scientific developments in FRM. Experts simply defined it as the use of simple to complex mathematical and statistical functions to recreate a historical flooding event and using these results to predict the future. It explores the diverse dimensions of flood risk using depth-damage curves, hydrographs, frequency analysis, stochastic, empirical, and physically based hydrodynamic models. A major reason flood modelling is a crucial science in FRM is because experts need to undertake flood risk assessment and produce flood risk maps to communicate the spatial and temporal dimensions of flood risk to non-scientific communities, who will then use this to make an informed decision about their lives and occupation of flood-prone areas. Therefore, it is important that we understand the overall theory and then pinpoint areas that can assist experts to delimit the issues with flood modelling in developing countries.

Research on the utility of hydraulic and hydrologic models for simulating flood hazards has grown over the years. However, flood modelling for developing countries, which includes development of new models or the application of existing ones, is still a major concern for hydrology and numerical mathematics research. A solution to the lingering inability of current science to do a holistic benchmarking, calibration, and validation of existing flood models on external locations has yet to attain a global outreach (Wing *et al.*, 2021). Although there are now global flood models, for example Ward *et al.* (2015), how they fit into the actual objectives of flood disaster management in many developing countries has yet to be justified. In developing countries, the lack of quality data to parametrise existing models, down-scaling of model computation cost, and unconditional stability of models are still being debated (Teng *et al.*, 2017). Therefore, Nkwunonwo *et al.* (2020), who examined flood modelling in data-poor areas, recognised the importance of bespoke flood simulation schema for developing countries. Given the current state of flooding and socio-economic vulnerabilities in developing countries, it is important to look at the scientific aspects of flood modelling.

In extant flood modelling research, approximate solutions to the full and simplified versions of the shallow water equations (SWEs) are the mathematical framework which underpins flood modelling. Equations (1) and (2) are respectively the continuity and momentum versions of the SWEs. This chapter presents only a summary of these equations and the rationale for bespoke flood modelling in developing countries. We can find comprehensive discussions on the SWEs and their simplifications in Chow *et al.* (1988) and Bates *et al.* (2010). Originally, the SWEs are partial differential equations (PDEs) for which no simple or exact solution exists. However, analytical solutions are likely by non-trivial techniques which have often involved numerical simplifications; advanced mathematical concepts, for example, the cellular automata; and some hybrid techniques. Apart from the mathematical formulation, another major science in flood modelling is the discretisation of numerical schemes on very fine topography, afforded by mostly digital elevations models (DEMs), for example, high-resolution light detection and ranging (LiDAR). Equation (3) is Manning's formula for discretising the topography and solving the continuity and momentum SWEs. Using top-end computing tools to optimise the flood simulation process and visualisation of results is also a major factor in flood modelling.

$$[1] \quad \frac{\partial Q}{\partial x} + \frac{\partial A}{\partial t} + q = 0$$

$$[2] \quad \frac{1}{A}\frac{\partial Q}{\partial t} + \frac{1}{A}\frac{\partial}{\partial x}\left(\frac{Q^2}{A}\right) + g\frac{\partial y}{\partial x} - g(S_o - S_f) = 0$$

$$[3] \quad V = \frac{h^{\frac{2}{3}} * \sqrt{S_f}}{n}$$

where Q (L^{3}T^{-1}) is the discharge, q (L^{2}T^{-1}) is the discharge per unit width, g (LT^{-2}) is the force of gravity, A (L^2) is the cross-sectional area, t (T) is the time, x and y (L) are spatial dimensions, S_o and S_f are bed slope and friction slope (-), V (LT^{-1}) is the velocity, h (L) is the water depth, and S_f (L) is the water surface slope.

We have made some progress in flood modelling research to accommodate the particularities of developing countries, for example, the pan-African flood forecasting system in Thiemig *et al.* (2015), which is a probabilistic model for medium- to large-scale African river basins, based on a hydrological model, a regional GIS database, meteorological ensemble predictions, and critical hydrological thresholds. Komi *et al.* (2017) showed that the LISFLOOD open-access hydrological model can simulate flooding in African catchments. A new flood simulation schema, which uses only LiDAR DEM, Manning's friction coefficients, and single-value rainfall intensity to simulate to urban flooding, was recently proposed in Nkwunonwo *et al.* (2019).

A major limitation in these model developments is that there are many assumptions, and trying to eliminate them resulted in a very unstable and unrealistic model solution. Data in developing countries were not able parameterise most of the underlying theories in flood modelling. However, with the growing availability of open-access geospatial data and software, there are prospects for improvement in flood modelling in developing countries. Flood risk experts in developing countries can rely on bespoke models which are fit for purpose. This will rely on simple scientific techniques that can deliver on the needs of developing societies to build coping capacities and flood-related resilience rather than relying on top-end global techniques which may never be realistic. These bespoke flood models should mainly promote flexibility, affordability, inclusiveness, and multi-stakeholder participation and should provide room for incremental improvement and upgrading. Thus, a combination of local indigenous approaches and western best practices can serve the purpose of FRM in developing countries.

Conclusion

Flooding is pervasive within a global context, but the concept of its management is a great divide between developed and developing societies. The main issues on the borderline are communities' flood-related resilience and adaptation, community participation in flood risk mitigation, knowledge of the hazards and their threats to society's socio-economic attributes, and the general science of flood risk management. The lack of quality data, transferable technology, and reproducible methods constrains efforts towards preparing for future flood risk and disaster scenarios. There is a lack of political capacity (to create positive realities in the social and economic life of the society) and knowledge of free and open-source geospatial infrastructure and how to utilise them based on relevant technical skills which play a critical role in the ability to exploit geospatial technology opportunities.

Rapid urbanisation and population growth mean that flood management attention should be consolidated, particularly in areas where the needs are most exigent in terms of societal exposure and vulnerabilities to flood risk. This is a scenario that describes the experience developing countries have with flooding and the management of its threats to people and developments. In tackling the threats of flooding, it is important to apply measures that improve the living standard of people and enable them to cope with the challenges of flooding. Management of flooding can either be to eliminate flooding itself, which is inevitable considering the history of flooding, or to minimise the consequences, which is what constitutes flood risk management. Therefore, ideal flood risk management in developing countries should aim to comprehend various issues that attend flood risk and disasters and to enable inclusive stakeholder engagement in tackling the consequences of flooding.

This chapter has reviewed flood risk and disaster management policies in developing countries. It explored their strengths and weaknesses against the background of global best practices and highlighted the potential grey areas, challenges, and opportunities associated with implementing fit-for-purpose techniques in developing countries. Using Nigeria as a reference for flooding and its management, this chapter has explored the significance of formulating policies based on indigenous approaches and solutions endogenous to flood disaster and risk in developing countries. Finally, it recommended developing bespoke models and approaches that will take advantage of free and open-source geospatial infrastructure, training local technical capacity, and building local infrastructure to reinforce sustainable development objectives in developing countries.

References

Acharya, I. (2012). Indigenous flood management techniques from gender perspective: A case of Kailali district. *Administration and Management Review*, 24(1), 103–119.

Addo, K. A., Jayson-Quashigah, P. N., Codjoe, S. N. A., & Martey, F. (2018). Drone as a tool for coastal flood monitoring in the Volta Delta, Ghana. *Geoenvironmental Disasters*, 5(1), 1–13.

Aggarwal, A. (2016). Exposure, hazard and risk mapping during a flood event using open-source geospatial technology. *Geomatics, Natural Hazards and Risk*, 7(4), 1426–1441.

Aitsi-Selmi, A., Egawa, S., Sasaki, H., Wannous, C., & Murray, V. (2015). The Sendai framework for disaster risk reduction: Renewing the global commitment to people's resilience, health, and well-being. *International Journal of Disaster Risk Science*, 6(2), 164–176.

Albano, R., Mancusi, L., Sole, A., & Adamowski, J. (2015). Collaborative strategies for sustainable EU flood risk management: FOSS and geospatial tools – challenges and opportunities for operative risk analysis. *ISPRS International Journal of Geo-Information*, 4(4), 2704–2727.

Albano, R., Mancusi, L., Sole, A., & Adamowski, J. (2017). FloodRisk: A collaborative, free and open-source software for flood risk analysis. *Geomatics, Natural Hazards and Risk*, 8(2), 1812–1832.

Albano, R., Sole, A., Adamowski, J., Perrone, A., & Inam, A. (2018). Using floodrisk GIS freeware for uncertainty analysis of direct economic flood damages in Italy. *International Journal of Applied Earth Observation and Geoinformation*, 73, 220–229.

Al Mamun, M. A., & Al Pavel, M. A. (2014). Climate change adaptation strategies through indigenous knowledge system: Aspect on agro-crop production in the flood prone areas of Bangladesh. *Asian Journal of Agriculture and Rural Development*, 4(393–2016–23936), 42–58.

Bates, P. D., Horritt, M. S., & Fewtrell, T. J. (2010). A simple inertial formulation of the shallow water equations for efficient two-dimensional flood inundation modelling. *Journal of Hydrology, 387*(1–2), 33–45.

Boulange, J., Hanasaki, N., Yamazaki, D., & Pokhrel, Y. (2021). Role of dams in reducing global flood exposure under climate change. *Nature Communications, 12*(1), 1–7.

Brovelli, M. A., Codrina, M. I., & Coetzee, S. (2019). Openness and community geospatial science for monitoring SDGs – an example from Tanzania. In *Sustainable development goals connectivity dilemma* (pp. 313–324). CRC Press.

Campbell-Lendrum, D., & Corvalán, C. (2007). Climate change and developing-country cities: Implications for environmental health and equity. *Journal of Urban Health, 84*(1), 109–117.

Carter, J. G., Handley, J., Butlin, T., & Gill, S. (2018). Adapting cities to climate change – exploring the flood risk management role of green infrastructure landscapes. *Journal of Environmental Planning and Management, 61*(9), 1535–1552.

Casado, M. R., & Leinster, P. (2020). Towards more effective strategies to reduce property level flood risk: Standardising the use of unmanned aerial vehicles. *Journal of Water Supply: Research and Technology-Aqua, 69*(8), 807–818.

Choi, J., Hwang, M. H., Kim, H., & Ahn, J. (2016). What drives developing countries to select free open-source software for national spatial data infrastructure?. *Spatial Information Research, 24*(5), 545–553.

Chow, V. T., Maidment, D. R., & Mays, L. W. (1988). *Applied hydrology*. McGraw-Hill: New York.

Cislaghi, A., Masseroni, D., Massari, C., Camici, S., & Brocca, L. (2020). Combining a rainfall – runoff model and a regionalization approach for flood and water resource assessment in the Western Po Valley, Italy. *Hydrological Sciences Journal, 65*(3), 348–370.

Coronese, M., Lamperti, F., Keller, K., Chiaromonte, F., & Roventini, A. (2019). Evidence for sharp increase in the economic damages of extreme natural disasters. *Proceedings of the National Academy of Sciences, 116*(43), 21450–21455.

Cvetkovic, V. M., & Martinović, J. (2020). Innovative solutions for flood risk management. *International Journal of Disaster Risk Management, 2*(2), 71–100.

Dedekorkut-Howes, A., Torabi, E., & Howes, M. (2021). Planning for a different kind of sea change: Lessons from Australia for sea level rise and coastal flooding. *Climate Policy, 21*(2), 152–170.

Dieperink, C., Hegger, D. T., Bakker, M. H. N., Kundzewicz, Z. W., Green, C., & Driessen, P. P. J. (2016). Recurrent governance challenges in the implementation and alignment of flood risk management strategies: A review. *Water Resources Management, 30*(13), 4467–4481.

Ebhuoma, E. E., & Simatele, D. M. (2019). 'We know our terrain': Indigenous knowledge preferred to scientific systems of weather forecasting in the Delta state of Nigeria. *Climate and Development, 11*(2), 112–123.

Egbinola, C. N., Olaniran, H. D., & Amanambu, A. C. (2017). Flood management in cities of developing countries: The example of Ibadan, Nigeria. *Journal of Flood Risk Management, 10*(4), 546–554.

Ekeu-wei, I. T., & Blackburn, G. A. (2018). Applications of open-access remotely sensed data for flood modelling and mapping in developing regions. *Hydrology, 5*(3), 39–76.

Fabiyi, O. O., & Oloukoi, J. (2013). Indigenous knowledge system and local adaptation strategies to flooding in coastal rural communities of Nigeria. *Journal of Indigenous Social Development, 2*(1).

Fisher, R. P., Hobgen, S. E., Haleberek, K., Sula, N., & Mandaya, I. (2018). Free satellite imagery and digital elevation model analyses enabling natural resource management in the developing world: Case studies from Eastern Indonesia. *Singapore Journal of Tropical Geography, 39*(1), 45–61.

Giustarini, L., Chini, M., Hostache, R., Pappenberger, F., & Matgen, P. (2015). Flood hazard mapping combining hydrodynamic modelling and multi annual remote sensing data. *Remote Sensing, 7*(10), 14200–14226.

Guler, D., & Yomralioglu, T. (2021). Bicycle station and lane location selection using open-source GIS technology. In *Open-source geospatial science for urban studies* (pp. 9–36). Springer.

Haer, T., Husby, T. G., Botzen, W. W., & Aerts, J. C. (2020). The safe development paradox: An agent-based model for flood risk under climate change in the European Union. *Global Environmental Change, 60*. https://doi.org/10.1016/j.gloenvcha.2019.102009

Henrico, S., Coetzee, S., Cooper, A., & Rautenbach, V. (2021). Acceptance of open-source geospatial software: Assessing QGIS in South Africa with the UTAUT2 model. *Transactions in GIS, 25*(1), 468–490.

Hooli, L. J. (2016). Resilience of the poorest: Coping strategies and indigenous knowledge of living with the floods in northern Namibia. *Regional Environmental Change, 16*(3), 695–707.

Huang, W., Hashimoto, S., Yoshida, T., Saito, O., & Taki, K. (2021). A nature-based approach to mitigate flood risk and improve ecosystem services in Shiga, Japan. *Ecosystem Services, 50*. https://doi.org/10.1016/j.ecoser.2021.101309.

Iloka, N. G. (2016). Indigenous knowledge for disaster risk reduction: An African perspective. *Jàmbá: Journal of Disaster Risk Studies, 8*(1), 1–7.

Juarez-Lucas, A. M., & Kibler, K. M. (2016). Integrated flood management in developing countries: Balancing flood risk, sustainable livelihoods, and ecosystem services. *International Journal of River Basin Management, 14*(1), 19–31.

Juarez-Lucas, A. M., Kibler, K. M., Ohara, M., & Sayama, T. (2016). Benefits of flood-prone land use and the role of coping capacity, Candaba floodplains, Philippines. *Natural Hazards, 84*(3), 2243–2264.

Juarez-Lucas, A. M., Kibler, K. M., Sayama, T., & Ohara, M. (2019). Flood risk-benefit assessment to support management of flood-prone lands. *Journal of Flood Risk Management, 12*(3), e12476. https://doi.org/10.1111/jfr3.12476

Komi, K., Neal, J., Trigg, M. A., & Diekkrüger, B. (2017). Modelling of flood hazard extent in data sparse areas: A case study of the Oti River basin, West Africa. *Journal of Hydrology: Regional Studies, 10*, 122–132.

Leal Filho, W., Balogun, A. L., Olayide, O. E., Azeiteiro, U. M., Ayal, D. Y., Muñoz, P. D. C., . . . Li, C. (2019). Assessing the impacts of climate change in cities and their adaptive capacity: Towards transformative approaches to climate change adaptation and poverty reduction in urban areas in a set of developing countries. *Science of the Total Environment, 692*, 1175–1190.

Leidig, M., & Teeuw, R. (2015). Free software: A review, in the context of disaster management. *International Journal of Applied Earth Observation and Geoinformation, 42*, 49–56.

Mai, T., Mushtaq, S., Reardon-Smith, K., Webb, P., Stone, R., Kath, J., & An-Vo, D. A. (2020). Defining flood risk management strategies: A systems approach. *International Journal of Disaster Risk Reduction, 47*. https://doi.org/10.1016/j.ijdrr.2020.101550.

Makondo, C. C., & Thomas, D. S. (2018). Climate change adaptation: Linking indigenous knowledge with western science for effective adaptation. *Environmental Science & Policy, 88*, 83–91.

Meadows, M., & Wilson, M. (2021). A comparison of machine learning approaches to improve free topography data for flood modelling. *Remote Sensing, 13*(2), 275.

Moore, S. (2018). The political economy of flood management reform in China. *International Journal of Water Resources Development, 34*(4), 566–577.

Morales-Hernández, M., Sharif, M. B., Kalyanapu, A., Ghafoor, S. K., Dullo, T. T., Gangrade, S., . . . Evans, K. J. (2021). TRITON: A multi-GPU open source 2D hydrodynamic flood model. *Environmental Modelling & Software, 141*, 105034.

Morrison, A., Westbrook, C. J., & Noble, B. F. (2018). A review of the flood risk management governance and resilience literature. *Journal of Flood Risk Management, 11*(3), 291–304.

Naughton, O., Johnston, P. M., & Gill, L. W. (2012). Groundwater flooding in Irish karst: the hydrological characterisation of ephemeral lakes (turloughs). *Journal of Hydrology, 470*, 82–97.

Neeta, N. (2016). Indigenous knowledge practices as mechanism for flood management and disaster risk reduction: The case of the Lozi people of Zambia. *Indilinga African Journal of Indigenous Knowledge Systems, 15*(2), 140–150.

Nkwunonwo, U. C. (2016). A review of flooding and flood risk reduction in Nigeria. *Global Journal of Human-Social Science B: Geography, Geo-Sciences, Environmental Science and Disaster Management, 16*(2), 22–42.

Nkwunonwo, U. C. (2018). Exploring the inadequacy of pertinent capacities for urban flood risk management in the developing countries. *Handbook of Disaster Risk Reduction & Management*, 563–587.

Nkwunonwo, U. C., Whitworth, M., & Baily, B. (2016). A review and critical analysis of the efforts towards urban flood risk management in the Lagos region of Nigeria. *Natural Hazards and Earth System Sciences, 16*(2), 349–369.

Nkwunonwo, U. C., Whitworth, M., & Baily, B. (2019). Urban flood modelling combining cellular automata framework with semi-implicit finite difference numerical formulation. *Journal of African Earth Sciences, 150*, 272–281.

Nkwunonwo, U. C., Whitworth, M., & Baily, B. (2020). A review of the current status of flood modelling for urban flood risk management in the developing countries. *Scientific African, 7*. https://doi.org/10.1016/j.sciaf.2020.e00269.

Noh, S. J., Lee, J. H., Lee, S., Kawaike, K., & Seo, D. J. (2018). Hyper-resolution 1D-2D urban flood modelling using LiDAR data and hybrid parallelization. *Environmental Modelling & Software, 103*, 131–145.

Norbury, M., Shaw, D., & Jones, P. (2018, May). Combining hydraulic modelling with partnership working: Towards practical natural flood management. In *Proceedings of the institution of civil engineers-engineering sustainability* (Vol. 172, No. 7, pp. 372–384). Thomas Telford Ltd.

Nur, I., & Shrestha, K. K. (2017). An integrative perspective on community vulnerability to flooding in cities of developing countries. *Procedia Engineering, 198*, 958–967.

Obi, R., Nwachukwu, M. U., Okeke, D. C., & Jiburum, U. (2021). Indigenous flood control and management knowledge and flood disaster risk reduction in Nigeria's coastal communities: An empirical analysis. *International Journal of Disaster Risk Reduction, 55*, 102079.

Oladokun, V., & Proverbs, D. (2016). Flood risk management in Nigeria: A review of the challenges and opportunities. *International Journal of Safety and Security Engineering, 6*(3), 485–497.

Parsons, M., Nalau, J., Fisher, K., & Brown, C. (2019). Disrupting path dependency: Making room for indigenous knowledge in river management. *Global Environmental Change, 56*, 95–113.

Pender, G., & Faulkner, H. (Eds.). (2010). *Flood risk science and management*. John Wiley & Sons.

Petzold, J., Andrews, N., Ford, J. D., Hedemann, C., & Postigo, J. C. (2020). Indigenous knowledge on climate change adaptation: A global evidence map of academic literature. *Environmental Research Letters, 15*(11), 113007.

Rehman, J., Sohaib, O., Asif, M., & Pradhan, B. (2019). Applying systems thinking to flood disaster management for a sustainable development. *International Journal of Disaster Risk Reduction, 36*, 101101.

Samela, C., Albano, R., Sole, A., & Manfreda, S. (2018). A GIS tool for cost-effective delineation of flood-prone areas. *Computers, Environment and Urban Systems, 70*, 43–52.

Schanze, J. (2006). Flood risk management – a basic framework. In *Flood risk management: Hazards, vulnerability, and mitigation measures* (pp. 1–20). Springer.

Schwanenberg, D., Natschke, M., Todini, E., & Reggiani, P. (2018). Scientific, technical, and institutional challenges towards next-generation operational flood risk management decision support systems. *International Journal of River Basin Management, 16*(3), 345–352.

Sillitoe, P. (1998). The development of indigenous knowledge: A new applied anthropology. *Current Anthropology, 39*(2), 223–252.

Singh, H., & Garg, R. D. (2016). Web 3D GIS application for flood simulation and querying through open-source technology. *Journal of the Indian Society of Remote Sensing, 44*(4), 485–494.

Strazzera, E., Atzori, R., Meleddu, D., & Statzu, V. (2021). Assessment of renaturation measures for improvements in ecosystem services and flood risk mitigation. *Journal of Environmental Management, 292*, 112743.

Tabari, H. (2021). Extreme value analysis dilemma for climate change impact assessment on global flood and extreme precipitation. *Journal of Hydrology, 593*, 125932.

Teeuw, R. M., Leidig, M., Saunders, C., & Morris, N. (2013). Free or low-cost geoinformatics for disaster management: Uses and availability issues. *Environmental Hazards, 12*(2), 112–131.

Teng, J., Jakeman, A. J., Vaze, J., Croke, B. F., Dutta, D., & Kim, S. (2017). Flood inundation modelling: A review of methods, recent advances and uncertainty analysis. *Environmental Modelling & Software, 90*, 201–216.

Thiemig, V., Bisselink, B., Pappenberger, F., & Thielen, J. (2015). A pan-African medium-range ensemble flood forecast system. *Hydrology and Earth System Sciences, 19*(8), 3365–3385.

UNDRR. (2020). *Global assessment report on disaster risk reduction.* Retrieved from United Nations Office for Disaster Reduction: https://www.undrr.org/global-assessment-report-disaster-risk-reduction-gar

Ward, P. J., Jongman, B., Salamon, P., Simpson, A., Bates, P., De Groeve, T., . . . Winsemius, H. C. (2015). Usefulness and limitations of global flood risk models. *Nature Climate Change, 5*(8), 712–715.

Wing, O. E., Smith, A. M., Marston, M. L., Porter, J. R., Amodeo, M. F., Sampson, C. C., & Bates, P. D. (2021). Simulating historical flood events at the continental scale: Observational validation of a large-scale hydrodynamic model. *Natural Hazards and Earth System Sciences, 21*(2), 559–575.

Wolff, E. (2021). Investigating the promise of a "people-centred" approach to floods in the context of the Sendai framework: Types of participation in the global literature of citizen science and community-based flood risk reduction. *Progress in Disaster Science*, https://doi.org/10.1016/j.pdisas.2021.100171

Section IV
Infrastructure systems, urban systems and their management

16

Integrated water resources management and flood risk management

Opportunities and challenges in developing countries

Rudresh Kumar Sugam, Md. Humayain Kabir, Sherin Shiny George and Mayuri Phukan

Introduction

Floods are natural events induced by a combination of climatic and edaphic conditions, river channel features, and anthropogenic influences (Akintola and Ikwuyatum, 2006). The number of documented flood incidents has been steadily increasing during the last two decades. The number of individuals killed or seriously harmed by flood disasters has increased dramatically around the world (United Nations-Water, 2011). Every year, floods affect an estimated 520 million people around the world, resulting in up to 25,000 deaths (Jha et al., 2012). Analysing high resolution global spatial flood event data (Global Flood Database), Tellman et al. (2021) found that between 2000 and 2015, there has been a 20%–24% increase in the proportion of the population exposed to floods (around 58–86 million more people at flood risk), which is nearly ten times higher than previous estimate (2.6%) between 1970 and 2010 (Jongman et al., 2012). Developing countries are the worst sufferers, as increased flood exposure was concentrated in Asia and sub-Saharan Africa, whereas developed countries are better prepared for flood risks (Figure 16.1).

As a flood is a hydrological extreme event, integrated water resources management (IWRM) can play a significant role in mitigating the risks and impacts. Many international institutions such as the Global Water Partnership (GWP), the World Water Council, the World Bank, and the United Nations, as well as national governments, have promoted IWRM as a key means of improving access to safe water supply and sanitation and, more broadly, alleviating poverty and improving people's lives in many developing countries. The GWP defines IWRM as 'a process that promotes the coordinated development and management of water, land, and related resources to maximise the resultant economic and social welfare in an equitable manner without compromising the sustainability of vital ecosystems' (GWP, 2000). IWRM also advocates for collaboration across all sectors of water management, such as water availability, quality, flood

DOI: 10.1201/9781003160823-20

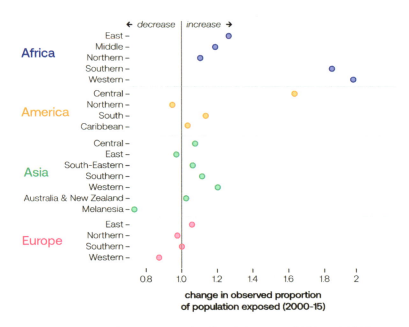

Figure 16.1 Increase in population exposed to floods between 2000 and 2015

Source: Global Flood database (a collaboration between Cloud to Street and the Flood Observatory [DFO]), accessed at Global Flood Database (cloudtostreet.ai) on 10 Nov 2021

risk, and ecosystem health. Integrated planning and management can enable a more comprehensive assessment of risks and uncertainties by considering river basin as a unit, as well as promoting learning and adaptation as new threats develop (Folke et al., 2005).

Generally, the flood risk management cycle encompasses a risk management plan (characterisation, risk perception and communication, risk assessment and mapping), flood mitigation (land use planning, flood zoning, regulations), flood protections (structural and non-structural measures), flood preparation (flood forecasting, early warning systems, information and warnings, flood emergency management plans), flood response (rescue, damage mitigation information), and flood recovery (short-term and long-term). Flood risk management involves the adoption of measures such as the construction of levees and embankments in order to reduce flood damage while also allowing for some flooding (Vis et al., 2003). Natural floods also bring several benefits to humans, especially in the agriculture sector; for example, they bring nutrients to agricultural fields and increase the agricultural output. The community needs to build adaptive capacity to minimise flood risks. Adaptive capacity is influenced by a variety of social, economic, technical, knowledge-related, institutional, and cultural mechanisms (Brouwer et al., 2007). IWRM as an approach talks about all these dimensions of water management. Floating homes and adapted interiors for houses (e.g., not putting electrical installations in the basement) are some examples of structural flood-resilient structures (McLeman and Smit, 2006).

Countries across the world such as the United States (US), the European Union (EU), India, Bangladesh, and several African countries have adopted IWRM concepts in managing flood risks and holistic management of water resources. However, there are some challenges to the realistic application of IWRM. These include a mismatch between needs and conditions in developing countries, the complexity of the IWRM approach, lack of skills and financial resources, institutional capacity, lack of sensitivity to traditional practices, genuine stakeholder participation, equity, and accountability, and rushed implementation (Beveridge and Monsees, 2012).

Literature review

Fragmented traditional approaches to water management are being phased out with an increasing focus on multi-dimensional approaches. The globally accepted concept of IWRM has evolved over multiple decades. The Mar del Plata Conference of 1977 laid the foundations of the IWRM concept wherein the importance of integrated management of resources was highlighted. The Dublin Principles, 1992, further outlined the importance of multi-level stakeholder participation for holistic resource management. The recommendations of the Dublin principles were presented in Agenda 21, which was developed from the 1992 Earth Summit in Rio de Janeiro (Ibisch et al., 2016). It emphasised that elements of water systems are connected through larger social and ecological processes and must not be neglected when taking economic development actions (Serra-Llobet et al., 2016b). IWRM has been adopted globally as a dedicated Sustainable Development Goal (SDG) 6.5, which aims that all countries "By 2030, implement IWRM at all levels, including through trans boundary cooperation as appropriate". The target is monitored by two indicators: degree of IWRM implementation and proportion of trans-boundary basin area with water cooperation arrangements.

The IWRM concept implies that the maximum benefits will accrue across economic, social, and environmental dimensions when the degree of coordination is optimised. This is closely tied to concepts of the 'nexus approach' and 'adaptive capacity' (Ibisch et al., 2016). Underlying principles of IWRM include institutional integration, level of institutional interaction, economic valuation of resources, environmental protection, stakeholder participation, and equity and efficiency (Benson and Lorenzoni, 2017). IWRM seeks to bring coordination across distinct aspects of management such as water supply, water quality, flood risk, and ecosystem health.

Flood risk management (FRM) is a vital component of IWRM (Verweij et al., 2021). It aims to increase the productivity of floodplains and coastal zones, achieve efficient usage of river basin resources, decrease negative impacts on livelihoods, and decrease existing flood risk levels while also increasing the resilience of the system (Associated Programme on Flood Management, 2017). FRM includes the following elements: adopting a suitable combination of structural and non-structural measures, integration of land and water management in planning, and adopting integrated management of hazards such as landslides. The aim is to arrive at an optimal combination of measures that decrease flood risk to acceptable environmental, social, and economic costs. (Topalović and Marković, 2018). Based on cost-benefit analysis and the capability of an intervention (structural or non-structural) to lower the flood risks/damage, each proposed intervention is assigned a different priority level. (Serra-Llobet et al., 2016a). A complete understanding of the risk necessitates a systems-based approach since it covers diverse sources of flooding, flooding pathways, exposure of people to flood events, and potential consequences at a basin level. With integration, there may be options to merge projects under different sectors to maximise benefits, for example, combining FRM projects with urban sewerage and waste disposal projects or rail and road development projects. Equitable sharing is also necessary when developing FRM plans, and a balance must be determined between social and economic benefits (Asian Development Bank, 2018).

Performance measures of IWRM and FRM include the level of development of river board institutions, poverty reduction, adaptive management, and social learning, among others. Monitoring of IWRM and FRM typically involves evaluating multiple dimensions such as degree of integration, the scale of management, institutionalisation across actors and levels, stakeholder participation, gender equity, ecology, and the environment. (Gain et al., 2017). Indicators such as assessment tools, management scale, water use integration, stakeholder consultation, awareness, capacity building, funding, regulatory aspects, technical means, availability of joint plans, consistency of timelines for monitoring, coherence of goals, and interventions across project boundaries (maximised synergy, managed trade-offs, maximised time and cost efficiencies) are used to assess the strength of policies for the purpose of integration with respect to FRM as well as to evaluate the implementation of FRM as a whole (Cumiskey et al., 2019).

Several developed countries have adopted IWRM; however, it is not clear whether it is being implemented in its totality. Analysis of Swiss flood risk management policies since the mid-19th century reveals that flood policies have not been framed as economic or voluntary instruments but rather as coercive requirements. Most flood risk–related policies were developed as reactions to flooding events rather than as pre-emptive guidance (Metz and Glaus, 2019). Implementation of flood risk–related prevention measures is not always straightforward; for example, a study based in the UK found that urban development continues to be allowed in flood-prone areas, and in Paris, building activities continue in flood-prone areas since financial mechanisms are available for compensating flood losses (Dieperink et al., 2016). In the latest UNEP monitoring report of SDG 6.5, it was found that around 52 countries made moderate progress, and 22 countries made substantial progress in IWRM (see Figure 16.2). Although the concept of IWRM has been around for almost three decades, 47% of the countries (87) report "low" or "medium-low" levels of IWRM implementation. Local and regional governments tend to be lax when implementing non-structural mitigation measures since these tend to be non-binding and more informal.

Case studies on adoption of IWRM for FRM

Bangladesh

Bangladesh stands as the sixth most vulnerable country in the world for flooding (Rosaidul Mawla et al., 2020). Seasonal monsoon rainfall, discharge from the upstream region, and sea-level rise cause severe flooding. Climatic change–induced extreme events, unplanned economic development, rapid urbanisation, land-use changes and poor governance are also responsible for the increased flood risk in the country (Gain et al., 2015). Furthermore, the country confronts issues in managing climate change impacts, water demand, and safe drinking water supply, as well as deteriorating water quality, reversing fishery decline, and maintaining natural habitats such as coastal wetlands and marshes.

Policy reforms

The first master plan for water management for Bangladesh was prepared after the floods of 1954 and 1955 with the objective of increasing agricultural production through engineering solutions to flooding control and drainage improvement, followed by irrigation facilities. The recommendations of the International Bank for Reconstruction and Development (IBRD)

Water resources management and flood risk management

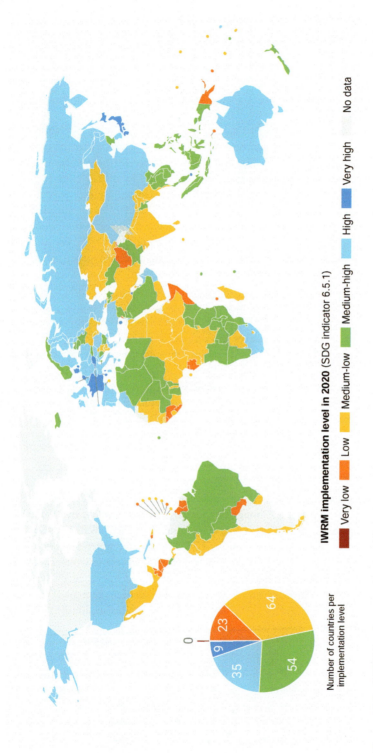

Figure 16.2 IWRM implementation level by country
Source: UNEP, 2021

report of 1972 shifted emphasis from flood control using purely structural engineering solutions to water management using a combination of structural and non-structural measures in Bangladesh. The "integrated management" approach of IWRM has been highly reflected in policy shifts. The adoption of the National Water Policy (NWPo) in 1999 served as a milestone towards the institutionalisation of IWRM in Bangladesh. It addressed the institutional, legal, and financial aspects with incentives, formulation of water rights, and water pricing for equitable water management. It also acknowledged transboundary cooperation among co-riparian countries.

The National Water Plan (NWP) Phase-I of 1986 assessed water availability from various sources and projected the future water demand by different sectors. After the consecutive floods of 1987 and 1988, the Flood Action Plan (FAP) was endorsed by the government of Bangladesh with the aim of stabilising food production in 1992. The Flood Plan Coordination Organisation published a set of guidelines that officially recognised community participation in flood management projects. The National Water Plan Phase-II of 1991 introduced catchment-scale planning and analytical tools for flood management by categorising the country into different hydrological regions. The National Water Management Plan (NWMP) of 2004 adopted a multi-sectoral approach to water management and emphasised non-structural (soft) approaches instead of only hard engineering approaches (Gain et al., 2017).

The Bangladesh Climate Change Strategy and Action Plan (BCCSAP) integrates issues of climate change in planning and designing processes to support economic growth and poverty reduction. The National Adaptation Programme of Action (NAPA) provides guidelines on the implementation of adaptation initiatives through building synergies with other programmes. The NWMP and BCCSAP have considered intersectoral integration for economic development, poverty reduction, food security, and protection of the natural environment.

Key policy documents such as NWPo, NWMP, and NAPA have recognised the involvement of diverse groups of stakeholders in water management projects. Water is also recognised as an economic and social good through the formulation of water rights allocation and water pricing tools in NWPo and the Bangladesh Water Act.

Institutional reforms

Institutional transition in Bangladesh through the establishment and reorganisation of various key institutions reflects the "integrated management" approach of IWRM. The Water Resources Planning Organisation (WARPO) and National Water Resources Council (NWRC) coordinate with different ministries for integrated water management. WARPO has transitioned from overlooking agricultural water management to overseeing water resource management of the country through the authority provided by the recent enactment of the Bangladesh Water Act. The Institute of Water and Flood Management (IWFM) undertakes multi-disciplinary research and capacity development programs in flood management with a focus on IWRM.

The Bangladesh Haor and Wetlands Development Board (BHWDB) and Barind Multipurpose Development Authority (BMDA) have adopted catchment-scale planning for water management. The institutional setup of WARPO and BWDB recognises the impact of ecological and environmental issues on water resources. Also, the Local Government Engineering Department (LGED) facilitates people's participation in project formulation and implementation. Key institutions such as LGED and BMDA have also adopted economic principles of cost recovery, pricing, and tariffs in water projects. The Bangladesh Water Development Board, (BWDB) with the original mandate of providing infrastructures for flood control, drainage, and irrigation, has undergone institutional reforms and has undertaken the partial implementation of cost recovery.

Transition in water development projects

Since the 1960s, there has been a gradual change in conceptualisation, design, operation, and management of water development projects (Gain et al., 2017). The Char Development and Settlement Project (CDSP) undertook sectoral integration at the project level through integrated management of land and water along with safe drinking water provision and sanitation, social forestry, livelihood enhancement, and disaster management. It prioritised social equity, human rights, and women's representation by working directly with the poor and marginalised communities. Projects such as the Coastal Embankment Rehabilitation Project (CERP) and Coastal Embankment Improvement Project (CEIP) took into consideration the hydrological environment during the project planning stage. The Southwest Area Integrated Water Resources Management Project (SWAIWRMP) and Integrated Planning for Sustainable Water Management (IPSWAM) added prominence to IWRM implementation in the country. Recent projects such as the Blue Gold programme have integrated participatory water management with the development of business and market linkages and consideration of potential impacts of climate change and upstream interventions.

Challenges and opportunities

The principles of IWRM are well reflected in the policies, institutional setups, and water project planning in Bangladesh. However, improvements are required in terms of capacity building of key stakeholders, periodic monitoring and evaluation of project performance, and legal framework for the enforcement of law and policies. Catchment-scale planning has been suggested across major policies and project frameworks in the country; however, basin-scale planning for transboundary rivers is not smoothly happening with co-riparian countries.

Institutional transition in Bangladesh is driven by various physical and socio-political factors such as water hazards, arsenic contamination, saline intrusion, population pressure, poverty, and vulnerability. Coordination among different institutions started due to the observed effect of large-scale water projects on other sectors. Various multinational donors and international agencies played a key role in the adoption of IWRM in Bangladesh.

Emphasis has been put on the participation of the local community after unsuccessful attempts at earlier project implementation due to lack of community support. Stakeholder participation also ensures the financial viability and long-term sustainability of public projects. Recent projects have also included gender and equity dimensions of community participation. However, studies have observed that at the field level, project implementation needs to address underlying socio-economic inequalities among communities and gender groups. While donor agencies emphasise community participation, participation is limited to public consultation. Significant improvement is required in mobilising decision making with respect to project outcomes between different stakeholder groups (Dewan et al., 2014).

India

Flood is a recurring disaster in India, causing large-scale loss of life, properties, and livelihood. According to National Disaster Management Authority (NDMA), out of the total geographical area of 329 million hectares (mha) of the country, more than 40 mha is prone to floods.

Institutional and policy framework for flood management

Following 1954, India made significant investments in riverine flood control, including the development of different structural measures such as embankments, detention reservoirs, and drainage improvements in river basins (Mohanty et al., 2020).

In India, water is a state topic; hence the central government often serves as a financial and advisory agency for water management. The Ministry of Jal Shakti, which is part of the central government, oversees disaster management in the country. The Central Water Commission (CWC) is the nodal agency for water resource management, providing technical advice to various agencies and states as well as promoting flood control practices through infrastructure development and maintenance, the development of flood forecasting systems, and information dissemination. India's two major river basins are Ganga and Brahmaputra; the Ganga Flood Control Commission (GFCC) and the Brahmaputra Board prepare flood control master plans for Ganga and Brahmaputra basin regions, respectively. For all disaster-related initiatives, the National Disaster Management Authority is the central agency. It is well supported by central agencies such as the India Meteorological Department (IMD), National Remote Sensing Centre (NRSC), and National Institute of Disaster Management (NIDM) in getting information and building capacity on floods in India. At the state level, there are disaster management authorities and departments such as irrigation and public works which support flood risk management. Thus, India has a well-established network of institutions working at various levels. However, the quality of data and overall response time during floods still need vast improvement.

The National Water Policy, 2012, establishes a clear intent of adopting IWRM principles for FRM as well as integrating flood forecasting technology backed by hydrological data management. It suggested morphological studies of the major rivers and the preparation of frequency-based flood inundation maps for flood-prone regions.

Transboundary river management

India shares water with Pakistan, Bangladesh, China, Nepal, and Bhutan. To date, there are not many information-sharing and joint efforts happening for flood management amongst the countries due to geo-political tensions.

Reservoir management

The Dam Safety Organization (DSO) was set up in 1979 to develop a dam safety procedure to assist the state governments in undertaking remedial measures in the event of dam failure. The National Committee on Dam Safety (NCDS) formulated the guidelines for the development and implementation of Emergency Action Plans (EAP) in 2005. The Dam Rehabilitation and Improvement Project (DRIP) was launched in 2012 to provide technological advances, rehabilitation material, capacity building, and technical regulations to the dam operating authority.

Flood forecasting network

The Central Water Commission is the government institution in charge of flood predicting data collection and analysis. The 325-station Flood Forecasting Network is dispersed over 20 river systems in 25 states. Flood forecasting and advance warning for 197 low-lying areas/towns and 128 reservoirs assist user agencies in determining mitigation steps such as persons being evacuated and their movable property being moved to safer sites. During floods, in a year, CWC regional offices across the country issue over 7000 flood forecasts and advance warnings to user agencies. The overall accuracy of CWC's forecasts over the last few years has been above 90% (Central Water Commission, GoI, 2021).

Opportunities for improving flood management through adoption of IWRM principles

It can be clearly observed that institutions and policies exist in the country for adopting IWRM for FRM. However, at the local level, participation of all stakeholders is still limited, and information dissemination in India usually follows a top-down approach. Capacity development of institutions at the local level and riparian communities is an essential step towards increasing adaptive capacity against flood risks. Also, the shift from reliance on structural measures to integrating non-structural measures such as flood forecasting, land-use planning, flood plain mapping, and increased coordination among government agencies in India needs a further push. The governance structure involved in flood management requires better coordination among the state and the centre to make up for the lack of efficient enforcement of projects and guidelines, with timely monitoring of project activities (MoWR, 2017). Empowering local communities to participate in flood management needs to be accelerated. While several policy recommendations on flood management taking an integrated approach towards water resource management have been developed, their implementation has been inadequate. The government should also explore micro-insurance and appropriate grants for post-disaster recovery. Further at the basin level, co-riparian countries will have to develop better relationships for achieving IWRM in a true sense.

European Union

The EU adopted the Water Framework Directive (WFD) in 2000 to ensure sustainable use of fresh water across Europe though conservation and restoration of water bodies. It tasked the EU Member States with preparing river basin management plans (RBMPs) to achieve the objective of "good ecological status" of the water bodies. WFD steered in the concept of IWRM in EU with its elements such as a basin-wide approach, public participation, and incorporation of precautionary principles for environmental protection (Global Water Partnership, 2015). The EU adopted the Floods Directive (FD) in 2007 after the devastating floods in Central Europe in 2002, which directed Member States to conduct assessment for flood risk and formulate catchment-based flood risk management plans. The FD refers to the WFD in its instruction to undertake coordinated application of the two directives, particularly in the formulation of RBMPs and FRMPs and active participation of relevant stakeholders in the planning process (European Commission, 2014; Hedelin, 2016).

This resulted in integration of flood risk management with the existing river basin management planning process by considering the impacts of flood control measures on river health, flow, and physico-chemical elements. In 2015, EU Member States presented river basin plans which incorporated flood risk management plans with a focus on flood prevention, protection, and preparedness. Various non-structural measures such as land-use management and river and floodplain restoration are promoted under the WFD. Flood hazard and risk maps were prepared after preliminary assessment of riverine and coastal regions prone to flooding. Funds for flood management activities were channelled through river basin authorities of state and regional and local governments, as well as local users. The EU also provides funds through various programs such as the LIFE programme, Special Accession Programme for Agricultural and Rural Development (SAPARD) or the EU Solidarity Funds; through structural and cohesion funds; or through the Common Agricultural Policy. Following is a case study of the Catalan River Basin in Spain representing typical FRM processes in the EU.

Catalan River Basin District, Catalonia, Spain

The Catalan River Basin region in Spain comprises an area of 16,428 km^2, which includes several smaller river basins that drain to the Mediterranean Sea. It has a high population density of 420 inhabitants/km^2. The rivers in the basin have highly irregular and variable flows; high urban and industrial water demand add to the low flow conditions and water stress in the region (Munné et al., 2021).

Institutions

Spain's framework for water resource management has been traditionally led by regional governments through the formation of river basin management districts since the 1920s. The integration of the WFD into Spanish legislation in 2003 brought about changes extending the decision-making process to stakeholders beyond water users (Hernández-Mora and Ballester, 2011).

The Catalan Water Agency (ACA), which is a public entity under the regional government, is responsible for implementation of WFD in the region. The Ministry of the Environment and Rural and Marine Affair Affairs is the central government agency which reports to the European Commission on the status and progress of WFD implementation. There are specified nodal agencies to look after various aspects of flood risk management, including land use planning and flood zoning (Directorate General of Territorial and Urban Planning), flood mitigation planning for inland waters (Catalan Water Agency) and coastal areas (State Directorate General of Coast and Sea Sustainability), emergency response (Directorate General of Civil Protection and the Meteorological Service), and post-disaster recovery – (the Insurance Compensation Consortium and the State Entity of Agricultural Insurance) (Ortega and Hernández-Mora, 2010; Serra-Llobet et al., 2016b)

Stakeholders' participation covering various aspects of IWRM

The Catalan Water Agency divided the basin into 16 sub-basins to improve public engagement, and each sub-basin did a shared diagnosis of flood risk assessment. Eventually, numerous players participated in multi-stakeholder consultations and workshops, which culminated in the creation of proposed disaster management methods. Information about management strategies was also provided to stakeholders through a system of feedback meetings (Hernández-Mora and Ballester, 2011). This exercise, which involved about 5000 actors, led to the development of catchment-based flood risk analysis: Flood Directive and flood hazard maps were developed for high-, medium-, and low-probability flood scenarios. These maps are used by the civil and municipal departments to define flood risk zones. While the Flood Directive requires the development of maps showing future flood risk, these maps are yet to be completely developed due to inadequate information availability about future socio-economic and climate changes.

Financing of implementation activities for river basin and flood management

The regional government, water consumers, municipal governments, and state governments all contribute to the budget for activity execution. Only 10% of the money goes to flood prevention and mitigation, with the rest going to water supply, water quality improvement, research innovation, and developmental activities. (Serra-Llobet et al., 2016b).

Thus, Europe has taken a top-down approach with a common obligatory flood-specific policy framework with the Floods Directive within the WFD. The requirements of the WFD led to integration of IWRM principles across flood risk management for Catalan and across the EU in general. The regional government of Catalan has attempted to integrate relevant national and regional laws, inter-institutional coordination, and encouraging stakeholder participation. Adoption of WFD has also improved the level of publicly available information as well as the relationship between different actors involved in water management.

Thus, the EU in a true sense has been able to implement IWRM for FRM by bringing co-riparian countries under one umbrella, developing institutional and financial mechanisms to support initiatives, and ensuring stakeholder participation at all levels.

United States

Various federal level initiatives such as the Water Resources Planning Act of 1965, Unified National Program (UNP) for Floodplain Management of 1994, and Flood Insurance Act of 1968 made attempts to bring concepts of IWRM for water resource management into the United States. However, policy implementation has been difficult due to the lack of coordination among agencies and disconnection with land-use decisions made at the local level. About 100,000 local agencies and 300 state-level agencies deal with distinct aspects of water management in the US. Following is a case study of the San Francisco Bay Area representing the adoption of IWRM for FRM in the US.

San Francisco Bay Area, California, US

The San Francisco Bay area in the US is an estuary covering a watershed of about 153,000 km^2 which receives runoff from the Sacramento-San Joaquin Rivers. It has a population of about 7 million. It is also a biodiversity hotspot, serving as a habitat for over 500 species of wildlife and a wintering and stopover area for migratory birds (McKee et al., 2013; Taylor and Kudela, 2021).

California created the Integrated Regional Water Management (IRWM) program in 2002 to encourage collaboration among local water agencies for the development of regional-level integrated water management plans. The nature of participation of regional agencies was voluntary to allow local governments to choose their jurisdictional boundaries and governance structure. However, this led to each of the 48 regions in the region developing individual IWRM plans with different approaches.

Institutions

The San Francisco Bay Area Integrated Regional Water Management Plan (IRWMP) was initiated in 2004 with regional and local governments, agencies and citizen groups signing a letter of mutual understanding to prepare a water management plan work by adopting IWRM framework. A coordinating committee consisting of representatives of agencies responsible for four different functional areas: water supply (the Bay Area Water Supply and Conservation Agency); wastewater management (the Bay Area Clean Water Agencies); water, flood, and storm water management (Bay Area Flood Protection Agencies Association); and catchment protection (public and non-profit agencies) served as the organising body for implementation of the IRWMP. However, the representation of agencies responsible for flood management was limited to land use planning divisions and flood control agencies. Multiple groups of key stakeholders were identified by the committee through information collected from water management agencies, public meetings, and public forums. Various community workshops, surveys, and regional and sub-regional meetings were conducted to discuss water management needs and to prioritise regional requirements (CCWater, 2019).

One of the five goals of the San Francisco Bay Area IRWMP was regional flood improvement, and it identified various strategies for effective flood risk management, including integrated land and water management, leveraging natural watersheds, and using a mix of structural and non-structural measures. It was not, however, integrated with the Bay Area Plan, which encourages the construction of transit hubs, many of which are in floodplains, or with local development plans. As a result, the San Francisco Bay Area IRWMP's land-use management contradicts the ideas in the Bay Area Plan (Serra-Llobet et al., 2016b).

Zimbabwe

Floods cause loss of livestock and human lives, crops, and infrastructure and lead to the outbreak of diseases such as malaria and cholera in flood-affected areas of Zimbabwe. For instance, the Mzarabani and Guruve districts located within the Zambezi basin in the northern part of Zimbabwe are affected by seasonal floods due to rainfall, occurring in January or February, and cyclone-induced floods. The districts are located between the Kariba and the Cabora Basa dams and at the confluence of two main tributaries of Kariba and Zambezi rivers. The release of water from the Kariba dam and rising water levels at Cabora Basa dam cause severe flooding in the region. The region covers an area of about 8,000 km^2 with a population of about 300,000 (Madamombe, 2004). The main economic activities in the region are commercial and subsistence agriculture, livestock rearing, and wildlife management.

Flood management practices

Zimbabwe adopts structural and non-structural measures for flood mitigation. Structural measures include dams and weirs which are put in place to improve water security through water storage and for flood mitigation. However, storage availability in these structures is inadequate and hence the flood control potential is limited. Water release from the dams also causes floods in the catchment area.

The non-structural flood mitigation measures include flood forecasting, clearly defined areas for settlement, and rescue operations. Based on meteorological forecasts, river flows are assessed for the probability of flooding. The information is disseminated by responsible agencies, and evacuation arrangements are made. However, there is a time lag between the flood forecast and the flood event, which reduces the time for flood preparation, and the accuracy of forecasts is also limited. In recent years, there has been an improvement in data collection of rainfall, information dissemination, and awareness generation. Multi-sectoral meetings on flood management are also coordinated by the civil protection agencies, which see attendance and involvement by relevant stakeholders.

Policy and institutional setup

The flood management process in Zimbabwe involves several agencies. Disaster preparedness is initiated by the central government, while local administration is responsible for its implementation (Gwimbi, 2004).

The Water Act (1998) played a key role in promoting IWRM in Zimbabwe, which has been adopted as a basis for water resources management. The country has been divided into seven catchments, each of which is managed by a Catchment Council with elected representatives from the different water users. The Zambezi Action Program (ZACPRO) brings the basin countries together for integrated management of the basin. However, at present, the member countries have different policies for the management of their portion of the Zambezi.

The Civil Protection Act presents the legal instruments and the powers vested in different agencies for disaster management. The Civil Protection Organization of Zimbabwe is responsible for the management of flood emergencies, which has a working party of representatives from the health, foreign affairs, water, mining, state security, and information departments of the government.

The National Policy for Disaster Management suggests the involvement of every citizen for efficient flood management. While the central government initiates hazard reduction measures through various sectorial ministries, the local administration is responsible for implementing and maintaining the activities. Existing government, private, and non-governmental organisations are adopted structurally, materially, and technically within the program, so that they can be involved in undertaking protective, relief, and rehabilitation measures during a disaster. The mandate for stakeholder involvement was adopted after cyclone-induced floods, and this change in policy reflects a major shift towards an integrated flood management approach.

Financial resources allocated annually by the government for flood management are extremely low, and in the case of larger resource requirements post-major disasters, the government allocates funds and assistance from the international community and the private sector.

Opportunities

The major issues observed in flood management framework of Zimbabwe are a fragmented approach to flood management, centralised decision making, inadequate training for rescue and relief, and lack of involvement of local community in the process of managing disasters (Gwimbi, 2004). In recent times, using a combination of hard engineering solutions and soft solutions and the involvement of different governmental agencies and stakeholder groups have improved flood management in Zimbabwe. Further improvements are required in the lead time and accuracy in forecasting for better flood preparedness. There is also a need to manage floods by considering the Zambesi basin as one unit in a coordinated manner by involving all eight basin countries.

Discussion

From the various case studies analysed, it can be clearly seen that IWRM principles have been accepted and adopted by countries across the globe for FRM. However, due to diverse levels of financial capability, technical skillsets, governance structure, cross-sectoral engagements, and stakeholder participation, the countries are at diverse levels of IWRM adoption. For example, in the EU countries where governance of all the river basins is done by WFD, it allows all the co-riparian countries to work together. Moreover, the EU has the requisite financial and technical skills to develop cross-sectoral aligned institutional structure and robust hydrological information system. In addition, since the stakeholders are contributing to flood management financially as well, they have a clear incentive in participating in any event on FRM.

In a financially restricted country such as Zimbabwe, there are several challenges when it comes to the implementation of the IWRM concept, as financial resources are limited. It is likely that the development of hydrological databases and institutional structures will be challenging and assistance from other countries or the private sector will be necessary to develop FRM plans. Also, in such cases, economic activity will be prioritised over ecological preservation (Tariq et al., 2020).

Countries like India and Bangladesh are well placed somewhere in between the EU, US, and Zimbabwe. While they have the policies and institutions in place at the central and state levels, at the ground level, there is a lack of long-term vision, financial independence, skilled workforce, and participation from stakeholders. In such cases, even when the funding is available, the projects are not implemented properly. Several capacity-building initiatives are required to build the adaptive capacity of the riparian communities and the departments associated with flood-related emergency measures in these countries. If we look at the case of the US, it is disintegrated due to its enormous size and lack of binding principle at the country level, though the country has been able to develop flood resilience due to the availability of a prominent level of technical skillsets, dedicated institutions, and finances. However, in terms of the adoption of IWRM for FRM, it lags far behind many countries.

IWRM is not an easy concept to understand and implement and to realise its full potential on the ground. The key challenges identified according to the UNEP report (2021) included lack of coordination and institutional collaboration, low policy coherence, low financing, weak institutional capacity, and lack of data on monitoring of IWRM. At a regional level, efforts were lagging in Southern and Central Asia, Caribbean, sub-Saharan Africa, and Oceania. The EU is seen as the region where IWRM has been achieved in true sense; however, there is no single country that has accomplished all four dimensions of IWRM (i.e., an enabling environment, institutional structure, multisectoral stakeholder participation and suitable management instruments, and sufficient financing). In the survey carried out by UNEP, it was found that financing for infrastructure and IWRM management is the dimension that performed the worst and remains a barrier to successful IWRM implementation. Around 135 countries mentioned financing as a major challenge. Basin-level management arrangements have also been found to be lagging. Around 70% of countries reported that the budget requirements for undertaking IWRM activities at the basin level are grossly insufficient. Considering finance options such as increasing revenue from water services, increasing central government investment, and improving implementation efficiency can improve the financial situation. Developing countries are dependent on international financing that forms a substantial portion of the finance for IWRM. In terms of disaster management, including that of flooding, about 50% of the countries reported that disaster management is mostly implemented on an ad hoc basis, and there are limited long-term programs (UNEP, 2021).

To execute policy to action, strategies on formation and mobilisation of organisations (civil societies, public and private sectors) across scales is often overlooked, although they play a vital role in catalysing change (World Water Council, 2014). Challenges also occur when integrating across sectors, since there may be competing sector demands and conflicts. Furthermore, developing balanced solutions and agreeing to risk sharing and trade-offs is not easy (Howarth, 2017). Participation of local people is important to make projects successful since it would bring in the social and equity contexts while also helping determine the context for adaptive capacity (Gain et al., 2017), thus setting the correct stage for IWRM adoption in the long term.

Conclusion

There is no doubt that IWRM is a long-term and sustainable method of FRM and increasing the adaptive capacity of institutions and communities. However, achieving all four elements of IWRM should be envisaged as a long-term process and not an easily achievable target. There are several things to learn from countries which have been successfully able to make progress in this direction. Specifically, for developing and under-developed countries, the transition from traditional structural measure–based FRM to IWRM would be a challenging task. They would require financial as well as technical support from other countries and the private sector. Last but not least, the role of information, communication, and education activities across various departments, educational institutes, and the local community level would be the main step towards the adoption of IWRM for FRM.

References

Akintola, F. O. and Ikwuyatum, G. O. (2006) 'Sustainability issues in flood management in Nigeria', in Ivbijaro, F. A. and Akintola, F. O. (eds.). *Sustainable environmental management in Nigeria*. Book Builders, pp. 197–207.

Asian Development Bank (2018) *Integrating flood and environmental risk management: Principles and practices*. ADB East Asia Working Paper Series 15. ADB. doi: http://dx.doi.org/10.22617/WPS189607-2

Associated Programme on Flood Management (2017) *Selecting measures and designing strategies for integrated flood management*. Policy and Tools Document Series 1. www.floodmanagement.info/publications/guidance%20-%20selecting%20measures%20and%20designing%20strategies_e_web.pdf

Benson, D. and Lorenzoni, I. (2017) 'Climate change adaptation, flood risks and policy coherence in integrated water resources management in England', *Regional Environmental Change*, 17, pp. 1921–1932. doi: https://doi.org/10.1007/s10113-016-0959-6

Beveridge, R. and Monsees, J. (2012) 'Bridging parallel discourses of integrated water resources management (IWRM): Institutional and political challenges in developing and developed countries', *Water International*, 37(7), pp. 727–743. doi: https://doi.org/10.1080/02508060.2012.742713

Brouwer, R. et al. (2007) 'Socioeconomic vulnerability and adaptation to environmental risk: A case study of climate change and flooding in Bangladesh', *Risk Analysis*, 27(2), pp. 313–326. doi: https://doi.org/10.1111/j.1539-6924.2007.00884

CCWater (2019) *San Francisco Bay Area integrated regional water management plan*. Contra Costa Water District. www.ccwater.com/DocumentCenter/View/8741/Bay-Area-IRWM-Plan-2019-Update-PDF

Cumiskey, L. et al. (2019) 'A framework to assess integration in flood risk management: Implications for governance, policy, and practice', *Ecology and Society*, 24(4), p. 17. doi: https://doi.org/10.5751/ES-11298-240417

Dewan, C., Buisson, M. C. and Mukherji, A. (2014) 'The imposition of participation? The case of participatory water management in coastal Bangladesh', *Water Alternatives*, 7(2), pp. 342–366.

Dieperink, C. et al. (2016) 'Recurrent governance challenges in the implementation and alignment of flood risk management strategies: A review', *Water Resources Management*, 30, pp. 4467–4481. doi: https://doi.org/10.1007/s11269-016-1491-7

European Commission (2014) *Links between the floods directive (FD 2007/60/EC) and water framework directive (WFD 2000/60/EC)*. European Commission.

Flood Forecasting/Hydrological Observation | Central Water Commission, Ministry of Jal Shakti, Department of Water Resources, River Development and Ganga Rejuvenation, GoI (2021). www.cwc.gov.in

Folke, C., Hahn, T., Olsson, P. and Norberg, J. (2005) 'Adaptive governance of social-ecological systems', *The Annual Review of Environment and Resources*, 30, pp. 441–473. doi: https://doi.org/10.1146/annurev.energy.30.050504.144511

Gain, A. K., Mondal, M. S. and Rahman, R. (2017) 'From flood control to water management: A journey of Bangladesh towards integrated water resources management', *Water*, 9(1), pp. 1–16, 55. doi: https://doi.org/10.3390/w9010055

Gain, A. K. et al. (2015) 'An integrated approach of flood risk assessment in the eastern part of Dhaka City', *Natural Hazards*, 79(3), pp. 1499–1530. doi: https://doi.org/10.1007/s11069-015-1911-7

Global Water Partnership (GWP) (2000) *Integrated water resources management*. TAC Background Paper No. 4. Global Water Partnership. www.gwpcacena.net/en/pdf/tec04.pdf

Global Water Partnership (GWP) (2015) *Integrated water resources management in Central and Eastern Europe: IWRM vs EU water framework directive*. Global Water Partnership.

Gwimbi, P. (2004) 'The effectiveness of early warning systems for the reduction of flood disasters: Some experiences from cyclone induced floods in Zimbabwe', *Journal of Sustainable Development in Africa*, 9(4).

Hernández-Mora, N. and Ballester, A. (2011) *Public participation and the role of social networks in the implementation of the water framework directive in Spain*. Ambientalia SPI. Retrieved 29 March 2022.

Hedelin, B. (2016) 'The EU floods directive trickling down: Tracing the ideas of integrated and participatory flood risk management in Sweden', *Water Policy*, 19(2), pp. 286–303. doi: https://doi.org/10.2166/wp.2016.092

Howarth, W. (2017) 'Integrated water resources management and reform of flood risk management in England', *Journal of Environmental Law*, 29(2), pp. 1–11. doi: https://doi.org/10.1093/jel/eqx015

Ibisch, R. B., Bogardi, J. J. and Borchardt, D. (2016) *Integrated water resources management: Concept, research and implementation*. Springer International Publishing, pp. 3–32. doi: https://doi.org/10.1007/978-3-319-25071-7

Jha, A. K., Bloch, R. and Lamond, J. (2012) *Cities and flooding: A guide to integrated flood risk management for the 21st century and a summary for policy makers*. The World Bank.

Jongman, B., Ward, P. J. and Aerts, J. C. J. H. (2012) 'Global exposure to river and coastal flooding: Long term trends and changes', *Global Environmental Change*, 22, pp. 823–835. doi: https://doi.org/10.1016/j.gloenvcha.2012.07.004

Madamombe, E. (2004) *Zimbabwe: Flood management practices – selected flood prone areas Zambezi basin*. WMO/GWP. www.floodmanagement.info/publications/casestudies/cs_zimbabwe_full.pdf

McKee, L., Lewicki, M., Schoellhamer, D. and Ganju, N. (2013) 'Comparison of sediment supply to San Francisco Bay from watersheds draining the Bay Area and the Central Valley of California', *Marine Geology*, 345, pp. 47–62. https://doi.org/10.1016/j.margeo.2013.03.003

McLeman, R. and Smit, B. (2006) 'Migration as an adaptation to climate change', *Climatic Change*, 76, pp. 31–53. doi: https://doi.org/10.1007/s10584-005-9000-7

Metz, F. and Glaus, A. (2019) 'Integrated water resources management and policy integration: Lessons from 169 years of flood policies in Switzerland', *Water*, 11(1173), pp. 1–27. doi: https://doi.org/10.3390/w11061173

Mohanty, M., Mudgil, S. and Karmakar, S. (2020) 'Flood management in India: A focussed review on the current status and future challenges', *International Journal of Disaster Risk Reduction*, 49, p. 101660. doi: https://doi.org/10.1016/j.ijdrr.2020.101660

MoWR (2017) *Report of the comptroller and auditor general of India on schemes for flood control and flood forecasting*. Ministry of Water Resources. https://cag.gov.in/webroot/uploads/download_audit_report/2017/Report_No.10_of_2017

Munné, A. et al. (2021) 'A proposal to classify and assess ecological status in Mediterranean temporary rivers: Research insights to solve management needs', *Water*, 13(6), p. 767. https://doi.org/10.3390/w13060767

Ortega, C. and Hernández-Mora, N. (2010) 'Institutions and institutional reform in the Spanish water sector: A historical perspective', in *Water policy in Spain*. CRC Press.

Rosaidul Mawla, M., Sultana, N. and Shiblee, M. (2020) 'Myths, safety and awareness of lightning protection in Bangladesh', *International Journal of Electrical Components and Energy Conversion*, 6(2), p. 7. https://doi.org/10.11648/j.ijecec.20200602.11

Serra-Llobet, A., Conrad, E. and Schaefer, K. (2016a) 'Governing for integrated water and flood risk management: Comparing top-down and bottom-up approaches in Spain and California', *Water*, 8(1), pp. 1–22, 445. doi: https://doi.org/10.3390/w8100445

Serra-Llobet, A., Conrad, E. and Schaefer, K. (2016b) 'Integrated water resource and flood risk management: Comparing the US and the EU', *3rd European Conference on Flood Risk Management*, 7, pp. 1–12. doi: https://doi.org/10.1051/e3sconf/20160720006

Tariq, M. A. U. R., Farooq, R. and van de Giesen, N. (2020) 'A critical review of flood risk management and the selection of suitable measures', *Applied Sciences*, 10(23), 8752, pp. 1–18. doi: https://doi.org/10.3390/app10238752

Taylor, N. and Kudela, R. (2021) 'Spatial variability of suspended sediments in San Francisco Bay, California', *Remote Sensing*, 13(22), p. 4625. https://doi.org/10.3390/rs13224625

Tellman, B. et al. (2021) 'Satellite imaging reveals increased proportion of population exposed to floods', *Nature*, 596, pp. 80–86. doi: https://doi.org/10.1038/s41586-021-03695-w

Topalović, Ž. and Marković, D. (2018) 'Integrated approach to flood management', in Anguillari, E. and Dimitrijević, B. (eds.). *Integrated urban planning: Directions, resources and territories*. TU Delft Open, pp. 143–169.

UNEP (2021) *Progress on integrated water resources management*. Tracking SDG 6 Series, Global Indicator 6.5.1 Updates and Acceleration Needs. www.unwater.org/publications/progress-on-integrated-water-resources-management-651-2021-update/

United Nations-Water (2011) *Cities coping with water uncertainties*. Media Brief, UN-Water Decade Programme on Advocacy and Communication. www.un.org/waterforlifedecade/swm_cities_zaragoza_2010/pdf/02_water_uncertainties.pdf

Verweij, S., Busscher, T. and van den Brink, M. (2021) 'Effective policy instrument mixes for implementing integrated flood risk management: An analysis of the "Room for the River" program', *Environmental Science and Policy*, 116, pp. 204–212. doi: https://doi.org/10.1016/j.envsci.2020.12.003

Vis, M. et al. (2003) 'Resilience strategies for flood risk management in the Netherlands', *International Journal of River Basin Management*, 1(1), pp. 33–40. doi: https://doi.org/10.1080/15715124.2003.9635190

World Water Council (2014) *Integrated water resource management: A new way forward*. A Discussion Paper of the World Water Council Task Force on IWRM. www.worldwatercouncil.org/sites/default/files/Initiatives/IWRM/Integrated_Water_Resource_Management-A_new_way_forward%20.pdf

17

Developing resilient cities in developing countries

Bolanle Wahab and Oluwasinaayomi Kasim

Introduction

The Sustainable Development Agenda, signed in 2015 by United Nations (UN) Member States, recognises cities as an integral part of the achievement of sustainability. City functionality and its physiological makeup exhibit distinguishable traits of living organisms (Egunjobi, 1999). Just as organisms are birthed and are subjected to the experience of survival, growth, happiness, ailment, agedness, and death, the same is attributed to the diverse anthropomorphic explanation of cities. In recent times, city managers and urban planners are more interested in the survival and resilient nature of cities to various natural and human-induced stress and shocks (Kasim, 2018; Wahab & Falola, 2017; Kasim et al., 2021). Increased urbanisation and climate change have pressured the hydrological cycle, thereby increasing the occurrence of natural hazards such as floods in recent years (Ghazali et al., 2018). Recognising that not all of these hazards can be avoided, the concept of resilience has gained popularity (Renschler et al., 2010).

In recent times, the city has accommodated more than half of the world's population, and this number is rising. As a result, more assets, people, facilities, and goods are concentrated in cities, putting them at risk of flooding (Barbaro et al., 2021). Major cities around the world have been impacted by floods (Adedeji et al., 2019). One of the key concepts used in many different ways and in many different fields underpinning the sustainability of cities as projected in the United Nations Sustainable Development Goals (SDGs) is the urban resilience concept. In the field of urban planning, urban resilience is becoming a huge priority for strategic planning at the local, national, and international levels (Bush & Doyon, 2019; Lalovic et al., 2020). However, there seems to be limited knowledge and applicability of the concept in the development of cities in developing countries, including those in Africa (Cilliers et al., 2013).

The major explanations for the limited application of the resilience concept in developing countries are rapid urbanisation and informality (Wahab & Agbola, 2017; Kasim et al., 2020b; World Bank, 2019). This rapid scale of urbanisation comes with numerous environmental challenges such as meeting the accelerated demand for affordable housing, uncontrolled pollution, increased demand for transport systems, the upsurge of informal settlement and urban expansion, poor waste management systems, and uncoordinated land use development. Climate change has also had tremendous negative impacts on city life, with the occurrence of intense urban temperature and flooding, especially in poorly planned neighbourhoods with substandard layouts found in most cities in developing countries, especially in Africa. In addition, urban flooding and erosion are becoming more and more frequent owing to several interrelated factors including changes in land use and land cover, increased rate of impervious surfaces, encroachment into floodplains, reduced capacity of drainage systems, and increasing population size in the absence of a well-coordinated urban management strategy (Kasim et al., 2020a). These environmental challenges have awakened the consciousness towards achieving especially flood-resilient cities within the developing countries of the world.

The term "resilience" has different understandings across disciplines. However, it has become increasingly popular in programming for sustainable development policy circles, disaster risk reduction, and climate change adaptation (Adewole et al., 2015). Recent disasters have emphasised the vulnerability of urban centres to natural hazards and human-induced disasters, emphasising the necessity to build disaster-resilient cities. Scholars have called attention to cities' vulnerabilities and the need to make them more robust to withstand the impact of disasters (Albrito, 2012). The need for flood-resilient cities cannot be overemphasised owing to the incessant occurrence of flood disasters in major cities of developing nations. The building of flood-resilient cities will not only afford cities the ability to absorb and recover from flood shock and stress, but it will also assist in maintaining cities' essential structures, functions, and identity as well as adapting to continual environmental changes (Agbola & Kasim, 2018). Based on statistics from the World Health Organisation (2019, 2020), in the last two decades, floods affected more than 2 billion people worldwide. Records have also shown the adverse impact of flood disasters in developing countries, including in Africa, in the last few decades. For example, as noted by Egbinola et al. (2017), over 1 million people were displaced in nations such as Burkina Faso, Ethiopia, Mali, Niger, Sudan, Togo, and Uganda in 2007 as a result of flood disasters.

Furthermore, floods in 2012, as a result of excessive rainfalls, affected millions of people in 13 countries in central and west Africa (United Nations Office for the Coordination of Humanitarian Affairs [OCHA], 2012). During this period, Nigeria experienced the most devastating flooding in more than 40 years, as 33 of the 36 states in the country were affected. Most affected areas were urban centres such as Lagos, Ibadan, Benin City, Kaduna, Kano, Warri, Calabar, and Port Harcourt (Agbola et al., 2012; Adedeji et al., 2019). According to Egbinola et al. (2017), about 370 persons were reported dead, more than 7 million people were affected, and more than 618,000 structures including residential houses were destroyed or damaged by flood, leading to the internal displacement of over 2 million people.

In the city of Ibadan, reputed as the largest indigenous city in West Africa, river flooding has been a recurrent disaster since 1902 when the first event occurred. Between 1902 and 2016, the city experienced 16 flood disasters (Wahab & Falola, 2018). The 1963, 1980, and 2011 flood disasters were the three most devastating flood events in Ibadan in terms of casualties and damage to properties. In 2011, Ibadan experienced the most devastating flood disaster that ravaged the entire city and its 11 local government areas, with serious negative impacts on human livelihoods and infrastructural development (Adewole et al., 2015). About 27,000 buildings were located within the statutory setbacks of various streams and rivers totalling 3,168.68 km in length; 2,105 buildings were flooded, over 100 people purportedly died, and properties worth billions of naira were destroyed.

The causes of the flooding were classified into hydrological, waste management, institutional, and awareness factors (Agbola et al., 2012). More detailed causes have been identified as excessive rainfall; uncoordinated urban development; rapid population growth; competing land uses; increased impervious surfaces; poor drainage facilities and weak/decaying hydraulic infrastructures; lack of proper environmental planning and management strategies; building on floodplains; ignorance; and human behaviour, especially haphazard municipal waste disposal and blockage of drains; lack of early warning information; and climate change coupled with inadequate preparedness (Agbola et al., 2012; Wahab, 2013; Osayomi & Oladosu, 2016; Wahab & Falola, 2017). With the concept of resilience, cities are readily prepared to deal with flood disasters. Flood predictions show that events of flooding will continue to increase in frequency and severity due to global climate change and extreme precipitation (WHO, 2020). Unless preventive steps are adopted, vulnerability to the devastating effects of flooding is expected to double owing to population increase and rapid urbanisation. Therefore, this chapter seeks to examine strategies for achieving resilient cities in developing countries of the world.

Conceptualising resilience and flood-resilient cities

Definitions of resilient cities

Cities are complex and interdependent systems with extreme vulnerability to a wide range of threats. These threats manifest in natural and human-induced hazards (Kasim, 2018). For sustainability, cities are expected to be resilient to numerous threats as a result of locational choice and dynamics of anthropogenic activities. The word "resilience" is derived from the Latin verb *resilire*, which means to jump back. It refers to the capacity to recover quickly from difficulties or unpleasant events (World Bank, 2019). From a broader perspective, resilience is a multidisciplinary concept to address issues relating to shocks and bounce-back mechanisms in the environment (Chelleri et al., 2015). It is also seen as the capacity of a system exposed to a hazard to resist, adjust, or adapt its functionality to meet the demand of the hazard, as shown in Figure 17.1.

Historically, resilience was originally a concept used in psychology and social ecology in the 1970s. However, its applicability to city management and urban planning has gained prominence since the 1990s as a result of several disasters (Shamsuddin, 2020). Since 2012, the resilient concept has focused on the city's ability to "spring back" from a disaster (Hofmann, 2021). This is expedient, as modern-day cities are required to respond, adapt, and absorb inner strength towards environmental shocks and stresses (Liao, 2012; Kasim et al., 2021).

The resilient city concept is based on the principles of past experiences of disaster in urban areas. According to Chelleri et al. (2015), resilient cities should be able to withstand shock and stress without necessarily being subjected to either chaos/damage or permanent distortion or breach. Also, a resilient city is prepared to anticipate, plan for, and mitigate risks while making the best use of opportunities associated with political, economic, social, and environmental changes (Agbola & Kasim, 2018). Since resilient cities respond to human-induced disasters such as urban flooding, a flood-resilient city is a city that is prepared for flooding while the urban system still retains its ability to function during and after a flood event (Batica & Gourbesville, 2016). Implicit in this definition is the need to intensify flood risk management systems in urban centres which are developed to prevent or reduce the impacts of flooding and to curtail the level of damage posed by flooding and associated impacts.

Attributes of resilient cities

Attributes of resilient cities are multifaceted owing to different contextual interpretations. It is also worth stating that while substantive progress has been made to achieve resilient cities, concrete agreements have not been reached by academics on specific characteristics of resilient

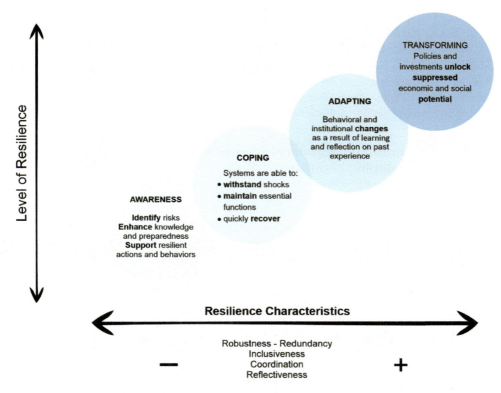

Figure 17.1 Systems-level urban resilience framework
Source: World Bank, 2019

cities. Therefore, there are various schools of thought as to the nature, attributes, and characteristics of resilient cities.

In a bid to explain resilience, Wardekker et al. (2020) considered three elements of urban resilience, "compartmentalization", "foresight and preparedness/planning", and "flexible planning/design". Ahern (2013) considered a resilient city to have eight major attributes, diversity, modularity, allowing for variability, innovation, tight feedback, overlap in governance, social capital, and ecosystem services. As noted by Godschalk (2003), resilient cities should exhibit physical components and human communities while responding to urban disasters. The physical system consists of the built environment of the city, while the social and institutional organisations represent the human communities and elements of the city. The physical systems act as the body of the city (arteries, bones, muscles) and may include physical infrastructure (roads, bridges, and buildings), communication facilities, and other natural features (soils, topography, geology, waterways). At the time of any disaster occurrence, the physical systems, as expected, should be able to function efficiently and survive extreme shock and stress. The human elements, which include informal and formal entities and stable and ad hoc human associations in an urban area such as agencies, schools, enterprises, neighbourhoods, organisations, and task forces, among others, should act as the mind of the city by directing its activities, responding to the city's needs, and deploying learning from past experiences. When a disaster occurs, the physical system and community networks within the city should be able to continue to function under unique conditions for the city to be resilient.

Given the complexity of cities and resilience, Newman et al. (2017) also established six key principles of a resilient city. According to the authors, a resilient city must create sustainable mobility and distributed energy systems, invest in renewable energy, shape disaster recovery for the future, foster inclusive and healthy cities, build biophilic cities within the city's bioregion, and produce a more cyclical and regenerative metabolism.

Resilient city index and approaches to making cities flood resilient in developing countries

Risk and uncertainty are becoming more prevalent in today's increasingly urbanised environment. One of the major aims of urban planning has always been to mitigate the risk of future uncertainties associated with climate change, population change, and economic shocks, among others (Coaffee & Lee, 2016). As observed in the United Nations urbanisation projections, 55% of the world's population would be living in cities in 2018, a 24% increase since 1950. Rapid urbanisation is expected to accelerate in the future, with 68% of the global population living in cities by 2050. In 40 years, the urban population of African countries is forecast to more than triple, from 395 million in 2010 to 1.339 billion in 2050 (UN-Habitat, 2013). With the projected population increase in the urban centres, in the event of climate change-induced hydrohazards, a greater number of people, property, and infrastructure would be impacted (Kasim et al., 2020a; Ramiaramanana & Jacques, 2021).

Damage to essential industrial buildings and residences; infrastructure, commercial, and residential items; and significant losses caused by general disruption, including business interruption to communities, are all repercussions of major flood disasters (Wahab & Falola, 2017; Kasim et al., 2021). These conditions have necessitated the development of a positive, proactive, and plan-led approach by the city to handle the problems provided by the rising flood risk. Recent flood disasters have shown flaws in the community flood risk management strategy due to poor planning and design of flood risk abatement measures (Birmingham City Council, 2018). As a result, it is becoming critical to delve deeper into the present flood risk management (FRM) strategy, as well as to contemplate additional research into developing more successful ways and structures to improve city resilience to flood disasters.

Identifying stakeholders in building urban resilience

Building resilience requires stakeholder participation (O'Brien et al., 2010). It allows for a variety of needs to be met and the sharing of knowledge and experience, as well as the adoption of proactive actions (Wildavsky, 1988). In the face of uncertainty, diversity, which is created through broadening the set of actors, is extremely important to resilience building. For example, a diverse group of individuals could help to tackle the unknown on time and allow for a larger pool of possible solutions and ideas for the development and implementation of plans that would be impossible to envision in a smaller group of people. Wider collaboration with experts in the built environment, academics, local non-governmental organisations (NGOs), and local populations enhances the possibility of identifying additional flood-related impacts. Individuals and local communities are critical actors in creating resilience, and community participation must be increased (Gero et al., 2011; Nelson, 2011). This is because they are an important source of information, and their demands and perspectives must be taken into account (O'Brien et al., 2010).

Financing resilience

With the frequency of flood-related disasters compounded by climate change, it is ironic that resilience is underfunded (GFDRR, 2010). Multi-year disaster resilience funding is frequently advocated, particularly for long-term crises like floods. It is never easy allocating scarce resources for proactive interventions in the face of growing competing and equally important developmental issues, especially in developing countries. Resilience programming naturally includes trade-offs between sectors and populations. Building the resilience of vulnerable individuals through diversification of small-scale livelihood enterprises may be a compromise for economic growth and larger agricultural and industrial enterprise development in long-term crises. Despite these obstacles, research has demonstrated that catastrophe-resilience development is more cost-effective than late humanitarian intervention (GFDRR, 2010; United Nations Office for the Coordination of Humanitarian Affairs [OCHA], 2021). Spending on flood disaster mitigation and preparedness strategies such as early warning systems, well-maintained essential infrastructure that works during and after disasters, and environmental buffers that act as physical protection has been commended by academics as some of the most effective strategies than responding in the aftermath of a disaster (GFDRR, 2010).

Investing in urban resilience

It is important to note that flood disasters have resulted in a significant increase in losses in recent years. As the global population increases and increasing urbanisation in underdeveloped countries threaten to undo hard-won development achievements, these trends are projected to become more pronounced (Kasim, 2018). By 2030, 325 million people will be severely impoverished in the 49 countries that are most vulnerable to disasters (Shepherd et al., 2013). Individuals and businesses concentrate assets in cities that have poor track records of infrastructural networks, communications systems, supply chains, and utility connections. Disruptions to these over-stressed, highly dependent, and precariously interrelated systems, by natural and human-induced perturbation, can have a disastrous impact on a city's ability to provide its residents' most basic requirements in the absence of a resilient strategy to absorb, rebound, and recover from the shock.

The global demand for urban infrastructure investment is between USD 4.5 and 5.4 trillion per year, with a premium of 9 to 27% as a prerequisite in making infrastructure low-carbon emission and climate change-resilient (Cities Climate Finance Leadership Alliance [CCFLA], 2015). Cities in emerging countries account for a large amount of this demand. Appropriate financing arrangements are critical to bolstering investments in and maintaining resilient urban infrastructure.

Insuring and shock-proof cities

Reduced and managed financial repercussions from climate-related disasters and extreme events are becoming increasingly crucial in developing countries as climate change evolves (Kasim, 2018; Wahab & Falola, 2018). The emergence of insurance instruments for transferring disaster risks and impacts to global financial markets and the severity of the post-disaster capital gap have prompted many development institutions, as well as developing country governments, non-governmental organisations, and other donor organisations, to adopt preventive disaster financial strategies as an important element in disaster risk management. Therefore, flood disaster insurance is becoming more widely recognised as a vital instrument for climate risk management, especially in developed countries (Kasim et al., 2021). This approach, if properly administered, in developing countries can assist in strengthening the resilience of individuals and the cities.

World Bank's City Resilience Programme

The World Bank Group's City Resilience Programme (CRP) was launched in June 2017 with the premise that a resilient future for cities is attainable. The CRP is aimed at strengthening cities and the required tools needed to develop complete resilience-building plans and access a variety of finance choices (UNDRR, 2019). Building urban resilience is a difficult process. However, the CRP is expected to effectively leverage the World Bank Group's broad range of sectoral expertise to assist cities to incorporate climate and risk scenarios into upstream urban planning. Relying on advancements in geospatial technology such as remote sensing, the programme is expected to break through sector divisions by using spatial planning to graphically communicate essential risk information, including climate change scenarios in investment planning. Within the provisions of the CRP's objective, cities are to be resilient, with the innate ability to plan for and mitigate climate change–induced disaster impacts by protecting lives and properties, reducing losses, and unlocking social and economic potential. The programme's goal, which is being implemented in 105 cities, is to accelerate a change toward longer-term, more robust multidisciplinary technical and financial services, creating a conduit for feasible and replicable city-wide projects that increase and enhance resilience to flood and other disasters (World Bank, 2019).

Best approaches for planning resilient cities

Under climate change scenarios, increasing vulnerabilities and severe environmental circumstances make it necessary to study urbanisation processes and spatial development policies to create resilient communities (Deshkar et al., 2011). According to Ahern (2013), there are five techniques for increasing city resilience capability. These are urban ecological networks and connection, biodiversity, multi-functionality, redundancy and modularisation, and adaptive design.

Biodiversity

Biodiversity is recognised by a wide range of stakeholders and decision-makers as a crucial feature of city and region resilience. Indeed, as the population of cities grows, the preservation and promotion of biodiversity in urban settings become increasingly important to protect the natural dynamics and promote the interconnectedness of humans and the natural environment. Resilience combines social communities and institutional solutions/strategies with physical infrastructure systems on a macro- to meso-scale. As a result, structural and engineering science is a crucial component of integrated resilience solutions. While traditional disaster management procedures and preparation are necessary, the ideal solution to reduce the escalating human and property losses caused by extreme flooding is to change the way we design and construct (Geis, 2000). The "last line of defence" of resilience techniques is micro-and meso-scale building and neighbourhood design. For example, in the case of floods, each construction site should be designed to either act as a component of a watershed problem or as a part of a watershed solution.

The multi-functionality of urban places and facilities increases diversity and efficiency, both of which are important for shock absorption and recovery. While parks and green spaces are mostly used for recreation, thermal comfort, and air pollution reduction, they can also give extra benefits in terms of flood prevention and evacuation. Policymakers should develop measures to understand how climate change affects urban biodiversity and how to describe and assess biodiversity functions in cities in a spatially explicit manner to improve city resilience to a perturbation such as flooding.

Urban ecological networks and connection

Ecological networks are defined as a framework of ecological components that provide the necessary physical circumstances for ecosystems and species populations to survive. For connectivity-dependent operations such as transportation, communication, and energy distribution, urban connectedness must be well-developed. Cities are intricate and interconnected systems that are very sensitive to natural hazards, human-induced disasters, and other forms of risk (Godschalk, 2003). Greenways, ecological networks, blue-green networks, riverways, and parkways are examples of sustainable multifunctional networks that can be used to connect urban environments. The connection of the ecological network should be developed within the confines of sustainable development and the integration of a disaster risk reduction strategy (Kasim, 2018). This will, to a larger extent, mitigate the impact of flooding and build the resilience of the city environment to recover from the shock associated with the prevailing disasters in the community.

Multi-functionality

Multi-functionality is essential in the context of resilience. For example, the efficient and effective use of space in urban planning as scarce and restricted spaces will enhance urban resilience and reduce vulnerability to disasters. Therefore, adopting multifunctional solutions associated with space and land use to satisfy the needs of rising populations globally can reduce the destructive effects of human activities on the urban environment. This is expected to make cities more resilient not only to flooding but also to other human-induced hazards and disasters.

Redundancy and modularisation

Contemporary urban planning geared toward resilience proposes a modular approach with redundant and regionalised features. The ability of cities and the residents to maintain the proper operation of the vital safety and critical infrastructure systems, particularly during periods of extreme environmental stress, is intimately related to their physical, social, and economic health. Therefore, the city cannot flourish in an unhealthy social space and poor environment. Hence the concept of redundancy and modularisation suggests a conscious planning process that allows human society to live within the provisions of the bio-physical environment (Basiago, 1999; Kasim et al., 2021). This entails proactive measures focusing on the involvement of all the essentials in the environment developed to guarantee ecosystem integrity and maintain the environment's carrying capacity and biodiversity in line with environmental issues and natural attributes such as opportunities and challenges inherent in the locality.

Adaptive design

Using available knowledge, it is important to innovate to improve the resilience of a city and to a larger extent a region. In terms of hazard and risk management, as well as climate change adaptation, it is critical to address immediate and long-term perspectives. The most effective and efficient adaptation and risk reduction initiatives are those that provide development advantages in the near term, as well as decreases in vulnerability over the long term, as shown in Box 1. The institutional capacity to analyse and anticipate future occurrences, organise and manage resources to address these difficulties, and strengthen institutional skills to deliver the objectives is at the heart of city resilience. This includes decisions on adaptation strategies to improve land use and resources, the quality of infrastructure and service coverage, the priority of new investments, and the integration of current and future community requirements into local plans and policies (Satterthwaite & Dodman, 2013; Field et al., 2014).

Box 1 Flood intervention for resilience in Ibadan

Connecting bridge at Cele Rainbow, Ibadan, Nigeria, before flood intervention

Connecting bridge at Cele Rainbow, Ibadan, Nigeria, after flood intervention

Challenges to resilient cities in developing countries: the example of Ibadan, Nigeria

Human activities play a significant role in developing the built environment with an adequate degree of resilience, to absorb, withstand, process, and respond to catastrophic threats.

A well-coordinated human activity is of paramount importance in the attainment of sustainable urbanisation and safer cities (Bosher, 2008; Agbola & Kasim, 2018).

However, creating a disaster-resistant built environment, particularly in developing cities, is a complicated task with numerous hurdles. Ibadan, in southwestern Nigeria, is the country's third-largest metropolis, after Lagos and Kano, with a population of more than 3.0 million people. The city's population has risen swiftly from roughly 60,000 in the early 1800s and is projected to increase to 5.6 million by 2033. The city's urban footprint has equally grown significantly to its current size in this regard. The main cause of sprawl is poor land-use planning, which results in low population densities, especially when compared to other large cities. Using Ibadan as an example, this section examines the difficulty of establishing a disaster-resilient city in developing countries.

Lack of regulatory frameworks

In Ibadan and the majority of the developing countries cities, the statutory law and regulations for planning, design, and construction do not effectively include disaster consequences and resilience, according to empirical research (Agbola et al., 2012; Wahab, 2013; Wahab & Falola, 2017). Nonetheless, at the state level, several measures, such as the World Bank-backed Ibadan Urban Flood Management Project (IUFMP), have developed risk and hazard maps, disaster resilience planning, and design and construction rules, including planning setback enforcement. However, many of these programmes have not been mainstreamed into local governments' construction, physical planning, and development approval processes, and as a result, community-level personnel who are involved in the development approval process are somewhat unaware of these new initiatives.

Therefore, a substantial portion of Ibadan city lacks a development plan to effectively control urban growth and physical development. The Ibadan City Masterplan prepared in 2018 has not been launched by the Oyo state government for implementation by November 2021. Also, disaster-resilient features are frequently overlooked while developing plans to manage urban development. Furthermore, owing to inadequate awareness, the majority of people did not obtain building plan approval before construction. This, in addition to poverty, harms and compromises the city's efforts to create a resilient-built environment. At the moment, the city is continuously undergoing fast urbanisation; however, the servicing and protective infrastructure are insufficient to service the growing population. Many residential buildings and other structures are constructed without due consideration for disaster risks, thereby compounding the existing vulnerabilities.

At-risk infrastructure and institutional bottlenecks

The core areas of cities in developing and developed countries grew organically and were characterised by a large number of historic building stocks and infrastructure that were constructed without proper consideration for disaster abatement measures and resilience. Regenerating these ageing structures and vulnerable infrastructure will cost a lot of money. It is difficult for the authorities to find the resources needed for thorough renovation and retrofitting in a developing metropolis. Furthermore, the majority of individuals who live in historic buildings have an emotional attachment to them because they have been inhabited for many years by numerous generations.

Many governmental organisations and agencies in the city designed, developed, operated, and maintained/managed the built environment. For example, urban planning in Ibadan works in close partnership with several government ministries, agencies, and organisations. These organisations collectively and independently have unique functions in housing and infrastructure planning, design, and approval. As a result, all of these departments are in charge of initiating disaster risk reduction and contributing to the construction of a safer city. However, the current system is seen to have some flaws, including a lack of clearly defined duties and tasks, overlapping obligations, lack of coordination and turf protection among organisations, and most importantly lack of political will to function. The sheer number of stakeholders and organisations involved has made the process of creating a resilient city complex and difficult.

Inadequate funding

The goal of building a disaster-resilient city environment is to ensure that constructed assets can withstand a hazardous event. It necessitates the relocation of existing vulnerable structures; the use of hazard resilient designs; the enforcement of resilient building standards and codes; adherence to specifications on materials, construction methods, and technologies; the protection of vital infrastructure and the development of urban planning that is sustainable; construction-protective infrastructure; and efficient land use practices (Godschalk, 2003; Haigh & Amaratunga, 2010; Malalgoda et al., 2013). Projects require a significant amount of capital and funding to be completed successfully. Accordingly, financial constraints are a major impediment to creating disaster-resistant cities in developing countries, such as Nigeria, where Ibadan is located.

Because qualified staff with an understanding of disaster risk reduction (DRR), particularly at the local government level, is scarce, several resilience measures are missed. It is difficult to establish a resilient city and successfully oversee development operations in the city without an adequate number of skilled employees who are educated about DRR. Developing the capacity of city managers to effectively address DRR and resilience issues requires (specialised) training. In the absence of finance, compounded by equally important competing activities jostling for the attention of the dwindling resources, it will be difficult to structurally position Ibadan and most of the cities in developing countries on the path of resilience and sustainable development.

The way forward: sustainability and resilience building

A disaster can be a major impediment to long-term development in highly exposed and vulnerable regions, nations, and cities to natural hazards and human-induced disasters (Bello, 2017). As reported by the World Risk Index, more than 60% of developing countries face a medium to a high level of disaster risk, with more than half of the inhabitants exposed to a higher level of risk (ECLAC, 2018). In nations with high disaster risks such as flooding, a single incident might result in a catastrophic situation, wiping out all previous progress and having a systemic influence on all aspects of the development process.

People who are unemployed and underemployed, those living in poverty, persons with disabilities, women and girls, young folks, aboriginal people, internally displaced persons, refugees and migrants, and elderly persons are among the most vulnerable groups. In the aftereffects of a disaster occurrence, these people may be trapped in protracted cycles of deprivation and low productivity, including low wages (UNDRR, 2019). The growing vulnerability of human activities to disasters raises the risk of environmental challenges and repercussions which may lead to difficult-to-predict vicious cycles of crippling cause-and-effect outcomes on human lives, economic projection, political stability, and environmental outlooks. Risks become increasingly systemic in a more complex settings like cities.

Urbanisation, poor urban governance, deteriorating ecosystems, poverty, and climate change are some of the major disaster risk factors globally. Aside from the broad effects of climate change, there are sector-specific development initiatives and patterns that increase risk and predispose humans to harm which include the building of infrastructure in flood-prone coastal areas. The occurrence of flooding in such areas, in the face of growing poverty, will exacerbate existing inequities through development patterns that could lead to increased poverty and exclusion from social and political processes that could further compound the existing vulnerability and disaster risk. Countries must develop creative ways of vulnerability analysis for the strengthening of urban resilience to achieve sustainable development. Globally, strengthening urban resilience is critical to the attainment of sustainable development and the dual aim of eradicating extreme poverty and increasing the share of prosperity in the urban environment.

Conclusion and recommendations

Droughts, flooding, and cyclones have increased in frequency and severity during the last decade. The effects of disasters on development, vulnerability, and poverty have necessitated the need to improve disaster resilience. Disaster resilience is the ability of a household, community, or country to cope with and adapt to disaster shocks and stressors. As the global population increases and increasing urbanisation in underdeveloped countries threatens to undo hard-won development achievements, these trends are projected to become more pronounced. There is mounting evidence that disaster resilience saves lives and protects social systems, infrastructure livelihoods, and the environment. Therefore, disaster resilience is more cost effective in both the short and long term than the current disaster relief and development aid combination. In practice, resilience is becoming an increasingly significant component of an all-inclusive approach to lessening the effects of disasters on the most vulnerable people. As climate changes and disasters become more common in cities that house a growing percentage of the world's poor, constructing resilient cities is becoming increasingly important. According to evidence, the following proposals can help developing countries establish disaster-resilient cities:

1. Assisting communities in recognising, reducing, and managing risk. Empowering communities to recognise risks and risk drivers as well as their capacity is essential to addressing risks at the community level. This could include ways to reduce the quantity of garbage generated in the home and providing instructions on the safe disposal of the rubbish. Establishing democratic institutions like waste pickers' associations, savings and investment clubs, art groups, community safety forums, and recycling co-ops could promote effective waste management.
2. Supporting local governments to reduce risk. All levels of government are responsible for providing services and infrastructure, as well as supporting and protecting vulnerable people and building resilience. Local governments, however, frequently lack adequate human, financial, and technical resources. Building resilience necessitates enhancing the ability of local governments to identify and address risk drivers in the short and long term. It is critical to enhancing the capacity and capability of local governments to offer services and infrastructure while also addressing challenges such as poor accountability and resource management.
3. Strengthening urban planning and regulatory frameworks. Strengthening urban and land use planning, as well as enforcing proper building standards, might go a long way toward boosting resilience. By limiting settlement and development in unsafe areas, better urban and land use planning can help to reduce exposure to dangers. These processes must be

supplemented by initiatives to improve government institutions and private developers' accountability for buildings that are poorly placed, illegal, or of poor quality.
4. Facilitating dialogue and collaboration to reduce risk. Building effective interactions between local governments, vulnerable populations, the business sector, NGOs, and community-based organisations (CBOs) can help with both immediate risk reduction and long-term adaptation– especially in areas where governmental capacity is limited. This includes enhancing government institutional capacity to identify and utilise crucial linkages that facilitate adaptation.
5. Supporting holistic risk reduction across sectors. Risk reduction will be most effective when NGOs, UN agencies, civil society organisations, and community groups work together to form strong bonds, identify gaps, and capitalise on shared interests. This should entail bringing together humanitarian, human rights, development, and disaster risk practitioners to discover best practices and innovation, as well as to encourage shared learning.

References

Adedeji, T., Proverbs, D., Xiao, H., Cobbing, P. and Oladokun, V. (2019). Making Birmingham a flood-resilient city: Challenges and opportunities. *Water*, 11(8), 1699.

Adewole, I. F., Agbola, S. B. and Kasim, O. F. (2015). Building resilience to climate change impact after the 2011 flood disaster at the University of Ibadan, Nigeria. *Journal of Urbanization and Environment*, 27(1), 199–216.

Agbola, S. B., Ajayi, O., Taiwo, O. J. and Wahab, W. B. (2012). The August 2011 flood in Ibadan, Nigeria: Anthropogenic causes and consequences. *International Journal of Disaster Risk Science*, 3(4), 207–217.

Agbola, T. and Kasim, O. F. (2018). *Urban development, climate change and disaster management Nexus in Africa*. Paper presented at the Sustainable African Cities Conference Organized by the Ghana Academy of Arts and Sciences (GAAS), the Network of African Science Academies (NASAC) and the German National Academy of Sciences Leopoldina, supported by the German Federal Ministry of Education and Research (BMBF) Held on 4–6 July, 2018 at Research Crescent, Hayford Road, Accra, Ghana.

Ahern, J. (2013). Urban landscape sustainability and resilience: The promise and challenges of integrating ecology with urban planning and design. *Landscape Ecology*, 28(6), 1203–1212.

Albrito, P. (2012). Making cities resilient: Increasing resilience to disasters at the local level. *Journal of Business Continuity and Emergency Planning*, 5(4), 291–297.

Barbaro, G., Miguez, M. G., de Sousa, M. M., Ribeiro da Cruz Franco, A. B., de Magalhães, P. M. C., Foti, G. . . . Occhiuto, I. (2021). Innovations in best practices: Approaches to managing urban areas and reducing flood risk in Reggio Calabria (Italy). *Sustainability*, 13(6), 34–63.

Basiago, A. (1999). Economic, social, and environmental sustainability in development theory and urban planning practice. *The Environmentalist*, 19(1), 145–161.

Batica, J. and Gourbesville, P. (2016). Resilience in flood risk management: A New communication tool. *Procedia Engineering*, 154, 811–817.

Bello, O. (2017). *Mainstreaming disaster risk management strategies in development instruments: Policy briefs for selected member countries of the Caribbean development and cooperation committee*. Studies and Perspectives series – ECLAC Sub-regional Headquarters for the Caribbean, No. 58 (LC/TS.2017/80; LC/CAR/TS.2017/6), Santiago, Economic Commission for Latin America and the Caribbean (ECLAC).

Birmingham City Council. (2018). *Managing the risk and response to flooding in Birmingham: Issues arising from May 2018 major flooding event*. file:///C:/Users/S16147278/Downloads/Managing_the_Risk_and_Response_to_May_2018_Flooding_FINAL%20(5).pdf (accessed 21 August 2021).

Bosher, L. (2008). Introduction: The need for built-in resilience. In Bosher, L. (ed) *Hazards and the built environment: Attaining built-in resilience*. London: Taylor & Francis, 3–19. https://doi.org/ 10.4324/9780203938720.

Bush, J. and Doyon, A. (2019). Building urban resilience with nature-based solutions: How can urban planning contribute? *Cities*, 95. https://doi.org/10.1016/j.cities.2019.102483.

Chelleri, L., Waters, J. J., Olazabal, M. and Minucci, G. (2015). Resilience trade-offs: Addressing multiple scales and temporal aspects of urban resilience. *Environment and Urbanization*, 27(3), 181–198.

Cilliers, S., Cilliers, J., Lubbe, R. and Siebert, S. (2013). Ecosystem services of urban green spaces in African countries – perspectives and challenges. *Urban Ecosystems*, 16(5), 681–701.

Cities Climate Finance Leadership Alliance (CCFLA). (2015). *The state of city climate finance, 2015*. www.citiesclimatefinance.org/2015/12/the-state-of-city-climate-finance-2015-2/.

Coaffee, J. and Lee, P. (2016). *Urban resilience: Planning for risk, crisis and uncertainty*. London: Palgrave Macmillan.

Deshkar, S., Hayashia, Y. and Mori, Y. (2011). An alternative approach for planning the resilient cities in developing countries. *International Journal of Urban Sciences*, 15(1), 1–14.

ECLAC (Economic Commission for Latin America and the Caribbean). (2018a). *Disaster assessment methodology exercise guide*. Project Documents (LC/TS.2018/64), Santiago.

Egbinola, C. N., Olaniran, H. D. and Amanambu, A. C. (2017). Flood management in cities of developing countries: The example of Ibadan, Nigeria. *Journal of Flood Risk Management*, 10(4), 546–554. https://doi.org/10.1111/jfr3.12157

Egunjobi, L. (1999). *Our gasping cities*. An Inaugural Lecture Delivered at the University of Ibadan on Thursday, 21 October.

Field, C., Barros, V., Dokken, D., Mach, K., Mastrandrea, M., Bilir, T., Chatterjee, M., Ebi, K. L., Estrada, Y. O. and Genova, R. C. (2014). *Climate change: Impacts, adaptation, and vulnerability, part A: Global and sectoral aspects; contribution of working group ii to the fifth assessment report of the intergovernmental panel on climate change*. Cambridge and New York: Cambridge University Press.

Geis, D. E. (2000). By design: The disaster-resistant and quality-of-life community. *Natural Hazards Review*, 1(3), 151–160.

Gero, A., Méheux, K. and Dominey-Howes, D. (2011). Integrating community-based disaster risk reduction and climate change adaptation: Examples from the Pacific. *Natural Hazards and Earth System Sciences*, 11(8), 101–113.

GFDRR. (2010). *Natural hazards, unnatural disasters: The economics of effective prevention*. United Nations, World Bank. www.gfdrr.org/node/281.

Ghazali, D. A., Guericolas, M., Thys, F., Sarasin, F., Arcos Gonzalez, P. and Casalino, E. (2018). Climate change impacts on disaster and emergency medicine focusing on mitigation disruptive effects: An international perspective. *International Journal of Environmental Research and Public Health*, 15(7), 1379.

Godschalk, D. R. (2003). Urban hazard mitigation: Creating resilient cities. *Natural Hazards Review*, 4(2), 136–143.

Haigh, R. and Amaratunga, D. (2010). An integrative review of the built environment discipline's role in the development of society's resilience to disasters. *International Journal of Disaster Resilience in the Built Environment*, 1(1), 11–24.

Hofmann, S. Z. (2021). 100 resilient cities program and the role of the Sendai framework and disaster risk reduction for resilient cities. *Progress in Disaster Science*, 11, 100189. https://doi.org/10.1016/j.pdisas.2021.100189.

Kasim, O. F. (2018). Wellness and illness: The aftermath of mass housing in Lagos, Nigeria. *Development in Practice*, 28(7), 952–963. https://doi.org/10.1080/09614524.2018.1487385.

Kasim, O. F., Agbola, S. B. and Oweniwe, M. F. (2020a) Land use land cover change and land surface emissivity in Ibadan, Nigeria. *Journal of Town and Regional Planning*, 77, 71–88. http://dx.doi.org/10.18820/2415-0495/trp77i1.6.

Kasim, O. F., Wahab, B. and Olayide, O. E. (2020b). Assessing urban liveability in Africa: Challenges and interventions. In: Leal Filho, W., Azul, A. M., Brandli, L., Lange Salvia, A. and Wall, T. (eds) *Industry, innovation and infrastructure: Encyclopedia of the UN sustainable development goals*. Cham: Springer. https://doi.org/10.1007/978-3-319-71059-4_70-1.

Kasim, O. F., Wahab, B. and Oweniwe, M. F. (2021). Urban expansion and enhanced flood risk in Africa: The example of Lagos. *Environmental Hazards*. https://doi.org/10.1080/17477891.2021.1932404

Lalovic, K., Sentic, I. and Zivojinovic, I. (2020). Urban and regional planning for sustainability. In: Leal Filho, W., Azul, A. M., Brandli, L., Özuyar, P. G. and Wall, T. (eds) *Climate action: Encyclopedia of the UN sustainable development goals*. Cham: Springer. https://doi.org/10.1007/978-3-319-95885-9_77.

Liao, K. (2012). A theory on urban resilience to floods – a basis for alternative planning practices. *Ecology and Society*, 17(4), 48. http://dx.doi.org/10.5751/ES-05231-170448.

Malalgoda, C., Amaratunga, D. and Haigh, R. (2013). Creating a disaster resilient built environment in urban cities: The role of local governments in Sri Lanka. *International Journal of Disaster Resilience in the Built Environment*, 4(1), 72–94.

Nelson, D. (2011). Adaptation and resilience: Responding to a changing climate. Wiley interdisciplinary reviews. *Climate Change*, 2(1), 113–120.

Newman, P., Beatley, T. and Boyer, H. (2017). *Resilient cities: Overcoming fossil fuel dependence*. Washington, DC: Island Press.

O'Brien, G., O'Keefe, P., Gadema, Z. and Swords, J. (2010). Approaching disaster management through social learning. *Disaster Prevention and Management*, 19(4), 498–508.

Osayomi, T. and Oladosu, O. S. (2016). "Expect more floods in 2013": An analysis of flood preparedness in the flood-prone city of Ibadan, Nigeria. *AJSD*, 6(2), 215–237.

Ramiaramanana, F. N. and Jacques, T. (2021). Urbanization and floods in Sub-Saharan Africa: Spatiotemporal study and analysis of vulnerability factors – case of Antananarivo Agglomeration (Madagascar). *Water*, 13(2), 149. https://doi.org/10.3390/w13020149

Renschler, C. S., Frazier, A. E., Arendt, L. A., Cimellaro, G. P., Reinhorn, A. M. and Bruneau, M. (2010). *Developing the 'PEOPLES' resilience framework for defining and measuring disaster resilience at the community scale*. Paper No 1827. Proceedings of the 9th U.S. National and 10th Canadian Conference on Earthquake Engineering, 25–29 July, 2010, Toronto, ON, Canada.

Satterthwaite, D. and Dodman, D. (2013). Towards resilience and transformation for cities within a finite planet. *Environment and Urbanization*, 25(4), 291–298.

Shamsuddin, S. (2020). Resilience resistance: The challenges and implications of urban resilience implementation. *Cities*, 103, 102763. https://doi.org/10.1016/j.cities.2020.102763.

Shepherd, A., Mitchell, T., Lewis, K., Lenhardt, A., Jones, L., Scott, L. and Muir-Wood, R. (2013). *The geography of poverty, disasters and climate extremes in 2030*. London: ODI.

UN-Habitat. (2013). *State of the world's cities 2012/2013: Prosperity of cities*. London: Routledge.

UNDRR (United Nations Office for Disaster Risk Reduction). (2019). *Global assessment report on disaster risk reduction*. Geneva: UNDRR.

United Nations Office for the Coordination of Humanitarian Affairs (OCHA). (2012) *Overview: Impact of floods – West and Central Africa*, 15 September 2012. https://reliefweb.int/map/niger/overview-impact-floods-west-and-central-africa-15-september-2012 (accessed 1 September 2021).

United Nations Office for the Coordination of Humanitarian Affairs (OCHA). (2021). *Global humanitarian overview 2021*. Geneva: OCHA.

Wahab, B. (2013). Disaster risk management in Nigerian human settlements. In: Wahab, B., Atebije, N. and Yunusa, I. (eds) *Disaster risk management in Nigerian rural and urban settlements*. Abuja: Nigerian Institute of Town Planners and Town Planners Registration Council, 1–37.

Wahab, B. and Agbola, B. (2017). The place of informality and illegality in planning education in Nigeria. *Planning Practice and Research*, 32(2), 212–225.

Wahab, B. and Falola, O. (2017). The consequences and policy implications of urban encroachment into flood-risk areas: The case of Ibadan. *Environmental Hazards*, 16(1), 1–20.

Wahab, B. and Falola, O. (2018). Vulnerable households and communities' responses to flood disasters in the Ibadan Metropolis, Nigeria. *Ibadan Journal of Sociology*, 7, 47–75.

Wardekker, A., Wilk, B., Brown, V., Uittenbroek, C., Mees, H., Driessen, P., Wassen, M., Molenaar, A., Walda, J. and Runhaa, H. (2020). A diagnostic tool for supporting policymaking on urban resilience. *Cities*, 101, 102691. https://doi.org/10.1016/j.cities.2020.102691.

WHO. (2019). *WHO aids flood-hit populations across Africa and the Eastern Mediterranean Region.* www.who.int/news/item/08-11-2019-who-aids-flood-hit-populations-across-africa-and-the-eastern-mediterranean-region (accessed 3 September 2021).

WHO. (2020). *Floods*. www.who.int/health-topics/floods#tab=tab_1 (accessed 2 September 2021).

Wildavsky, A. (1988). *Searching for safety*. New Brunswick, NJ: Transaction Books.

World Bank. (2019). *Building urban resilience: An evaluation of the World Bank Group's evolving experience (2007–17)*. Independent Evaluation Group. Washington, DC: World Bank.

18

Flood risk and urban infrastructure sustainability in a developing country

A case study of Central Java Province, Indonesia

Purwanti Sri Pudyastuti and Isnugroho

Introduction

Central Java is a province situated in Java Island, Indonesia, between latitudes 6°–7°30' S and longitudes 108°30'–112°00' E (Marfai et al., 2008). The province's area is about 32,800.69 km^2, which spans 29 regencies and 6 municipalities and is inhabited by more than 34 million people (Central Bureau of Statistics Republic of Indonesia – Central Java Province, 2020). The administrative map of Central Java Province (Figure 18.1) shows the major rivers and river basins traversing the province. There are tens of rivers in the province spanning nine river basins in Central Java Province: Jratunseluna, Bengawan Solo, Wiso Gelis, Progo Opak Serang, Bodri Kuto, Serayu Bogowonto, Pemali Comal, Citanduy, and Cimanuk Cisanggarung. Some of the river basins are transboundary rivers which belong to more than one province and are more complicated to manage. Some regions in the province have high flood risk, such as Semarang, Banyumas, Demak, Kendal, Kudus, Cilacap, Solo, and Sukoharjo (Dinas Pusdataru Provinsi Jawa Tengah, 2008).

According to Isnugroho (2010), drainage issues, in terms of coverage, capacity, operation and maintenance, and irregular solid waste disposal were the causes of flood in Java Island's urban areas. In addition, flood risk in Central Java Province could be triggered by coastal flooding, heavy rainfall, and overflow from big rivers in the regions. The average annual rainfall in Central Java Province based on the rainfall data in 2000–2015 was 2,147 mm. The annual rainfall isohyet map in Central Java Province is shown in Figure 18.2.

Semarang City, the capital of Central Java Province, is located in a lowland area and has a high risk of coastal flooding. Flood events in Central Java Province capital (Semarang) have caused disruption to the international airport and railway station repeatedly. During 2013–2018, the flood frequency in Semarang City was more than 25 times a year (Wismana Putra et al., 2020). Recently, according to Dartmouth Flood Observatory, from 19–22 February 2021, heavy rain in some of Indonesia's regions triggered floods which affected about a 13,179.26 km^2 area (Brakenridge, 2021).

Flood risk and urban sustainability in a developing country

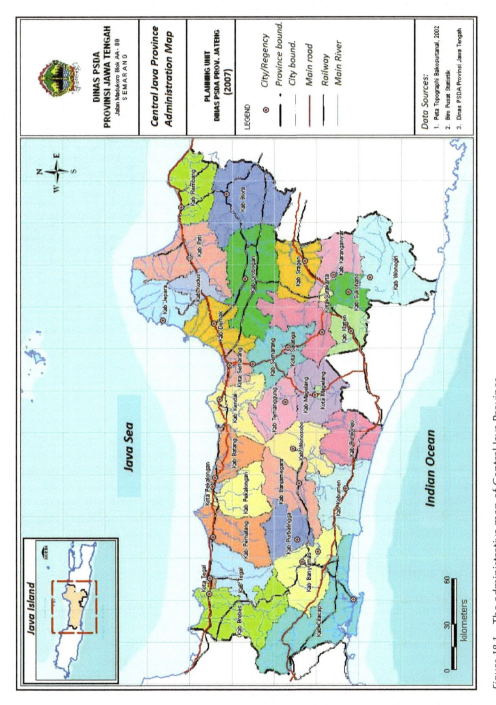

Figure 18.1 The administrative map of Central Java Province
Source: Modified from Pudastaru Propinsi Jawa Tengah, 2008

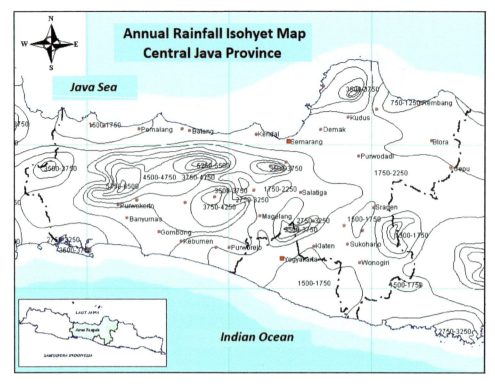

Figure 18.2 Annual rainfall isohyet map of Central Java Province
Source: Modified from Pudastaru Propinsi Jawa Tengah, 2008

Flood event can damage and disturb transportation infrastructure, which can affect other sectors, such as the economic and health sectors (Pudyastuti and Nugraha, 2018). A study conducted by (Hatmoko et al., 2017) reported that the contribution of flooding to road damage along the East Kendal Highway on the northern coast of Central Java was significant (Hatmoko et al., 2017). The flood event in Kendal Regency at that time was suspected to have been caused by the poorly maintained drainage system (Hatmoko et al., 2017). In addition, the total loss due to flooding in the province in 2014 was about $69 million (Beritasatu, 2014).

Historical context

Indonesia has a high frequency of floods, and Java Island has the highest frequency. Several cities in Central Java Province experienced flooding in the past. For example, in 1966, Surakarta City, one of the most populated cities in Central Java Province, was flooded due to high rainfall intensity which triggered the break of an embankment along the Solo River. Most of the area in Surakarta City was inundated during the flood event. A big flood struck the city again at the end of 2007 and the beginning of 2008. The flood that occurred in Surakarta City was caused by high rainfall intensity causing river overflow and embankment breaks. In 1990, a flash flood occurred in Semarang City (the capital city of Central Java Province) and damaged the housing, transportation, and water infrastructures in the city (Priyanto and Nawiyanto, 2014).

Table 18.1 Flooded area in Central Java Province in 2007

Regency/city	Flooded Area (hectares)	Regency/city	Flooded Area (hectares)	Regency/city	Flooded Area (hectares)
Banyumas	1,488.0	Kudus	22,500.0	Sukoharjo	1,367.0
Batang	3,065.0	Pati	7,500.0	Surakarta	2,105.0
Blora	5,000.0	Pekalongan City	10.0	Tegal City	1,250.0
Brebes	15,504.0	Pekalongan Regency	3,925.0	Tegal Regency	4,757.0
Cilacap	32,500.0	Pemalang	8,295.0	Wonogiri	480.0
Demak	15,061.0	Purbalingga	750.0		
Grobogan	12,500.0	Purworejo	195.0		
Jepara	5,000.0	Rembang	11,000.0		
Kebumen	10,000.0	Semarang	15,000.0		
Kendal	9,875.0	Sragen	10,300.0		

In 2007, there were 91 flood events in Central Java Province (Marfai et al., 2008). The flood events in that year caused 1,337 deaths and damaged 211,159 houses, 635 km of road, 16 bridges, and thousands of hectares of agricultural land. Table 18.1 and Figure 18.3 show the flooded area in Central Java Province in 2007 (Dinas Pusdataru Provinsi Jawa Tengah, 2008; Marfai et al., 2008). Furthermore, based on a study conducted by (Wisnu et al., 2017), most of the regencies and cities in Central Java Province have high flood risk and vulnerability.

In 2016, Central Java Province experienced 521 flood events, which was the highest number of flood events in Indonesia at that time (Purwaningsih et al., 2018). In addition, according to the head of the water resources management board of Central Java Province, from late 2019 to mid-January 2020, there were 143 flood events in the province (Suwiknyo, 2020). During the flood events, there were 52 damaged embankments, and 21,633 people had to be evacuated. Figure 18.4 shows damaged roads due to inundation caused by flooding in Surakarta City and Semarang City. Figure 18.5 shows the flooding at Tawang Railway Station and Terboyo Bus Station in Semarang City.

The floods not only damaged transportation infrastructures and housing but also damaged irrigation system and agricultural land. Recent flood events at the beginning of 2021 were reportedly caused by extreme rainfall, land-use changes, and land subsidence, particularly on the northern coast of the province (Kompas, 2021; Kontan, 2021). The neighboring countries of Indonesia in South-East Asia, such as Malaysia and Vietnam, are affected by disasters such as flooding and landslides. Flood disasters in Malaysia have occurred frequently and caused great damage annually (Chan, 2012). In Vietnam, flood disasters also cause severe economic losses and damage to infrastructures (Nguyen et al., 2021).

Flood risk management

A previous study conducted by Japan International Cooperation Agency (JICA, 2004) reported that urban areas in Java Island became inundated due to an insufficient drainage system and poor maintenance work. Furthermore, according to Stanton-Geddes and Vun (2019), the regions which have high flood risk in the province have experienced pluvial flooding, fluvial flooding, and coastal flooding. Fluvial flooding can be triggered by upstream rainfall, flash flood, and river

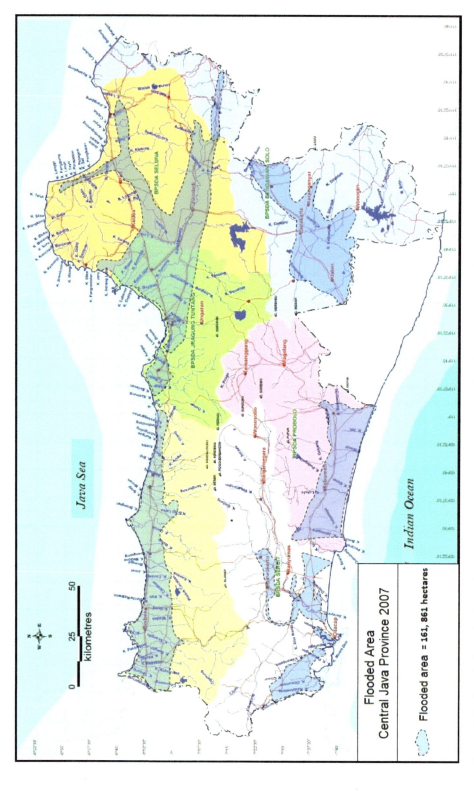

Figure 18.3 Flooded area in Central Java Province 2007 (Dinas Pusdataru Provinsi Jawa Tengah, 2008; Marfai et al., 2008)

Flood risk and urban sustainability in a developing country

Figure 18.4　Damaged roads in Surakarta City and Semarang City (Pos Jateng, 2018; Solopos, 2016a)

Figure 18.5 Flooding at Tawang Railway Station (top) and Terboyo Bus Station (bottom), Semarang City

Source: Tempo online, 2021; Detiknews, 2017

overflow, whereas coastal flooding is linked to extreme offshore weather (Stanton-Geddes and Vun, 2019).

Semarang City and Surakarta City are the most populated cities in Central Java Province which have the highest flood risk in the province (Stanton-Geddes and Vun, 2019). Surakarta City experienced pluvial flooding, which is linked to extreme rainfall, runoff, and drainage system issues. In October and November 2016, some areas in Surakarta City were flooded, as

Flood risk and urban sustainability in a developing country

can be seen in Figure 18.4. According to the head of the Disaster Research Center of Sebelas Maret University of Surakarta, the flood was caused by poor function of drainage system, low soil infiltration capacity in the city, high sedimentation in the rivers, and poor solid waste management (Solopos, 2016b).

Figure 18.6 Flooding in Surakarta City in October and November 2016
Source: Okezone News, 2016; Solopos, 2016a

Flood risk management is defined as the management of floodplain using a systematic process requiring risk assessment, strategic planning, development of risk reduction measures, and implementing activities which involve multiple stakeholder and sector cooperation to reduce flood risks in a sustainable way (ADCP and UNDP, 2006). In managing flood risk, a holistic approach considering structural and non-structural measures involving all the stakeholders is required. These approaches include reducing the runoff in the city, for example, by applying rainwater harvesting, vegetated rooftops, infiltration wells, and absorbent gardens. According to Pudyastuti and Saputra (2018), the application of green rooftops in selected flat-roofed buildings in Surakarta City can reduce the runoff by 18%. The most common intervention applied in Central Java Province to reduce runoff is constructing infiltration wells at housing and other buildings. An infiltration well (*sumur resapan*, in Indonesian) is a porous well constructed underground to catch and infiltrate rainwater into the aquifer. Some hotels in the province also have applied green rooftops. Furthermore, green infrastructure application can increase the infiltration and attenuation of storm water as well as decreasing the flowrate capacity and runoff coefficient, which can reduce flood risk (Ertan and Çelik, 2021).

River basins in the province are managed by the River Basin Organization, which cooperates and coordinates with other related institutions. For example, Solo River, which flows from Wonogiri Regency in the upstream area to Java Sea, crossing some cities and regencies in two provinces, Central Java Province and East Java Province, is a transboundary river such that conditions in the upstream area could impact the downstream area. Therefore, the river basins must be managed jointly by all relevant stakeholders in the upstream and downstream areas (Pudyastuti, 2008). The government of the Republic of Indonesia has spent a huge amount of funds to construct flood protection infrastructures to reduce flood risk in Central Java Province. Dams, embankments, and storage reservoirs were built to reduce the flood risk technically. According to Kundzewicz (1999), structural flood protection like dams, embankments, and storage reservoirs has been criticized in the context of sustainable development because they close options for future generations and introduce unacceptable disturbances into the ecosystem. Therefore, a paradigm shift to manage flood risk in the sustainability context is required (Kundzewicz, 1999). Moreover, the different aspects of flood risk, such as hydrological, hydraulic, economic, social, and ecological aspects, should be considered in flood risk management to obtain effective and efficient risk reduction measures (Tariq et al., 2020).

Sustainable urban infrastructure and flood risk management in central Java Province, Indonesia

Urban sustainability has been defined in several ways, with different criteria and emphases. Most of the definitions are derived from those of sustainability with concentration on enhancement of long-term human wellbeing. Urban sustainability harmonizes the three dimensions of sustainability, namely minimizing resource consumption and environmental damage, maximizing resource use efficiency, and ensuring equity and democracy (Huang et al., 2015). In addition, urban sustainability is "an adaptive process of facilitating and maintaining a virtual cycle between ecosystem services and human wellbeing through concerted ecological, economic, and social actions in response to changes within and beyond the urban landscape" (Wu, 2014). There are various urban sustainability indicators which depend on how sustainability is defined. Such indicators can include CO_2 emissions, energy, buildings, land use, transport, water and sanitation, waste management, air quality, and environmental governance (Huang et al., 2015).

In an urban system, there are many infrastructures such as transportation, energy, water supply, wastewater, communication, hospital, schools, and others. Urban infrastructure is essential to the urban system; therefore, sustainable urban infrastructure is required in achieving sustainability of an urban system across all dimensions (environmental, economic, and social). However, there has not been any significant study on the impact of flooding on urban sustainability indicators in Central Java Province. Some of the challenges to ensuring urban infrastructure sustainability in Central Java Province from an environmental, political, and social point of view include the following.

Environmental aspects

To ensure the sustainability of urban infrastructure from an environmental viewpoint, flood mitigation is an important effort. However, determining an appropriate flood mitigation approach requires availability of data. The challenge in Central Java Province is insufficient data, for example, hydrological data and hydraulic data. While there are many rain gauging stations in Central Java Province, many of them do not work well. Hydraulic data are also very limited. There are some automatic water level recorder applied in some rivers in the province; however, some of the rating curves resulting from the measurement are rarely calibrated. Other challenges in the environmental aspect include main rivers that are already polluted and have sedimentation problems, poorly managed drainage networks, and less green open space.

Political aspect

In Indonesia, when there is a change of elected leaders involving political party change, there is usually a change of policies. This instability of policy causes difficulty in ensuring sustainability of programs.

Social aspect

As mentioned previously, Central Java Province is densely populated. A region with various races, religions, education levels, and economic levels will have complicated social problems, which have become constraints to ensuring sustainability in an urban system.

Flood risk management in Indonesia

According to Deputi Bidang Sarana dan Prasarana (2008), in Indonesia, there are 5,590 main rivers, and 600 of them have potential risk of flooding. The flood-prone areas covered by these main rivers reach about 1.4 million hectares. From previous studies, floods occurred in vulnerable areas due to the following three factors:

1) human activities that cause changes in spatial planning and have an impact on natural changes;
2) natural events such as very high rainfall, sea level rise, storms, and so on;
3) environmental degradation such as loss of vegetation cover in the catchment area, river silting due to sedimentation, and narrowing of riverbanks.

Flood management in Indonesia is basically carried out in stages: prevention (before flooding), response/intervention (treatment during floods events), and recovery (after flood events). Flood management is conducted by applying structural interventions such as the construction of flood control in river areas (in-stream) to floodplain areas (off-stream) and non-structural interventions such as land use management and flood early warning systems.

Table 18.2 shows a list of related stakeholders in flood management in Indonesia according to the Ministry of Public Work and Housing (2017). In the governmental sector (national and regional levels), there are many boards/agencies under the institutions mentioned in the table which participate in flood management. Table 18.3 presents some examples of agencies under government institutions that take part in flood management in Indonesia and their action plans.

Table 18.4 presents a summary of common interventions applied in flood risk management in Indonesia and related stakeholders involved in each intervention in Central Java Province.

Table 18.2 List of related stakeholders in flood management in Indonesia

Private sector, community, and other institutions	Government (national level)	Government (regional level)

i. Investors
ii. Contractors
iii. Consultants
iv. Volunteer
v. Corporates
vi. Community
vii. NGO
viii. Universities
ix. Research centers
x. Professional association
	i. Parliament (DPR)
	ii. Ministry of:
	  a) Agriculture
	  b) Health
	  c) Social Affairs
	  d) Forestry
	  e) Marine and Fisheries
	  f) Finance
	  g) Communication and Information
	  h) Environment
	  i) Transportation and Telecommunication
	  j) Energy and Natural Resources
	  k) Public Work and Housing
		i. Regional Parliament (DPRD)
		ii. Government:
		  a) Provincial
		  b) Regency/City
		  c) Bappeda
		  d) Irrigation Agency
		  e) Districts
		  f) Jasa Tirta
		  g) Local government
		iii. National army
		iv. Rescue team
		v. Hospitals
		vi. Infrastructure projects
		vii. Red Cross

Table 18.3 Examples of agencies under government institutions in flood management in Indonesia and action plans

Agencies	Action Plans	Timeline
Research and Development Agencies (Balitbang)	• Water resources management plan • Institutional arrangement for water resources management plan • Financial assessment of water resources management plan • Flood control assessment as a part of water resources management plan	• Medium term, long term • Medium term, long term • Medium term, long term • Short term, long term
Regional Planning and Development Agencies (Bappeda)	• Make comprehensive and integrated plan • Master plans for the development of each service infrastructure system • Organizational and institutional arrangement • Plans to improve operation services	• Long term • Long term • Short term • Short term

Agencies	Action Plans	Timeline
Water Resources Management Agencies/ Irrigation Board	• Evaluation and review of river basin action plans • Water resources management and flood management • Evaluation and review of flood management plan for every river basin • Flood-prone area maps • Landslide-prone area maps	• Medium term, long term • Medium term, long term • Medium term, long term • Short term • Short term

Source: Ministry of Public Work and Housing, 2017

Table 18.4 Summary of interventions applied in flood risk management in Indonesia

Phase	Interventions
Prevention (before flooding)	*Structural*: • in-stream interventions • off-stream interventions *Non-structural*: • long-term mitigation • short-term mitigation • land use management • early warning system
Response (during flooding)	• flood alert • flood aid • evacuation
Recovery (after flooding)	• infrastructure reconstruction • physical and non-physical rehabilitation • compensation • study to find the cause of flooding

In Central Java Province, according to Dinas Pusdataru (2008), flood management is typically conducted by applying interventions as follows:

1) Store the runoff as much as possible by constructing dams and small reservoirs (*embung*, in Indonesia), and check dam.
2) Increase rainwater infiltration rate by constructing infiltration wells, provide open green space, and control rainwater of middle catchment areas by storing runoff in retention ponds.
3) Drain runoff in downstream catchment areas to the ocean as quickly as possible.
4) Educate the community to support catchment area conservation through GNKPA (*Gerakan Nasional Kemitraan Penyelamatan Air* or National Water Conservation Partnership Movement).

Strengths, weaknesses, opportunities, and threats analysis of current flood risk management in Central Java Province

Flood risk management has been applied in Central Java Province for decades, with different types of interventions which depend on the era, the leaders, and the advances of science and technology. Table 18.5 shows the strengths, weaknesses, opportunities, and threats (SWOT) analysis of current flood risk management in Central Java Province.

Table 18.5 SWOT analysis of current flood risk management in Central Java Province

Strength	Weaknesses
1. Disaster management agencies are available at national and regional level, for example, www.bnpb.go.id/ for the national level and http://bpbdjateng.info/ for Central Java Province. 2. There have been some community-based disaster reduction movements, for example, *Forum Peduli Banjir* (Flood Care forum) and *TAGANA* (*TAruna siaGA bencaNA* – Youth Disaster Preparedness and Quick Response). 3. Community solidarity during flood events. 4. Capacity building through regularly conducted workshops or training on flood risk management for staff of related agencies and universities. 5. Massive infrastructure project to reduce flood risk during the government of President Joko Widodo (since 2014), who constructed many new dams and small earth dams.	1. Too many agencies with lack of coordination and cooperation. 2. Lack of hydrology and hydraulic data in some catchment areas. 3. Dense population in flood-prone areas, for example, in Semarang and other northern coastal areas in Central Java Province. 4. Poor solid waste management causing some rivers to become garbage and solid waste collection places, thus reducing their capacities. 5. Sedimentation problems in some reservoirs which have function as flood control. 6. Less awareness and lack of adaptation to climate change impact.
Opportunities	*Threats*
1. Many agencies participate in flood risk management. 2. The government and parliament support activities to reduce flood risks. 3. Some other countries, such as Japan and the Netherlands, give support in funding and technology. 4. Local communities and students participate in flood management, particularly during and after flood events.	1. Lack of awareness in environmental preservation and conservation among some communities. 2. The flooding area is increasing. 3. Uncertainty of climate.

It should be an opportunity when many agencies get involved in flood risk management; however, the lack of coordination and cooperation can be a constraint to ensuring successful flood risk management planning.

Current condition of urban infrastructures in Central Java Province

According to the Indonesia Ministry of Public Work (Human Resources Development Agency, 2015), urban infrastructures can consist of water supply infrastructure, wastewater treatment, drainage systems, solid waste management plants, transportation infrastructure, electricity infrastructure, telecommunication infrastructure, and public building infrastructure (government office, business centers, hospitals, schools, etc.). The total number of hospitals and local medical centers in Central Java Province is 6,880, while there are 16 dams/reservoirs situated in the province. The dams, mostly located in rural areas in some regencies, are mostly multi purposes dams for flood control, fisheries, and irrigation water supply, as well as hydroelectric power plants. Beside dams, the government has planned to build hundreds of small reservoirs (*embung*, in Indonesia), with some of them still under construction. In all cities and regencies in Central Java Province, all the available infrastructures are typically similar; however, some small towns have no railway station. Table 18.6 presents the types of road surfaces in all regencies and cities

Table 18.6 Lengths and types of road surfaces in Central Java Province (Central Bureau of Statistics Republic of Indonesia – Central Java Province, 2020)

Regency	Type of Road Surface			
	Paved	Not paved	Others	Total (km)
Cilacap	1229	40	0	1269
Banyumas	805	0	0	805
Purbalingga	864	24	0	888
Banjarnegara	840	99	0	939
Kebumen	681	151	129	960
Purworejo	721	33	16	770
Wonosobo	781	123	95	999
Magelang	1001	0	0	1001
Boyolali	574	18	86	678
Klaten	725	3	41	770
Sukoharjo	574	7	24	605
Wonogiri	847	12	180	1038
Karanganyar	919	128	0	1047
Sragen	1011	9	0	1020
Grobogan	155	662	101	918
Blora	888	259	64	1211
Rembang	304	38	287	629
Pati	812	0	25	837
Kudus	601	4	34	639
Jepara	872	0	0	872
Demak	6	1	23	30
Semarang	638	98	0	736
Temanggung	591	11	48	650
Kendal	569	27	174	770
Batang	552	0	0	552
Pekalongan	595	24	51	670
Pemalang	710	21	34	766
Tegal	860	0	0	860
Brebes	316	0	394	710
Total (km)	20041	1792	1806	23639

Municipality/city	Type of Road Surface			
	Paved	Not paved	Others	Total (km)
Magelang	116	0	0	116
Surakarta	528	72	69	669
Salatiga	310	27	0	337
Semarang	527	180	134	840
Pekalongan	124	18	12	153
Tegal	225	2	4	231
Total (km)	1830	299	219	2346

in Central Java Province (Central Bureau of Statistics Republic of Indonesia – Central Java Province, 2020).

Moreover, the road network in Central Java province can be categorized into six road types (Central Bureau of Statistics Republic of Indonesia – Central Java Province, 2020), as follows:

1) National roads: 1,518.09 km.
2) Main artery roads: 979.76 km
3) Main collector roads: 538.33 km
4) Provincial roads: 2,404.74 km
5) Regency and city roads: 25,985 km
6) Toll roads: 338.42 km

The road network, particularly the regency and city roads, are prone to damage, as previously shown in Figure 18.4, due to inundation during wet season and yearly flood events. This means the government must spend more of its budget on repair of damaged road networks. In the capital city of Central Java Province, Semarang, the airport, train station, and bus station have been disrupted many times due to flooding, causing huge financial losses.

Solid waste management in Central Java Province is still poor. Most solid wastes are collected in open-air landfills, and some communities still throw garbage into rivers and drainage channels. During wet season, the pollutants and leachates from the landfill washed away by runoff are dangerous for the environment and human health.

Drainage systems are essential infrastructure in urban areas, needed for the discharge of storm water and reduction of inundation during the wet season. The insufficiency of drainage network capacity is amplified by extreme rainfall and poor maintenance that trigger high sedimentation in drainage channels.

In addition, during high rainfall intensity, electricity supplies in some regions are cut off by service providers. This condition is inconvenient, particularly for students who need electricity for lighting during their study activity at night. Flood event impact on electricity supply also affects other sectors, such as health services and communications.

Community-based development and disaster risk reduction in Central Java Province, Indonesia

Indonesia is located in disaster-prone areas and has had disasters such as the tsunami in Aceh at the end of 2004; the Mount Merapi eruption in November 2010; and floods in Jakarta in 2002, 2007, and 2013. There are three philosophies in disaster management: moving people from disaster, moving the disaster from people, and living in harmony with the disaster. Shifting the disaster away from people involves constructing disaster countermeasure systems such as dikes, levees, dams, river training, and coastal protection. The construction of disaster countermeasure systems is very costly. On the other hand, moving people away from disaster requires relocating people from hazard-prone areas, which could be very difficult due to the social problems involved. For various reasons, people are often reluctant to be relocated. In addition, for Indonesia, which is located in a disaster-prone area, it is very difficult to find areas that are really safe from disaster. Therefore, a new paradigm in disaster management is living in harmony with disaster. This means that while people still live in hazard-prone areas, they are capable of responding when disaster strikes so that losses and risk can be reduced. The idea of living in harmony with disaster can be achieved by implementing community-based development. Community-based development seeks to empower individuals and groups of people by providing them with the skills they need to effect change in their own communities. These skills are often created through the formation of large social groups working for a common agenda.

People live and survive in hazard areas by understanding that they live in a disaster-prone area and being aware of the effects of disasters. They should be capable of response and evacuation when disasters hit so as to minimize the damage and reduce the risk to the victims. Therefore, public empowerment and awareness should be built and encouraged in order to take into account the possibility of impending floods and prepare for evacuation and rescue.

Community-based development is concerned with the involvement of local stakeholders in decision making. If people in communities are to take initiative, be creative, learn, and assume responsibility for their own development, they must be actively encouraged to participate. This requires building into policies and projects features which enable people's participation. In order to encourage community-based development on a large scale, it is important to first understand the dynamics at the household, group, or community levels. Based on this understanding, what needs to happen to support community action can be defined at successively higher and more distant levels. Community-based development can be arranged through many kinds of community-based activities.

Through these community-based activities, people should be able to participate alongside government officials and experts' groups as the direct stakeholders of these activities. While people should own the problems, consequences, and challenges of any mitigation and/or preparedness initiative, it is necessary to take people's involvement further, into policy and strategy. This process induces a sense of ownership in people which results in their continuous engagement and long-term commitment to these activities. Involvement of communities is important for both pre-disaster mitigation and post-disaster response, as well as for recovery process.

Some of the activities that should be done in CBD are as follows:

1) The establishment of an organization to serve as a driving force. This organization is very important in training and motivating the people. Usually, such an organization is composed of community leaders and local youth.
2) Inventory of local wisdom in disaster risk reduction and raising it for use in the DRR program.
3) Teach people to make a simple early warning system (if possible, based on local wisdom they have) and train them how to use it in a community connection system. Community connections are the relationships necessary to develop, implement, and maintain an effective end-to-end early warning system. A multi-hazard warning center can only be successful if the warnings it produces reach individuals at risk and are easy to understand, resulting in appropriate responses. To ensure warnings are most effective, the staff at a center must establish trusted partnerships among international organizations, governmental agencies, community leaders and organizations, businesses, and local citizens prior to issuing a warning
4) Village mapping as well as evacuation routes: creating large-scale maps of the ground for everyone to see. The map contains houses, local shelters and low-lying areas, and hazard-prone areas, as well as the evacuation route. Such a map enables villagers to see and identify risks and plan accordingly. This mapping exercise is carried out by the villagers every year before the disaster.
5) Learning to rescue: different disaster-based lifesaving techniques of rescuing people, carrying the rescued/injured, and first aid.
6) Preparing family kits – family survival kits (FSKs) and child survival kits (CSKs) – using plastic bags as water proof kits and creating a safe storage space where families can keep and

preserve dry food, important documents, medical kits, children's toys, and school books in case of floods.
7) Child protection: education of young children in the event of a disaster. In case of separation, children will know their own names, names of their parents, name of their village, and so on.
8) The task forces formed through the project were also described to the delegates. Task forces were based on the felt need and aspiration of the community; special emphasis was provided for inclusion of the most vulnerable segments of society. These task forces were named after their specific roles, such as early warning and village protection team, search and rescue teams, temporary shelter and camp management, first aid and sanitation, child protection and school safety, coordination, and so on.

The main objective of the activity of community-based development for disaster risk reduction is to encourage participation and community preparedness in preparing themselves when a flood occurs up to the evacuation process. The stages are as follows:

1) The establishment of a community organization called *Forum Peduli Banjir* (Flood care forum) or *TAGANA* (*TA*runa sia GA benca NA – Youth Disaster Preparedness and Quick Response). The members of this organization come from non-governmental organizations, young citizens, Boy Scout members, and so on. This team is formed and trained to be ready for disasters (to take into account the possibility of a flood, so they can thus prepare for evacuation and rescue) as well as to serve as a motivator and mover for other populations.
2) To train people to understand that floods will occur. Based on their own initiative with support and guidance from the team, they created a very simple early flood warning system instrument. The instrument is a serial lamp, and the indicators are: green indicates safe conditions, blue shows that river water level is increasing, yellow is a warning phase meaning people should prepare to evacuate, and red combined with the alert sign informs of dangerous condition, so people should be evacuated promptly. The instrument is located in strategic places, such as a mosque, a gathering place (sports/entertainment yard), and so on. This instrument is connected with floating equipment that corresponds to the water level in the river a few kilometers upstream from the village, so there is a time lag of flood propagation in order to have time enough for evacuation.

The important things in this activity are: a danger alert code agreed on by the people, the assembly area, publishing the evacuation route, defining the evacuation hall/area map, and socialization to train people to use it. The danger alert code uses common signals for inhabitant habits such as mosque information, *kenthongan* (traditional equipment made of bamboo which can create a loud sound when hit by bamboo or a wooden stick), and so on.

The evacuation route is made by them people supported by the team. The most important thing is how to inform and train the people so they can apply this scenario. The community organization is responsible for the entire implementation of the event.

3) Village mapping as well as the evacuation route. The map contains houses, local shelters, and hazard-prone areas, as well as the evacuation route
4) Learning and practicing rescue. Learning and practicing rescue was carried out periodically by doing landslide, evacuation, and rescue simulations.

The people's training as well as the evacuation drills were carried out seriously. First, the community organization made several public announcements and gave the people training.

Then, after everything was well prepared, an evacuation drill was carried out by involving all the related stakeholders.
5) The last step is doing an evaluation and conducting practice periodically. Based on this activity, people are expected to be aware of risk conditions, and then they could understand when a flood will happen and know what they should do.

Research questions

To find solutions to these challenges, more studies and efforts need to be done using sufficient data and appropriate methods. Some pertinent research questions regarding flood risk management and urban infrastructure sustainability in Central Java Province that such studies should address include:

1) What are the appropriate sustainability indicators for urban infrastructures in Central Java Province?
2) How does flood risk management impact urban infrastructure sustainability in Central Java Province?
3) What are the impacts of flood risk management on urban infrastructure sustainability in Central Java Province?
4) What efforts are needed to improve the hydrology, hydraulic, and other data availability for flood mitigation in Central Java Province?
5) How can comprehensive evaluation and assessment of drainage network performance be done in Central Java Province?
6) What are the appropriate actions to raise community awareness to preserve the environment and awareness of the impact of climate change in Central Java Province?
7) What are the appropriate approaches to persuade political leaders to address environmental, sustainability, and climate change issues in Central Java Province?

Conclusions

Central Java Province in Indonesia has experienced flooding over the past decades. These flood events have impacted infrastructural systems in the transportation, housing, electricity, and irrigation sectors. Repeated flood occurrences have damaged roads, disrupted train stations and airports, and caused loss of human lives. Flood events in Central Java Province can be triggered by coastal flooding in the northern coastal area, extreme rainfall, poor drainage networks, poor solid waste management, degraded river systems, and less green open space. As a part of urban systems, urban infrastructure sustainability should be achieved to ensure urban system sustainability. Flood risk can be a constraint to achieve this goal. Therefore, the government and all stakeholders should work together to achieve successful flood risk management.

There are further research opportunities regarding flood risk and urban infrastructure sustainability in Central Java Province to answer the research questions raised in this chapter.

References

ADCP, UNDP, 2006. *Integrated flood risk management in Asia: A primer 332.* http://redac.eng.usm.my/EAD/EAD511/floodprimer2005.pdf
Beritasatu, 2014. *Kerugian Banjir di Jateng Capai Rp 1 Triliun.* www.beritasatu.com/nasional/164939/ganjar-kerugian-banjir-di-jateng-capai-rp-1-triliun (accessed 7.2.2021).

Brakenridge, G. R., 2021. *Global active archive of large flood events*. Dartmouth Flood Observatory, University of Colorado. https://floodobservatory.colorado.edu/Archives/index.html (accessed 22.2.2021).

Central Bureau of Statistics Republic of Indonesia – Central Java Province, 2020. *Provinsi Jawa Tengah Dalam Angka 2020*. https://bappeda.jatengprov.go.id/wp-content/uploads/2020/07/Provinsi-Jawa-Tengah-Dalam-Angka-2020.pdf.

Chan, N. W., 2012. Impacts of disasters and disasters risk management in Malaysia: The case of floods. *ERIA Research Project Report*, 503–551, December.

Deputi Bidang Sarana dan Prasarana, 2008. *Kebijakan Penanggulangan Banjir di Indonesia*. https://adoc.tips/download/kebijakan-penanggulangan-banjir-di-indonesia.html.

Dinas Pusdataru, J. T. 2008. *Konsep Pengendalian Banjir Jawa Tengah*. http://pusdataru.jatengprov.go.id/iNEWS/konsep-pengendalian-banjir/.

Dinas Pusdataru Provinsi Jawa Tengah, 2008. *Peta Rawan Banjir Jawa Tengah*. http://pusdataru.jatengprov.go.id/iNEWS/peta-rawan-banjir-jawa-tengah/#:~:text=Daerah rawan banjir antara lain, seperti Karanganyar%2C Solo dan Sukoharjo (accessed 28.1.2021).

Ertan, S., Çelik, R. N., 2021. The assessment of urbanization effect and sustainable drainage solutions on flood hazard by GIS. *Sustainability (Switzerland)*, 13, 1–19. https://doi.org/10.3390/su13042293.

Hatmoko, J. U. D., Setiadji, B. H., Wibowo, M. A., 2017. Evaluasi Pengaruh Banjir, Beban Berlebih dan Mutu Konstruksi pada Kondisi Jalan. *Jurnal Transportasi*, 17, 89–98.

Huang, L., Wu, J., Yan, L., 2015. Defining and measuring urban sustainability: A review of indicators. *Landscape Ecology*, 30, 1175–1193. https://doi.org/10.1007/s10980-015-0208-2.

Human Resources Development Agency, 2015. *Modul Penyusunan Rencana Terpadu Infrastruktur Kawasan Perkotaan*. https://bpsdm.pu.go.id/center/pelatihan/uploads/edok/2018/05/8b7cb_Modul_2_-_Penyusunan_Rencana_Terpadu_Kawasamn_Perkotaan.pdf.

Isnugroho, 2010. *CRBOM small publications series No. 25 Java's water security*. https://simantu.pu.go.id/personal/img-post/superman/post/20181128142118__F__KMS_BOOK_20180727110302.pdf.

JICA, 2004. *Country report Indonesia: Natural disaster risk assessment and area business continuity plan formulation for industrial agglomerated areas in the ASEAN region*. Japan International Cooperation Agency OYO International Corp. Mitsubishi Research Institute, Inc. CTI Engineering International Co. Ltd., 70–74.

Kompas, 2021. *Dampak Cuaca Ekstrem, 6 Daerah di Jateng Terendam Banjir*. https://regional.kompas.com/read/2021/02/08/18412371/dampak-cuaca-ekstrem-6-daerah-di-jateng-terendam-banjir (accessed 2.7.2021).

Kontan, 2021. *Ganjar ungkap penyebab banjir di Jawa Tengah, apa katanya?* https://regional.kontan.co.id/news/ganjar-ungkap-penyebab-banjir-di-jawa-tengah-apa-katanya?page=all (accessed 6.3.2021).

Kundzewicz, Z. W., 1999. Flood protection – sustainability issues. *Hydrological Sciences Journal*, 44, 559–571. https://doi.org/10.1080/02626669909492252.

Marfai, M. A., King, L., Singh, L. P., Mardiatno, D., Sartohadi, J., Hadmoko, D. S., Dewi, A., 2008. Natural hazards in Central Java Province, Indonesia: An overview. *Environmental Geology*, 56, 335–351. https://doi.org/10.1007/s00254-007-1169-9

Ministry of Public Work and Housing, 2017. *Modul Kelembagaan dan Koordinasi Pengendalian Banjir Pelatihan Pengendalian Banjir*. Ministry of Public Work and Housing.

Nguyen, M. T. et al., 2021. Understanding and assessing flood risk in Vietnam: Current status, persisting gaps, and future directions. *Journal of Flood Risk Management*, 14(2), 1–24. https://doi.org/10.1111/jfr3.12689.

Okezone News, 2016. *Sungai Bengawan Solo Meluap, Soloraya Dikepung Banjir*. https://news.okezone.com/read/2016/11/29/512/1553818/sungai-bengawan-solo-meluap-soloraya-dikepung-banjir (accessed 6.1.2021).

Pos Jateng, 2018. *Pemerintah Bisa Dituntut Gegara Jalan Rusak Akibat Banjir*. www.posjateng.id/warta/pemerintah-bisa-dituntut-gegara-jalan-rusak-akibat-banjir-b1UB59xH (accessed 15.6.2021).

Priyanto, E. H., Nawiyanto, N., 2014. Banjir Bandang di Kodya Semarang tahun 1990. *Publika Budaya*, 3, 9–17.

Pudastaru Propinsi Jawa Tengah, 2008. *Pusdataru open data.* https://pusdataru.jatengprov.go.id/data-/OPENDATA=/index.html.

Pudyastuti, P. S., 2008. *Banjir dan Kemitraan Hulu Hilir.* Harian SoloPos.

Pudyastuti, P. S., Nugraha, N. A., 2018. Climate change risks to infrastructures: A general perspective. *AIP Conference Proceedings*, 1977. https://doi.org/10.1063/1.5043000.

Pudyastuti, P. S., Saputra, M. M., 2018. *The effect of green rooftop to reduce runoff (a case study at selected buildings in Surakarta city).* The 12 Th SEATUC Symposium, pp. 1–5.

Purwaningsih, T., Prajaningrum, C. S., Anugrahwati, M., 2018. Building model of flood cases in central java province using geographically weighted regression (GWR). *International Journal of Applied Business and Information Systems*, 2, 14–27.

Solopos, 2016a. *Jalan Rusak Solo: Nyaris 67% Kondisi Jalan Hancur.* www.solopos.com/jalan-rusak-solo-nyaris-67-kondisi-jalan-hancur-707908 (accessed 15.6.2021).

Solopos, 2016b. *Banjir Soloraya: Genangan Cenderung Merata dan Sulit Diprediksi.* www.solopos.com/banjir-soloraya-genangan-cenderung-merata-dan-sulit-diprediksi-758629 (accessed 1.6.2021).

Stanton-Geddes, Z., Vun, Y. J., 2019. Strengthening the disaster resilience of Indonesian cities. *Time to ACT: Realizing Indonesia's Urban Potential*, 161–171. https://doi.org/10.1596/978-1-4648-1389-4_spotlight1.

Suwiknyo, E., 2020. *143 Kejadian Banjir di Jateng Timbulkan Kerusakan.* https://semarang.bisnis.com/read/20200119/535/1191721/143-kejadian-banjir-di-jateng-timbulkan-kerusakan (accessed 20.6.2021).

Tariq, M. A. U. R., Farooq, R., van de Giesen, N., 2020. A critical review of flood risk management and the selection of suitable measures. *Applied Sciences (Switzerland)*, 10, 1–18. https://doi.org/10.3390/app10238752.

Wismana Putra, I. S., Hermawan, F., Dwi Hatmoko, J. U., 2020. Penilaian Kerusakan Dan Kerugian Infrastruktur Publik Akibat Dampak Bencana Banjir Di Kota Semarang. *Wahana Teknik Sipil: Jurnal Pengembangan Teknik Sipil*, 25, 86. https://doi.org/10.32497/wahanats.v25i2.2154.

Wisnu, R. S., Pamungkas, L., Kurniawan, A., 2017. Kajian Risiko Bencana Jawa Tengah 2016–2020. *Deputi Bidang Pencegahan dan Kesiapsiagaan Badan Nasional Penanggulangan Bencana*, 1, 1–63.

Wu, J., 2014. Urban ecology and sustainability: The state-of-the-science and future directions. *Landscape and Urban Planning*, 125, 209–221.

19

A review of flood management in South Asia

Approaches, challenges, and opportunities

Md. Arif Chowdhury, Shahpara Nawaz, Md. Nazmul Hossen, Syed Labib Ul Islam, Sayeda Umme Habiba, Mir Enamul Karim, Md. Lokman Hossain and Md. Humayain Kabir

Introduction

South Asia is one of the most flood-prone regions of the world (Ziegler *et al.*, 2020). The region consists of mighty river systems (Mirza, 2011). There are eight countries with more than one-fifth of the world's population (Sivakumar and Stefanski, 2010). This region is one of the most disaster prone, with climate change, high density of population, degradation of natural resources, and unplanned urbanization and industrialization exacerbating the situation (Heinen, 2005; Sivakumar and Stefanski, 2010). The socioeconomically vulnerable population in South Asia has to cope with disasters that are getting worse every year due to climate extremes (Matheswaran *et al.*, 2019). Heavy rainfall, cyclonic disturbances, and increased glacial melting in the Himalayas are increasing the risk of flooding in this region (Kale, 2002; Dorji, 2021). The Ganges-Brahmaputra-Meghna (GBM) basin is the most significant area for flood risk management in Bangladesh and India (Nandargi *et al.*, 2010). The basins of the rivers Bagmati and Indus are also prone to severe flooding in Nepal, India, and Pakistan (Nandargi *et al.*, 2010). Heavy rainfall during the long monsoon and glacial lake outburst floods cause extensive flooding in Bhutan: there are at least 17 glacial lakes prone to severe flooding (NCHM, 2019).

The impacts of floods in South Asia are multi-dimensional. The loss of lives and economic damage are the direct and comprehensible impacts. However, there are long-term impacts that create underlying issues in society. Repeated recovery and rebuilding efforts after each flooding event cause economic stagnation and halt socio-economic development of already resource-constrained developing countries. Vulnerable communities live with the risks and the full knowledge of impending disasters, as they cannot find any alternatives (Shrestha and Takara, 2008). Being mainly agricultural countries, floods cause massive crop damage in South Asia. Due to climate change, these disruptions to food production are happening more frequently and more severely than ever before, destabilizing food security (Douglas, 2009). Floods also drive many social problems in this region, as the recurrent flooding forces the affected people into internal displacement and conflicts (Coelho, 2012). These issues exacerbate poverty and create a cycle that is very difficult to break.

On the other hand, being historically flood prone, the South Asian countries have long-standing experience in living with floods. Different measures have been tried in this region with varying degrees of success. Both structural and non-structural measures have been implemented for flood management in India and Bangladesh (Bhattacharya, 2019). Particularly in Bangladesh, an effective governance structure and early warning systems have been developed which have reduced the loss of lives due to floods in recent years (Abbas et al., 2016). The non-structural measures and early warning systems have generally been very successful at preventing losses. This is of particular importance to the resource-stricken countries, as the structural measures often require a large initial investment and continuous expenditure on maintenance.

Flood management strategies formulated through scientific research and policies informed by science perform significantly well in dealing with crises (Abbas et al., 2016). Community participation and local knowledge are other areas of crucial importance in flood management, which have been somewhat overlooked in the past (Kaushik and Sharma, 2012; Smith et al., 2017; Uddin, 2018; Nur et al., 2021). Bridging the knowledge gaps in these areas can play a huge role in achieving effective management of floods in South Asia. This chapter attempts to cover the issues and recent scientific findings related to natural flood management with a focus on the experiences of developing countries of South Asia. Therefore, this study examines three countries, Bangladesh, India, and Nepal, in order to generate insights for understanding the different aspects of flood management in South Asia. The three countries were selected purposively to understand the natural flood management issues, including the policies, successes, and failures, based on extensive literature review from online resources.

Overview: floods in South Asia

Historical floods

The South Asian region is drained by some of the mightiest rivers in the world, which causes frequent floods (Kale, 2002). The number of hydro-meteorological disasters in this region has increased significantly. In the previous decades, over 90% of those killed by natural disasters lost their lives in hydro-meteorological events such as droughts and floods. Floods affected the majority of disaster-affected people between 1991 and 2000 (WDR, 2001). Almost 140,000 Bangladeshis were killed by cyclones in the 1990s, while 2.5 million people were evacuated (WDR, 2002). The major floods in 1987, 1988, and 1998 affected 40%, 60%, and 68% of the total area of Bangladesh as well as killing 1800, 2379, and 918 people, respectively.

According to FFWC (2019), the Himalayan mountain range is one of the highest precipitation zones in the world. During the monsoon, a huge volume of water flows through the rivers. This often surpasses the carrying capacity of the rivers due to natural and anthropogenic reasons, resulting in floods. In India, nearly 8754 million people have been affected by floods in the past 50 years (CRED, 2021). However, the frequency and intensity of flooding in this region have been predicted to increase even more (IPCC, 2012; Turner and Annamalai, 2012).

Impacts of floods

Flood events cause enormous damage to human life, with deaths and injuries (Du et al., 2010). With time, the number of direct deaths is declining as preventative measures are being taken, but the economic loss is increasing due to the changes in land use. Other impacts of floods include displacement of the affected people and widespread damage of crops, infrastructure, and property (IPCC, 2007; Doocy et al., 2013).

Also, agricultural lands are becoming more vulnerable to floods, with sea-level rise causing salinity intrusion and internal displacement among people living around the Bay of Bengal (Revi, 2008; Douglas, 2009). Food insecurity, unemployment crisis, and famines are other common concerns after floods (Alamgir, 1980). Future floods are expected to increase the dimensions of such crises (Douglas, 2009).

Challenges to natural flood management in South Asia

The major challenge for flood management in South Asia is that, with the river basins crossing international borders, it is an issue that involves multiple countries. The past approaches of using structural measures such as embankments and reservoirs are insufficient and require collaboration between national and international stakeholders, which is very difficult to achieve (Mirza et al., 2003). Moreover, major institutional and legal system changes are required for the success of structural measures. Misinformation and lack of access to facts and knowledge are other constraints for South Asian countries (Dixit, 2003). However, proper formulation and implementation of policies can be achieved with evidence-focused research and sufficient monitoring (Abbas et al., 2016).

Governance/policy-level aspects regarding natural flood management

After starting with a strategy of flood control and irrigation for agriculture in the 1960s while overlooking ecological and societal needs, the national policies of Bangladesh have gradually shifted to an integrated management focus since the 1980s (Gain et al., 2017). The success of structural and non-structural measures depends on local conditions and robust planning (Shah et al., 2018). Management strategies have been slowly shifting from a reactive to a proactive approach (Tarrant and Sayers, 2012). In coastal Bangladesh, polders were utilized for preventing coastal flooding. After being initially successful, the polders are now causing drainage congestion by hampering natural drainage. A new approach, "tidal river management", is being tried to solve the problems created by polders (BWDB, 2013).

Furthermore, standing orders on disaster (SOD) by the Disaster Management Bureau (DMB) have been effective in reducing the damages caused by natural disasters in recent decades (Chowdhury, 2003). There is a National Disaster Management Council (NDMC) which comprises the prime minister and the ministers from different ministries. Additionally, local government-level committees work for disaster mitigation and post-disaster recovery with the participation of the local community as guided by the SOD (MoDMR, 2019). Also, the National Water Policy (NWPo) guides activities in the water sector. A National Water Management Plan (NWMP) considers long-term needs cross-cutting all the sectors and water-induced disasters. There is also a Comprehensive Disaster Management Plan (CDMP) for natural disasters.

In India, the National Commission on Floods made 207 recommendations in 1980 for flood management, following which an Integrated Water Resources Development and Management Plan was formulated in 1999. The National Water Policy of India was formulated in 1987 and revised in 2002 with the recommendation for an integrated and multi-disciplinary approach and prioritizing non-structural measures.

In Nepal, disaster management, through the Natural Disaster Relief/Calamities Act 1982, is relief focused. A new act was drafted in 2006, but the ratification was not deemed a priority by the Constituent Assembly (Jones et al., 2016). As such, most stakeholders are working according to the National Strategy for Disaster Risk Management, which was prepared through a multi-stakeholder process in 2009; however, there are some conflicts among different authorities regarding jurisdictions due to the absence of a clear act (Jones et al., 2016; Vij et al., 2020).

Flood impact mitigation approaches

Bangladesh

Flood management strategies (i.e. structural, and non-structural) in Bangladesh are aimed at minimizing the adverse socioeconomic impacts while creating an environment conducive to economic growth. Structural measures directly intervene in the physical process of flooding. Flood-protected tube wells, drainage infrastructure, and dams are all examples of such structural measures (Nur et al., 2021) (Table 19.1). Non-structural measures are flexible; less intrusive; and incorporate nature-based engineering elements, social awareness, forecasting, evacuation, shelter management, flood insurance, floodplain zoning, changes in cropping patterns, and so on (Rahman et al., 2014; Baten et al., 2018). Non-structural flood mitigation techniques in Bangladesh have enabled rapid adaptation to dynamic river basins, socioeconomic conditions, and climatic scenarios and are consistent with the purpose of ecologically sound and sustainable development (Shah et al., 2018).

The Bangladesh Water Development Board (BWDB) has built numerous dams, embankments, and canals to decrease flood damage and develop irrigation. Additionally, for early flood forecasting and warning dissemination, a network of hydrological stations has been linked to the Flood Forecasting & Warning Centre (FFWC) to develop and transmit flood forecast and warning information (Baten et al., 2018).

India

Flood protection initiatives in India invested heavily in structural measures to improve the drainage conditions in river basins (Mohanty et al., 2020) (Table 19.2). As non-structural measures, the government of India and the local authority established 226 flood predicting stations providing water level and discharge estimates to different local and reservoir operating organizations for flood prevention (Mohanty et al., 2020).

Nepal

The government of Nepal, with the assistance of donor agencies, is working towards increasing the network of hydrological and meteorological stations and designing experimental forecasting systems (Smith et al., 2017). Additionally, various scientific and participatory flood mitigation measures are being adopted (Table 19.3). Dhital and Tang (2015) found soil bioengineering, using vegetative check dams and wire nets, rapidly stabilized stream banks. Pandeya et al. (2021) recommended the application of citizen science (CS) for flood risk mitigation and monitoring system with local low-cost sensors. The use of community-based early warning systems (CBEWSs) is also prioritized in Nepal (Smith et al., 2017).

Table 19.1 Different issues related to natural flood management in Bangladesh

Study area	Flood management approach/mitigation measures		Success	Challenges	Opportunities	Reference
	Structural	Non-structural				
Kurigram	– Embankment and dam – Raising the plinth level – Sand-filled high-strength geotextile bags – Flood-protected tube wells – Drainage improvements	– Indigenous housing patterns – Locals cultivated grasses and plants along the stream bank – Prepared dried fish and grown vegetables – Safeguarding of household valuables and stockpiling food and gasoline – Increase floor of tube wells and houses – Homestead gardening – Utilization of banana trees to build rafts for use during floods – Application of indigenous tactics to safeguard crops	– Homestead gardening, planting trees, and using nets to protect fish – Crops cultivating the varieties that are appropriate for the time period of floods – Cultivation of grasses and plants along the stream bank		– Homes should be built with durable materials, away from flood-prone areas considering the height of the flood level – Flood shelter construction with proper design – Consider building dams to store extra water for agriculture – The government and key stakeholders should educate communities and local governments about flood danger due to climate change – Ministry of Agriculture (MoA) and cooperatives should encourage communities to promote upland farming and tree plantation – Raise tube wells and toilets – Develop and disseminate early warning signal distribution system – Improve the drainage system	(Nur et al., 2021)
Coastal areas	– Embankments, dikes, polders, diversion channels, and dams	– Application of tidal river management (TRM) – Land use changes upstream – Early warnings	– TRM decreased flood vulnerability – Non-structural flood mitigation techniques facilitate rapid adaptation – Improved socioeconomic conditions	– Operational disruptions due to conflict and management cost – Population expansion – Unplanned land use changes – Sedimentation in river channels	– New design of coastal polders regarding sea level rise – TRM	(Shah et al., 2018; Adnan et al., 2020)

Region	Adaptation measures	Benefits	Limitations	Recommendations	Reference
Haors basin in Sylhet	– Earthen submergible embankments – Dredging of rivers and canals	– Provision of free seeds and fertilizer after a flood – Training for farmers on agricultural technology – Developing an early warning flood forecasting system – Floating beds for vegetable cultivation – Introduce high-value crops – Credit agricultural tools – Cultivation of homestead vegetables and spices	– Cultivation of vegetables, spices – Development of flood early warning system	– Comprehensive action plan with effective implementation strategies – Reformation of current Department of Agriculture Extension (DAE) strategies – Collaboration among universities, government, and foreign sponsors	(Kamruzzaman and Shaw, 2018)
Coastal areas	– Approximately 800 water development projects	– Afforestation on the embankment and surroundings – Educate water management groups and associations – Ensure participation of women in earthwork, monitoring, and meetings – Water management improvement project (WMIP) promotes an integrated process to water resource management	– Increased food grain productivity	– Slowly transmitted downwards, and full adoption of top-down method at all levels – Single discipline–focused engineering education system	– Modifications are necessary to completely implement integrated water resources management (IWRM) concept at the project level – Capacity building of stakeholders is required – Proper enforcement of law and policy	(Gain et al., 2017)
Northeastern part	– Raising of national and regional roadways and railways Dredging	– Flood forecasting and warning system Construction of school buildings to use as flood shelters	– Forecasting, preparation, and post-disaster relief activities have minimized flood impacts	– Siltation and increased bed level due to steep embankments – Obstruction of drainage – Slowed tidal prism movement and hampered gravity changes in polders	– Flood zoning demarcation – Coordination among different stakeholders – Avoid costly expenditure in management – Prepare appropriate land development guidelines	(Rahman et al., 2014)

Table 19.2 Different issues related to natural flood management in India

Study area	Flood management approach/mitigation measures		Success	Challenges	Opportunities	Reference
	Structural	Non-structural				
Brahmani-Baitarani river basin in eastern India	– Embankments	– Flood forecasting and warning – Community-based preparedness – Evacuation and relief	More than 90% of the total flood risk would be reduced	– Inadequate or non-existent embankments – Insufficient government funds	–	(Marchand et al., 2021)
India	– Embankments – Detention reservoirs – Improvement of the drainage conditions in river basins	– Flood predicting stations	–	– Unproven effectiveness of structural interventions – Expansion of flood-prone locations – More emphasis on the structural approach – Undervalued flood preparedness – Less extensive existing forecasting network	– Committees and task groups created by the central government to understand effective flood management plans – Repeatedly address flood forecasting importance as a non-structural approach – Continuous development of pertinent plans	(Mohanty et al., 2020)
Cities of India	– Elimination of encroachments in flood plains – Construction of a flood wall – Airport runway detention ponds – Rain gardens	– Establishment of automatic weather stations – Flood risk mapping – Assessment of flood risk and resilience techniques	–	– Inadequate flood resilience measures in urban areas – The drainage system is under the control of the city authority – Changes in institutional approaches – Population growth – Sediments and garbage in drains disturbs the flow – Absence of relevant data for drainage design	– More focus on provision of water, sewerage, stormwater drains – Development of drainage systems for rainfall	(Gupta, 2020)
Chennai, India	– Strom drainage network	–	–	– Incapable storm drain network	– Inlet width may extend to reduce flood depth	(Andimuthu et al., 2019)
Eastern Indian state of West Bengal	–	– Forest cover	Forest cover acts as a buffer against occurrences and impacts of flood	–	– Combined application of ecosystem-based adaption and standard hard structure flood protection mechanisms	(Bhattacharjee and Behera, 2018)

Table 19.3 Different issues related to natural flood management in Nepal

Study area	Flood management approach/ mitigation measures		Success	Challenges	Opportunities	Reference
	Structural	Non-structural				
Karnali River Basin Nepal	–	– Low-cost sensors for hydrological data collection monitoring systems	– Produce river-level data – Community-based participatory monitoring network	– Inadequate hydroclimatic data – Limited network of hydrometeorological stations – Less capacity of national and local institutions – Insufficient funds to establish proper flood forecasting system	Collaboration of people from different backgrounds may ensure the effective operation of the system	(Pandeya et al., 2021)
Siwaliks, Nepal	–	– Soil bioengineering techniques like vegetative check dams	– Flow channels of the river were narrowed to reduce soil erosion – Bamboo plantation and vegetation check dams are useful for flood control	–	–	(Dhital and Tang, 2015)
Karnali River Basin in western Nepal	–	– Deployment of community-based early warning systems	– Increased integration of CBEWSs with national government suggested EWSs including community and governed components	– Low-lying inland delta with large, shallow, naturally movable channels – Extensive sediment movement – Difficult to predict geographical situation and uncertainty	–	(Smith et al., 2017)

Success and failure of flood management measures

Bangladesh

Bangladesh has developed a reasonable flood protection and disaster management plan based on previous disasters and has legislated climate change adaptation on a national level (Abbas *et al.*, 2016). Stronger institutional commitment and collaboration, as well as an increased trust among residents in the efficacy of the recommended efforts, have enabled this achievement. Controlled flooding regulated by water management groups has demonstrated positive results, allowing for restricted flooding in different units defined by hydrology and land topography. After the 1987–1988 floods, the Flood Action Plan (FAP) shifted the infrastructure development policy such that strategically important facilities were constructed above 100-year flood levels, and all the schools were built with temporary shelter arrangements in flood-prone locations. The community-based approach to flood risk management has demonstrated encouraging results. Forecasting, preparation, and post-disaster relief activities have minimized flood impacts in the country (Rahman *et al.*, 2014). Furthermore, improving the capacity of rivers has ensured easy conveyance to the Bay of Bengal while improving land and water management (Abbas *et al.*, 2016). On the other hand, increasing development in certain flood-prone cities has exacerbated susceptibility to floods, even though community-based hazard mitigation programs were incorporated in raising awareness and enhancing the resilience of threatened people. In comparison, the ineffectiveness of legislation in preventing deforestation, hill destruction for development, and overfishing have resulted in certain problematic issues with the ecosystem and flow regimes. Overreliance on structural flood protection techniques such as embankments and polders, along with inadequate maintenance, has resulted in internal drainage problems, waterlogging, and siltation (Abbas *et al.*, 2016).

To protect crops and homesteads, people are using indigenous tactics, such as cultivating appropriate varieties of crops to adapt to the physiographic and soil conditions of an area in rural Bangladesh (Nur *et al.*, 2021). Kamruzzaman and Shaw (2018) reviewed the association between floods and sustainable agriculture in the Haor basin of Bangladesh, which encounters distinctive impacts of floods. They noted that the government of Bangladesh (GoB) focused on structural mitigation measures, while non-governmental organizations (NGOs) focused on non-structural measures such as providing free seeds and fertilizer to the affected farmers following a flood, training farmers in the usage of modern agricultural technology, developing and testing an early warning system for flood forecasting, motivating the local people to adopt floating beds for vegetable cultivation, introducing high-value crops with a short maturation period, crediting agricultural tools and inputs, adapting plants for wetland harvesting, homestead vegetables, and spice cultivation and preserving crop seeds.

Adnan *et al.* (2020) assessed tidal river management (TRM) for flood alleviation and found that flood vulnerability was decreased in 35% of the chosen sites. Gain *et al.* (2017) found that projects under the umbrella of integrated water resources management (IWRM) helped to increase crop production by around 10 million tons per year and contributed to self-sufficiency in rice.

Initiatives of the BWDB, such as the Coastal Embankment Rehabilitation Project (CERP), Coastal Embankment Improvement Project (CEIP), Blue Gold project, and Water Management Improvement Project (WMIP), combined participatory water management with economic and market growth successfully. Local communities were included in all stages of the participatory scheme cycle management (PSM), from identification through monitoring and assessment (Gain *et al.*, 2017).

India

The importance of flood forecasting as a non-structural solution has been repeatedly acknowledged, and its extension is being developed continually in India (Abbas *et al.*, 2016). Abbas *et al.* (2016) discussed the challenges of the flood hazard zoning process and flood preparedness at the state and national level of India. The government of India has been generating and maintaining flood hazard maps at the state and district levels to aid in flood mitigation efforts since 1997. However, there has been a failure in developing floodplain zoning for the country despite being mandatory according to the 2012 National Water Policy. A few states, like Manipur and Rajasthan, have previously enacted legislation establishing floodplain zoning, but these measures lack enforcement mechanisms. In comparison, Uttar Pradesh, Bihar, and West Bengal are still in the enactment phase, while the remaining states have not even begun this process (Abbas *et al.*, 2016).

Inadequacy of the drainage network is a problem for India, as major cities like Chennai and Mumbai are incapable of withstanding intense rainfall for even a two-year return period. This is mainly due to the accumulation of debris and encroachment in flood plain areas, along with inadequate space for the natural flow of water (Gupta, 2020). To mitigate this situation, extensive monitoring systems are being developed. Flood risk mapping has been completed for some major cities as well (Gupta, 2020).

Marchand *et al.* (2021) noted the construction of embankments on the Brahmani–Baitarani river basin, as 90% of the total flood risk could be mitigated in this way. Following this structural strategy, embankments protecting rural regions are being planned with a 25-year return period and those protecting urban or industrial areas with a 100-year return period, with mixed success.

Nepal

Dhital and Tang (2015) found the application of vegetation to flood-damaged bare ground effective in Nepal. Several plant species have been found to grow at about the same rate on flood-affected and unaffected land; therefore, soil bioengineering techniques (SBTs) such as active vegetative check dams are considered one of the finest options for reducing flood hazards in Nepal. When paired with vegetative check dams, *I. fistulosa* plants were highly effective in flood control. Bamboo mixtures used to construct check dams and bamboo planting behind check dams were both exceedingly beneficial for streambank stability and flood risk reduction. Vegetation restoration has also been found efficient for the protection of flood-affected regions (Dhital and Tang, 2015).

Pandeya *et al.* (2021) highlight the fact that the robust hydrological and meteorological data needed for accurate forecast systems are particularly difficult to obtain in areas around the Himalayas. Community-driven methods and low-cost sensing technology are being applied in an attempt to improve this situation in data-scarce locations in Nepal (Pandeya *et al.*, 2021).

The first community-based early warning system (CBEWS) in Nepal was piloted in 2002 and was further developed and expanded to cover eight river basins across Nepal over the next decade (Smith *et al.*, 2017). The CBEWSs were progressively integrated with the national EWSs while retaining their community-driven characteristics. CBEWSs provide about two to three hours of preparation time before the occurrence of a flood and functioned efficiently during the 2014 floods in West Nepal. However, the CBEWS failed in the Babai River basin, resulting in loss of lives, due to the washing away of the water level measuring station meant to generate alarms. This is why the communities downstream could not receive information about when water levels exceeded warning and danger limits (Smith *et al.*, 2017).

Challenges

Bangladesh

There are many complicated issues regarding structural measures for flood management in Bangladesh. In addition to being ineffective during catastrophic floods (Shah *et al.*, 2018), the steep embankments along riverbanks induce siltation and raise riverbed levels, limiting the conveyance capacity of the river. Polders prevent salinity intrusion to some degree but obstruct the natural drainage and tidal prism movement and induce sedimentations in tidal rivers. To overcome these problems, proper flood zoning is of crucial importance (Rahman *et al.*, 2014). Social unrest with conflicts between the authorities and local people have interrupted the implementation of TRM with significantly less accretion (Shah *et al.*, 2018; Adnan *et al.*, 2020).

One of the most significant challenges was the use of a top-down method for implementing IWRM for flood management in Bangladesh. As a result, the concepts have slowly been transmitted to stakeholders at local levels, and complete dissemination is expected to take a longer time. In addition, stakeholder capacity building is required, as well as proper monitoring and evaluation of IWRM initiatives. Transparency and accountability need to be ensured in all relevant projects (Gain *et al.*, 2017).

Tropical cyclones annually form in the Bay of Bengal after and before the monsoon season. Bangladesh has developed disaster preparedness drawing from the experiences of catastrophic disasters, achieving increased adaptive capacity and reducing flood mortality (Abbas *et al.*, 2016). The disaster-affected people use indigenous knowledge to cope with extreme weather events (EWEs) to hold out until relief assistance arrives. However, several policy gaps persist for effective flood management in Bangladesh. Community involvement is not embedded in the decision-making process coming from a top-down approach. Additionally, indigenous knowledge is not taken into account with seriousness when formulating management plans. Also, unplanned urbanization is increasing the vulnerability of urban flooding, and poor maintenance is causing drainage problems.

Major focused parts and challenges to natural flood management in South Asia from the Bangladesh, India, and Nepal perspectives are presented in Table 19.4.

India

Different types of floods afflict different areas of India due to the varied topography. The arid areas of Rajasthan are prone to flash floods, while Himachal Pradesh is threatened by glacial lake outbursts. Pluvial flooding is the major challenge in the Ganges basin, while uncontrolled sand mining, drainage congestion, and river erosion drive the flood risk. The coastal states face tidal flooding and cyclones which are becoming more severe due to climate change. Numerous dams throughout the upstream rivers also often cause flash floods (Mohapatra and Singh, 2003). Monsoon rainfall often triggers landslides and mudslides which cause a high number of deaths as well.

India prioritized structural measures for flood mitigation while undervaluing preparedness and non-structural strategies, resulting in an expansion of flood-prone regions and ineffective drainage networks (Andimuthu *et al.*, 2019). Inadequate or non-existent embankments in low-lying rural areas exacerbate flood severity in India. There is also the financial constraint, as the country lacks funds to repair and maintain an extensive embankments network – hence the country is aiming to improve and implement more non-structural measures (Marchand *et al.*, 2021).

Table 19.4 Major focused parts and challenges to natural flood management in South Asia from Bangladesh, India, and Nepal perspectives

Countries	Main focused part	Challenges	Reference
Bangladesh	— Planning of post-disaster rehabilitation and relief — Effective evacuation and shelter management of people during coastal flooding — Significant changes in development policy for various infrastructures to mitigate flood risk	— Inadequate relief, rehabilitation, and rescue operations considering the needs of communities — Lack of community participation in practical implementation and execution phase — Lack of coordination among departments, budgetary scrapes, and rigorous institutional frameworks — Absence of coordination to generate disaster-related indicator data	(Saleem Ashraf *et al.*, 2017; Sadique and Kamruzzaman, 2021)
India	— Focus of disaster reduction programs on safety and protection of vulnerable communities and strengthening coordination and partnerships with local communities to enhance flood risk mitigation — State government in association with district governments executes rescue, disaster relief, and rehabilitation functions with the support of the national government	— Poor coordination with local communities — Failure in making floodplain zoning — Poor governance behind the sluggish implementation of integrated water resource management	(Mohanty *et al.*, 2020)
Nepal	— Successful implementation of disaster management policy and acts, disaster risk reduction program — Promote disaster risk reduction–related education program — Improvement in institutional structures and their collaboration	— Lack of data and limited understanding of upstream-downstream linkages of the river basin in flood management	(Panday, 2012; Saleem Ashraf *et al.*, 2017)

Lack of coordination between the Central Water Commission and local administration is a major concern in India (Mohanty *et al.*, 2020). Lack of transparency among the local authorities is highly detrimental to the attempt of risk mitigation. Moreover, disaster preparedness is a state obligation in India which creates an imbalance in national preparedness (Abbas *et al.*, 2016). Urban flood management is another major challenge for India, as the drainage systems in most of the major cities are inadequate and not properly maintained, along with encroachment and ignorance of building codes (Weinstein et al., 2019). The India Meteorological Department does not have intensity-duration-frequency (IDF) curves for India's largest cities – which also hampers the designing of drainage networks to address the issue of urban flooding (Gupta, 2020).

Nepal

Most river basins along the Himalayan foothills become unpredictable during the monsoon and are highly vulnerable to flooding (Dewan, 2015). In addition, the potential risks of glacial lake outburst floods have significantly increased over the last few decades. The government of Nepal has taken several steps to implement effective disaster risk reduction measures, but the major challenge for Nepal lies in the inadequacy of data and lack of preparedness. Efficient early warning systems are essential for flood management in Nepal, which cannot be addressed reliably while having a lack of data and resources (Pandeya et al., 2021). Implementation of effective CBEWS also faces a challenge, as it cannot capture all the scenarios and time-varying effects of sedimentation and bedload transport (Smith et al., 2017). In such a setting, reliable prediction is extremely difficult and fraught with uncertainty. Lack of coordination among various wings of the government and insufficient financial support also hinder the preparedness of Nepal (Devkota et al., 2014).

Policy-level gaps or barriers to implementing sustainable natural flood management

Bangladesh

There is a need for multi-sector, multi-agency, participatory, and collaborative approaches for the complex water management issues – for which IWRM can be a practical pathway. There do not appear to be any major barriers to the adoption of IWRM in Bangladesh (Gain et al., 2017). However, improved coordination is needed among different stakeholders (Rahman et al., 2014). The local government needs to take proper management initiatives with a long-term vision and community involvement. Awareness building is highly important in this regard. Livelihood diversification and opportunities should be enhanced for people in disaster-prone areas (Nur et al., 2021).

India

The major barrier for India lies in harmonizing the state-level initiatives across the country. Decentralized disaster governance is a major challenge with the balkanization of institutions and governance mechanisms. There are also a lack of coordination between different agencies and disagreements over strategies (Parthasarathy, 2016). State-level plans are customized to the needs of each state. There is a need for enforcing and monitoring current policies uniformly. Stakeholder participation and cross-state coordination need to be included in policy planning and execution. Oversight responsibilities need to be delegated properly to ensure coordination at the state and local district levels (Bezboruah et al., 2021).

Nepal

Institutions in Nepal need to ensure the implementation of feasible strategies for flood risk reduction. Institutional and governance issues need to be addressed to assign clear roles and responsibilities to stakeholders for disaster management (Delalay et al., 2018). Political leadership and legal enforcement of the laws related to disaster mitigation are essential in Nepal, along with good governance (Panday, 2012; Jones et al., 2016). Transparency and accountability need to be ensured for national resources and foreign aid (Panday, 2012). Gender inclusiveness also needs to be ensured while strengthening governance (UNISDR, 2015). Policies and projects need to be designed for the benefit of the Nepali people and the local contexts (Shakya, 2012). The lack of coordination between different governmental agencies needs to be addressed for effective disaster mitigation in Nepal as well (Vij et al., 2020).

Discussion

South Asia is one of the most susceptible regions to flood in the world, and floods occur very frequently due to high rainfall during the monsoon season (Mirza, 2011). To combat the impacts of floods in this region, different countries are applying various measures, and there are significant challenges to attaining sustainable natural flood management. This study assessed the natural flood management of Bangladesh, India, and Nepal in order to understand the requirements that need to be implemented in the future to reduce the impacts of floods on life and resources.

Both structural and non-structural approaches have been applied for flood mitigation in Bangladesh, India, and Nepal. Construction of embankments, raising plinth levels, protection of tube wells from floodwater, application of indigenous or local knowledge, homestead gardening, and so on are commonly practiced in Bangladesh (Shah et al., 2018; Nur et al., 2021). TRM is also being applied in Bangladesh, which can prove a comprehensive strategy for managing part of the GBM Basin (Gain et al., 2017). The GoB has taken 800 water development projects, and IWRM, as well as participatory approaches, are being integrated with the projects (Gain et al., 2017).

In India, embankments are mainly being used as structural flood control measures. Flood forecasting, establishing weather stations, and flood risk mapping are being used as non-structural measures in both India and Bangladesh (Rahman et al., 2014; Mohanty et al., 2020; Marchand et al., 2021). The application of non-structural measures is more common in Nepal, where low-cost sensors are being used for collecting hydrological data. Vegetation check dams and community-based early warning systems are being widely used to improve flood resilience (Dhital and Tang, 2015; Smith et al., 2017; Pandeya et al., 2021).

Flood hazard mapping and zoning along with homestead gardening, cultivation, and TRM are effective in reducing flood susceptibility. Although an insufficient drainage network is considered a problem in some cases for India, initiatives are being taken to improve the situation. Restoration of vegetation is effective in minimizing the flood risk in Nepal and Bangladesh. Stromberg (2001) stated that the restoration of vegetation should be conducted with community participation to ensure efficient management.

Lack of community involvement, insufficient data, and poor governance have been identified as the major challenges to flood management. More research is needed to address the issues of flood risk management, and the management practices should be informed by the latest research findings (Messner and Meyer, 2006).

Although Bangladesh, India, and Nepal are working on policy formulation, effective implementation is needed to ensure sustainable flood management. Capacity-building programs should be created, and sufficiently skilled professionals are needed at the administration and local levels. Lack of coordination among different stakeholders needs to be addressed rapidly (Hossen et al., 2019). Besides, gender inclusiveness and stakeholder engagement should be ensured (Hasan et al., 2019).

Summary/conclusion

Every year South Asia suffers from the impacts of devastating flood events, while socio-economic conditions deteriorate. Bangladesh, India, and Nepal are working to ensure effective natural flood management. Structural (e.g. embankments) and non-structural (e.g. restoration of vegetation) measures are being applied to address flood management. Despite applying different measures, several challenges, like the lack of participation, funding, and skilled people, exist. Effective governance is needed with the formulation of frameworks and implementation of projects. It is important to understand the patterns of floods and socio-economic conditions that are changing over time. Regional cooperation with neighbouring countries can also help address the basin-wide management of floods and the issue of robust data availability. Innovative solutions need to be developed with a concerted effort among the government and non-government sectors and local communities to develop effective flood management strategies in South Asia.

References

Abbas, A. et al. (2016) 'An overview of flood mitigation strategy and research support in South Asia: Implications for sustainable flood risk management', *International Journal of Sustainable Development & World Ecology*, 23(1), pp. 98–111. doi: 10.1080/13504509.2015.1111954.

Adnan, M. S. G. et al. (2020) 'The potential of tidal river management for flood alleviation in south western Bangladesh', *Science of The Total Environment*, 731, p. 138747.

Alamgir, M. (1980) *Famine in South Asia: Political economy of mass starvation.* Oelgeschlager, Gunn & Hain, Publishers, Inc.

Andimuthu, R. et al. (2019) 'Performance of urban storm drainage network under changing climate scenarios: Flood mitigation in Indian coastal city', *Scientific Reports*, 9(1), pp. 1–10.

Baten, A., González, P. A. and Delgado, R. C. (2018) 'Natural disasters and management systems of Bangladesh from 1972 to 2017: Special focus on flood', *OmniScience: A Multi-Disciplinary Journal*, 8(3), pp. 35–47.

Bezboruah, K., Sattler, M. and Bhatt, A. (2021) 'Flooded cities: A comparative analysis of flood management policies in Indian states', *International Journal of Water Governance*, 8.

Bhattacharjee, K. and Behera, B. (2018) 'Does forest cover help prevent flood damage? Empirical evidence from India', *Global Environmental Change*, 53, pp. 78–89.

Bhattacharya, A. K. (2019) 'An analysis of flood control in Eastern South Asia', in *Wastewater reuse and watershed management.* Apple Academic Press, pp. 201–216.

BWDB (2013) *Feasibility study of Coastal Embankment Improvement Project, Phase-I.* BWDB.

Chowdhury, J. R. (2003) Technical Paper presented in the 47th Annual Convention of the Institution of Engineers Bangladesh (IEB), 5–7 January, Chittagong, Bangladesh.

Coelho, S. (2012) 'Assam and the Brahmaputra: Recurrent flooding and internal displacement'. Available at: http://labos.ulg.ac.be/hugo/wp-content/uploads/sites/38/2017/11/The-State-of-Environmental-Migration-2013-63-73.pdf.

CRED (2021) *EM-DAT: International disaster database.* Centre for Research on the Epidemiology of Disasters, Universite Catholique de Louvain.

Delalay, M. et al. (2018) 'Towards improved flood disaster governance in Nepal: A case study in Sindhupalchok district', *International Journal of Disaster Risk Reduction*, 31, pp. 354–366.

Devkota, R. P., Cockfield, G. and Maraseni, T. N. (2014) 'Perceived community-based flood adaptation strategies under climate change in Nepal', *International Journal of Global Warming*, 6(1), pp. 113–124.

Dewan, T. H. (2015) 'Societal impacts and vulnerability to floods in Bangladesh and Nepal', *Weather and Climate Extremes*, 7, pp. 36–42. doi: 10.1016/j.wace.2014.11.001.

Dhital, Y. P. and Tang, Q. (2015) 'Soil bioengineering application for flood hazard minimization in the foothills of Siwaliks, Nepal', *Ecological Engineering*, 74, pp. 458–462.

Dixit, A. (2003) 'Floods and vulnerability: Need to rethink flood management', *Flood Problem and Management in South Asia*, pp. 155–179.

Doocy, S. et al. (2013) 'The human impact of floods: A historical review of events 1980–2009 and systematic literature review', *PLoS Currents*, 5.

Dorji, K. (2021) 'Adaptive human settlement planning for glacier lake outburst floods in Bhutan', *Social Science Asia*, 7(4), pp. 9–16.

Douglas, I. (2009) 'Climate change, flooding and food security in south Asia', *Food Security*, 1(2), pp. 127–136. doi: 10.1007/s12571-009-0015-1.

Du, W. et al. (2010) 'Health impacts of floods', *Prehospital and Disaster Medicine*, 25(3), pp. 265–272.

FFWC (2019) 'Annual flood report 2019: 126'. Available at: www.ffwc.gov.bd/images/annual19.pdf.

Gain, A. K. et al. (2017) 'Tidal river management in the South West Ganges-Brahmaputra delta in Bangladesh: Moving towards a transdisciplinary approach?', *Environmental Science & Policy*, 75, pp. 111–120.

Gain, A. K., Mondal, M. and Rahman, R. (2017) 'From flood control to water management: A journey of Bangladesh towards integrated water resources management', *Water*, 9(1), p. 55. doi: 10.3390/w9010055.

Gupta, K. (2020) 'Challenges in developing urban flood resilience in India', *Philosophical Transactions of the Royal Society A*, 378(2168), p. 20190211.

Hasan, M. R., Nasreen, M. and Chowdhury, M. A. (2019) 'Gender-inclusive disaster management policy in Bangladesh: A content analysis of national and international regulatory frameworks', *International Journal of Disaster Risk Reduction*, 41, p. 101324. doi: 10.1016/j.ijdrr.2019.101324.

Heinen, J. T. (2005) 'Book review: Geo yearbook 2003', *Environmental Practice*, 7(1), pp. 58–59.

Hossen, M. A. et al. (2019) 'Governance challenges in addressing climatic concerns in coastal Asia and Africa', *Sustainability*, 11(7), p. 2148. doi: 10.3390/su11072148.

IPCC (2007) *Contribution of working group II to the fourth assessment report of the intergovernmental*. Edited by Parry, M. L., Canziani, O. F., Palutikof, J. P., van der Linden, P. J. and Hanson, C. E. Cambridge and New York: Cambridge University Press, Panel on Climate Change.

IPCC (2012) *Managing the risks of extreme events and disasters to advance climate change adaptation: A special report of working groups I and II of the intergovernmental panel on climate change*. IPCC.

Jones, S., Oven, K. J. and Wisner, B. (2016) 'A comparison of the governance landscape of earthquake risk reduction in Nepal and the Indian State of Bihar', *International Journal of Disaster Risk Reduction*, 15, pp. 29–42.

Kale, V. S. (2002) 'Fluvial geomorphology of Indian rivers: An overview', *Progress in Physical Geography*, 26(3), pp. 400–433.

Kamruzzaman, M. and Shaw, R. (2018) 'Flood and sustainable agriculture in the Haor basin of Bangladesh: A review paper', *Universal Journal of Agricultural Research*, 6(1), pp. 40–49.

Kaushik, A. D. and Sharma, V. K. (2012) 'Flood management in India', *Indian Journal of Public Administration*, 58(1), pp. 119–136. doi: 10.1177/0019556120120109.

Marchand, M. et al. (2021) 'Flood protection by embankments in the Brahmani–Baitarani river basin, India: A risk-based approach', *International Journal of Water Resources Development*, pp. 1–20.

Matheswaran, K. et al. (2019) 'Flood risk assessment in South Asia to prioritize flood index insurance applications in Bihar, India', *Geomatics, Natural Hazards and Risk*, 10(1), pp. 26–48. doi: 10.1080/19475705.2018.1500495.

Messner, F. and Meyer, V. (2006) 'Flood damage, vulnerability and risk perception – challenges for flood damage research', in *Flood risk management: Hazards, vulnerability and mitigation measures*. Springer, pp. 149–167.

Mirza, M. M. Q. (2011) 'Climate change, flooding in South Asia and implications', *Regional Environmental Change*, 11(1), pp. 95–107.

Mirza, M. M. Q., Dixit, A. and Nishat, A. (2003) *Flood problem and management in South Asia*. Springer.

MoDMR (2019) *Standing orders on disaster 2019*. MoDMR.

Mohanty, M. P., Mudgil, S. and Karmakar, S. (2020) 'Flood management in India: A focussed review on the current status and future challenges', *International Journal of Disaster Risk Reduction*, 49, p. 101660.

Mohapatra, P. K. and Singh, R. D. (2003) 'Flood management in India', in *Flood problem and management in South Asia*. Springer, pp. 131–143.

Nandargi, S. *et al.* (2010) 'Hydrometeorology of floods and droughts in South Asia – a brief appraisal', in *Global environmental changes in South Asia*. Springer, pp. 244–257.

NCHM (2019) *Reassessment of potentially dangerous glacial lakes in Bhutan*. NCHM.

Nur, M. N. B., Rahim, M. A. and Rasheduzzaman, M. (2021) 'Flood impacts analysis and mitigation approach towards community resiliency at Nageshwari Upazila, Kurigram', *Asian Journal of Social Sciences and Legal Studies*, 3(5), pp. 178–192.

Panday, D. R. (2012) 'The legacy of Nepal's failed development', *Nepal in Transition: From People's War to Fragile Peace*, p. 81.

Pandeya, B. *et al.* (2021) 'Mitigating flood risk using low-cost sensors and citizen science: A proof-of-concept study from western Nepal', *Journal of Flood Risk Management*, 14(1), p. e12675.

Parthasarathy, D. (2016) 'Decentralization, pluralization, balkanization? Challenges for disaster mitigation and governance in Mumbai', *Habitat International*, 52, pp. 26–34.

Rahman, M. M., Hossain, M. A. and Bhattacharya, D. A. K. (2014) 'An analytical study of flood management in Bangladesh', *IOSR Journal of Engineering (IOSR-JEN)*, 4(1), pp. 1–6.

Revi, A. (2008) 'Climate change risk: An adaptation and mitigation agenda for Indian cities', *Environment and Urbanization*, 20(1), pp. 207–229.

Sadique, M. Z. and Kamruzzaman, M. (2021) 'Flood management perspectives in Bangladesh: An assessment of monsoon flood in 2020'. Available at: https://think-asia.org/handle/11540/13498.

Saleem Ashraf, M. L. *et al.* (2017) 'Understanding flood risk management in Asia: Concepts and challenges', *Flood Risk Management*, p. 177.

Shah, M. A. R., Rahman, A. and Chowdhury, S. H. (2018) 'Challenges for achieving sustainable flood risk management', *Journal of Flood Risk Management*, 11, pp. S352–S358.

Shakya, S. (2012) 'Unleashing Nepal's economic potential: A business perspective', *Nepal in Transition: From People's War to Fragile Peace*, pp. 114–128.

Shrestha, M. S. and Takara, K. (2008) 'Impacts of floods in South Asia', *Journal of South Asia Disaster Study*, 1(1), pp. 85–106.

Sivakumar, M. V. K. and Stefanski, R. (2010) 'Climate change in South Asia', in *Climate change and food security in South Asia*. Springer, pp. 13–30.

Smith, P. J., Brown, S. and Dugar, S. (2017) 'Community-based early warning systems for flood risk mitigation in Nepal', *Natural Hazards and Earth System Sciences*, 17(3), pp. 423–437.

Stromberg, J. C. (2001) 'Restoration of riparian vegetation in the south-western United States: Importance of flow regimes and fluvial dynamism', *Journal of Arid Environments*, 49(1), pp. 17–34.

Tarrant, O. and Sayers, P. B. (2012) 'Managing flood risk in the Thames Estuary – the development of a long-term robust and flexible strategy', in *Flood risk: Planning, design and management of flood defence infrastructure*. ICE Publishing, pp. 303–326.

Turner, A. G. and Annamalai, H. (2012) 'Climate change and the South Asian summer monsoon', *Nature Climate Change*, 2(8), pp. 587–595.

Uddin, K. N. (2018) 'Health hazard after natural disasters in Bangladesh', *Bangladesh Journal of Medicine*, 28(2), pp. 81–90. doi: 10.3329/bjmed.v28i2.33357.

UNISDR (2015) 'Global assessment report (GAR) on disaster risk reduction 2015 – making development sustainable: The future of disaster risk management'. Available at: www.preventionweb.net/english/hyogo/gar/2015/en/gar-pdf/GAR2015_EN.pdf%0A.

Vij, S. *et al.* (2020) 'Evolving disaster governance paradigms in Nepal', *International Journal of Disaster Risk Reduction*, 50, p. 101911.

WDR (2001) *World disaster report, an annual publication of the international federation of red cross and red crescent societies*. WDR.

WDR (2002) *World disaster report, an annual publication of the international federation of red cross and red crescent societies*. WDR.

Weinstein, L., Rumbach, A. and Sinha, S. (2019) 'Resilient growth: Fantasy plans and unplanned developments in India's flood-prone coastal cities', *International Journal of Urban and Regional Research*, 43(2), pp. 273–291.

Ziegler, A. D. *et al.* (2020) 'Flood mortality in SE Asia: Can palaeo-historical information help save lives?', *Hydrological Processes*, 35(1), p. 1.

Section V
Community perspectives, resilience and adaptation

20

The role of education in flood risk management

Building a resilient generation in developing countries

Edson Munsaka

Introduction

The climatic conditions in the African continent are incredibly diverse. Various scholars (e.g. Conway et al., 2009; Hamandawana, 2007; Laraque et al., 2001) have argued that the rainfall and rivers flow in this continent show high levels of variability across a range of spatial and temporal scales. This poses several complex challenges for the management of floods on the continent. Based on the number of people affected over the past 30 years, floods and droughts are the two natural hazards that have the largest humanitarian impacts in Africa (Lumbroso et al., 2016). However, in the past decade, floods have overtaken droughts in terms of the number of people they have impacted globally. In developing countries, floods have also caused damage to critical infrastructure. For instance, Hillard (2019) noted that the costs of damage to infrastructure caused by Cyclone Idai in Mozambique amounted to about US$1 billion. It is, therefore, not surprising that flood risk management has emerged as a key concern, posing several significant complex societal challenges in all communities around the world, though more pronounced in developing countries. Most, if not all, the commonly flood-affected places are not new, and it does not imply that flood risk management measures have not been put in place. Rather, continued community and infrastructure vulnerability to flood risk in developing countries is testimony to the existence of gaps in the measures that have been put in place by concerned countries. This chapter explores education as the missing ingredient in flood risk management.

The rest of the chapter is divided into different sections and will include the following: understanding the concept of education and forms of education, the effects of flooding on communities, the concept of flood risk management and the role of education in flood risk management, as well as its value in building community resilience to flood risk.

Understanding the concept of education

There are various viewpoints on what defines education. The term "education" has been used in different ways; as such, it is difficult to capture in any precise definition. According to Bartlett and Buron (2020), "Education is seen along with other social institutions as working to create and maintain a stable society" (p. 16). For Dewey (1916), education entails the continuous reconstruction as well as reorganization of experience which adds to the meaning of experience and increases the ability to direct the course of subsequent experience. Furthermore, education can also be construed as the process where an individual acquires or imparts basic knowledge to another. It is a deliberately initiated process with the intention of developing a person's skills essential to daily living along the way. Such an understanding of education resonates with the work of Sifuna and Otiende (2006). The authors define it as a process by which people are prepared to live well and efficiently in their environment or locality (Sifuna and Otiende, 2006).

It is also not far fetched to understand education as the transmission of the values and accumulated knowledge of a society on a particular event. It is a deeply practical activity – something that communities can do themselves and with others. In its broadest sense, the term "education" is normally thought to be about "acquiring knowledge and developing skills and understanding – cognitive capabilities" (Bartlett and Buron, 2020: 16). Having the experience of being impacted by flooding year in and year out, communities consciously or unconsciously develop knowledge on how to live in flood-prone areas, and that knowledge is transmitted inter-generationally. In this chapter, the notion of transmission is equated to what social scientists term enculturation. For instance, children are born without knowing the nitty-gritty of their culture but ultimately become knowledgeable of their culture. Perhaps the question to ask is "How do they learn that which they learn?" This demonstrates education's functional role. It guides them in learning a culture, moulding their behaviour in the ways of adulthood and directing them towards their eventual role in society. Learning to live in harmony with hazards such as floods and managing the risk of flooding is a culturally learnt behaviour by communities living in flood-prone areas. This ties in well with Hegel's conceptualisation of the role of education (Hegel, 1986). In a nutshell, Hegel conceives the role of education as that of elevating the human being as an individual into conformity with his universal nature. Literature (e.g. Lumbroso et al., 2008) has it on record that over half of Africa's population lives in rural areas. Many of these people depend on floodplains and rivers for their livelihoods and survival. Perhaps what may need further consideration is exploring the type of education they learn that enables harmonious coexistence with flood risk.

Forms of education and learning theory

The previous section focused on defining the term "education". This section focuses on the forms of education. However, prior to going into the discussion on forms of education, this section delves into the "concept of learning" and tries to unpack what constitutes learning.

What "learning" refers to always generates debate among scholars given its broad and abstract nature. Most textbook definitions of learning refer to learning as a change in behaviour due to experience (Lachman, 1997). This is a very basic functional definition of learning. It implies that learning is seen as a function that maps experience onto behaviour. In other words, learning is an effect of experience on behaviour. Some researchers (e.g. Domjan, 2010; Lachman, 1997; Ormrod, 2008) have insisted that such a simple functional definition of learning is narrow and unsatisfactory. They argue that changes in behaviour are neither necessary nor sufficient for learning to occur. On the contrary, De Houwer et al. (2013) find value in the definition. They argue that the definition does not imply that changes in behaviour are sufficient to infer the presence of learning. Rather, a change in behaviour is an instance of learning only if it is caused by some experience of the organism (De Houwer et al., 2013). Hence, in their argument, they contend that there are other factors besides experience that may cause changes in behaviour.

Therefore, behaviour changes that are due to factors other than experience (e.g. genetic factors) do not count as instances of learning. It therefore implies that learning refers to the individual acquisition of information, knowledge, behaviours, competencies or skills, including flood risk management skills through experience, practice, study or instruction. Globally, the acquisition of knowledge is provided through three recognised forms of education: formal, non-formal and/or informal. The state, local authorities and private sector play a critical role in the provision of education in developing countries.

Formal education

One known and recognised route through which disaster risk knowledge, including flood risk management, is acquired is formal education. Formal education is institutionalised. It is also intentional and planned. According to UNESCO (2012), formal education also consists mostly of initial education, vocational education and special needs education. On the other hand, initial education takes place when individuals are yet to enter the labour market.

Some countries offer education on disaster risks such as flood risk through formal learning. It is actually a component of their national school curriculum. In Zimbabwe, the Department of Civil Protection developed resource books on disaster education for school teachers and learners. Embracing hazard concepts including floods as a hazard as part of teaching content in view of the introduction of continuous assessment learning activities (CALA) as a teaching and learning strategy in Zimbabwe offers an opportunity to introduce learners to flood risk management issues through formal learning. CALA is based on "inquiry" learning that incorporates activity-based instruction across the curriculum.

Another example of the operationalisation of formal education can be found in China. Following the 2008 Wenchuan earthquake, the government of China integrated disaster education into the education system (Zhu and Zhang, 2017). Furthermore, in memory of this earthquake, 12 May is set aside as the National Disaster Prevention and Reduction Day in China. In their study covering four regions in China, Zhu and Zhang (2017) found that the majority of students and teachers view school-based disaster education as an effective way to deepen students' awareness of disaster risk reduction (DRR), including flood risk reduction. Therefore, including DRR and flood risk management in the school curriculum as new subjects is desirable. However, this approach could end up crowding school curricula. Sharpe and Kelman (2011) opine that there are many creative ways DRR including flood risk management can be integrated into the school curriculum. Luna (2012) concurs with Sharpe and Kelman (2011) by suggesting the use of lectures, research and school projects on specific subject matter as well as art as various creative ways flood risk management issues could be addressed in the formal education system. Equally, the relationship between flooding and other hazards relating to food security, water, energy and infrastructure as well as health could be explored using an integrated approach. Integration of subjects enables in-depth discussion without necessarily creating separate subjects.

Non-formal education

Non-formal education is also institutionalised, intentional and planned by an education provider. It is an addition, alternative and/or complement to formal education, and it caters to people of all ages. However, non-formal learning does not necessarily apply a continuous pathway structure when compared to formal education. It may be short in duration and/or low intensity and can be provided in the form of short courses, workshops or seminars. According to UNESCO (2012), non-formal education can cover programmes contributing to adult and youth literacy and education for out-of-school children, as well as programmes on life skills, work skills and social or cultural development.

While this chapter on the role of education in flood risk management largely focuses on developing countries, it is worthwhile to highlight that non-formal education, like formal education, is a globally recognised route through which DRR, including flood risk management knowledge and skills, can be acquired. In Germany, for instance, the German Federal Agency for Technical Relief and the Volunteer Fire brigade (Chadderton, 2015) are credited with providing non-formal adult learning services. A similar non-formal adult learning arrangement also exists in Japan and is overseen by the Japan Bousaisi Organization (Japan Bousaisi Organization, 2021). Bousaisi promotes disaster prevention and resilience abilities in communities (Kitagawa, 2016). Currently, nearly 205,000 Bousaisi exist in the country, mainly engaged in proactive building of community-based DRR (Japan Bousaisi Organization, 2021).

In developing countries, civil society organisations' role in providing non-formal disaster risk education, including flood risk–related education, cannot be overemphasised. In Zimbabwe, such organisations provide training and awareness programmes to communities on DRR, including flood risk management. Lumbroso et al. (2008) outline non-structural ways by which community members can reduce their vulnerability to floods. Some of the suggested non-structural solutions include raising their awareness of floods and improving preparedness. These initiatives, which also apply to formal education, come in different variations. One such initiative is the use of games. The use of games allows students to gain and broaden their understanding of the key issues in DRR, including flood risk management. Once capacitated, school children would normally make available to their parents or guardians what they have learnt, thereby reducing the risks of being a casualty in the case of a disaster, including a flood disaster, occurring. Communities can also engage in other adaptation actions. Creating and adopting a rainwater harvesting infrastructure is one such valuable activity which is useful to the communities not only in times of flood but also on a day-to-day basis. The rainwater harvesting infrastructure would be useful in providing a source of clean water during floods. In this way, non-formal education becomes a viable route through which communities in flood-prone areas can acquire skills on how to survive and live with a flood hazard.

Informal education

Informal education is another route by which flood risk management skills can be developed and inculcated. According to various literature, this type of education can be classified as deliberate and not institutionalised. It is consequently less organized and less structured than either formal or non-formal education. In this form of education, learning may include learning activities that occur in the family, workplace, local community and daily life or on a self-directed basis (UNESCO, 2012). Informal education to a large extent is presumed to account for most examples showing the application of disaster education in disasters, including flooding disasters. In the case of communities living in earthquake-prone areas, they are expected to have learnt and applied the "drop, cover and hold (on)" technique at some point. "Drop, cover and hold" takes the form of drilling (Kitagawa, 2021), and it establishes a stimulus/response relationship. This relationship enables individual community members to get used to their actions prior to the actual event. Many authors (e.g. Mahdavifar et al., 2009; Arlikatti et al., 2019) have noted the effectiveness of the "drop, cover and hold" action. It helps in reducing injury and death in the case of earthquakes. Tsunami evacuation drills hold the same principle. It is therefore imperative to note that engaging in drills as an activity constitutes a critical component of the informal form of learning for disaster education. Informal learning may occur during visiting flood-prone areas such as Muzarabani and Tsholotsho in Zimbabwe. Furthermore, photographs of communities taken during flooding can be displayed at community centres for future reference and learning. Such learning to a certain extent also takes place unintentionally. Acknowledging the existence of flood gauges in the neighbourhood as well as reading flood gauges constitutes a form of informal learning. It also applies to examining and being conscious of hazard maps, including flood risk maps distributed in the community.

Causes and effects of flooding on communities

Flooding has not been confined to rural areas alone in developing countries. In fact, urban flooding is a pressing issue on the African continent due to various reasons. According to Baker (2012), most cities and urban centres in Africa are now regarded as flood disaster risk hotspots. For instance, the blockage of water courses and drainage systems in Ibadan, a town in Nigeria, is a result of the dumping of refuse by residents (Egbinola et al., 2017). Such a practice is a significant factor in causing flooding. Adelekan (2016) also raises the same issue with regard to flooding in Lagos. However, the participation of communities in engineering flood risk is not a Nigerian issue alone. It has been observed in other countries, too. For instance, Tazen et al. (2019) also found that the significant increase in flood risk in Ouagadougou in Burkina Faso was largely due to human and environmental factors. There was evidence of unplanned urbanisation as well as significant levels of construction of infrastructure in flood-prone areas (Tazen et al., 2019). Thus, unplanned urbanisation in most developing countries and the associated increase of people living in floodplains have led to an increase in the number of fatalities related to floods in African cities (Di Baldassarre et al., 2010). When such flooding occurs, research has shown that much of the property, road infrastructure and at times life is lost, too. In the ensuing self-forced evacuation, the majority of the affected community members fail to remove their possessions to safety. Their houses and huts become flooded (Egbinola et al., 2017). Commenting on the 2018–2019 southwest Indian Ocean cyclone season, Masters (2019) opined that the cyclones resulted in a level of flood damage previously unseen in Africa. Cyclone Idai commenced in March 2019 as a tropical depression over Malawi primarily affecting Mozambique and Zimbabwe. It resulted in at least 602 deaths in Mozambique and 299 fatalities in Zimbabwe (Centre for Research on the Epidemiology of Disasters, 2019). Furthermore, the worst effects of flooding were recorded on the bridge components of road transport networks in Mozambique, where approximately 29% of the national road network was damaged. The damage included 20 bridges (World Bank, 2019). The destruction of road infrastructure has a devastating impact on the local economy. A loss of an individual road link as a result of a flood event, for instance, often leaves areas without a viable road connection to markets or essential services. Thus, the impacts of flooding on communities and their assets create the need to explore the importance of education to flood risk management as well as its role in creating communities resilient to flooding.

Education and flood risk management

The concept of flood risk management cannot be discussed and fully understood outside the broader concept of risk management. Risk management has been established as a well-defined procedure for handling risks due to natural, environmental or humanmade hazards, of which floods are representative.

As a process, flood risk management has been discussed extensively (UNDRO, 1991; Plate, 1997). It is the process of managing an existing flood risk situation. However, in a wider sense, flood risk management includes the planning of a system that will reduce the flood risk. Therefore, risk management for the operation of an existing flood protection system should be seen as the sum of actions for a rational approach to flood disaster mitigation. Its purpose is the control of flood disasters by way of being prepared for a flood and to minimize its impact should it occur.

Risk management is the process of risk analysis which provides the basis for long-term management decisions for the existing flood protection system. Risk analysis in the context of flood risk management helps to identify communities at risk as well as risk factors. Continuous improvement of the system is based on the reassessment of the existing risks and an evaluation of the hazards depending on the newest information available. Such an analysis process yields hazard risk maps, which can be drawn by means of geographical information systems (GIS) based on extensive surveys of vulnerability. The maps may serve either to identify or show weak points of the flood defence system or indicate a need for action. Maps may even show areas of vulnerability to flooding. This knowledge requires sharing or passing on to the next generation. This is done through formal, informal and/or non-formal means. For instance, a very different response may be required to floods in alluvial plains of large rivers, where velocities are comparatively low and the main danger to life is from the wide lateral extent of inundated areas, as was the case during the floods in Mozambique in February 2000. In the earliest days, people responded to such floods by moving their cities and villages out of reach of the highest flood they experienced. If an extremely rare flood was experienced, which reached even higher, then people had no choice but to live with the flood damage. Thus, in other areas, people learned to live with frequent floods. The learning was either formal, informal or non-formal. Therefore, the role of education, which in the context of disaster discourse is called disaster education, which entails what ought to be learnt, cannot be overemphasised in flood risk management. Education has a critical role.

Education of and for flood disasters

The driver and appetite for the education of and for flood disaster, commonly known as disaster education, can be traced from the Yokohama Strategy for a Safer World: Guidelines for Natural Disaster Prevention, Preparedness and Mitigation and its plan of action adopted in 1994. This was the output of the World Conference on Natural Disaster Reduction held in Yokohama, Japan. Education was identified as key in the strategy (UNISDR, 2005).

In 2005, the Yokohama Strategy was revised and gave birth to the Hyogo Framework of Action 2005–2015: Building the Resilience of Nations and Communities to Disasters (UNISDR, 2005). At that time, the United Nations Office for Disaster Risk Reduction, which was established in 1999, was playing a leading role in the field of disaster risk reduction. It is also important to note that the aim of DRR is two pronged. It aims to prevent new and reduce existing disaster risks. DRR also aims to manage residual risks. All these aims essentially contribute to strengthening community resilience to all kinds of disasters, including flooding. The concept of DRR, as put forth in the Hyogo Framework of Action, implies a stronger focus on risk preparedness and prevention when compared to response and recovery (DasGupta and Shaw, 2017). The role of education in flood risk management, as it embodies preparedness or prevention activities and measures, is implied in the priorities and therefore cannot be overemphasised.

In disaster-prone regions and countries, education of and for disasters is promoted as disaster mitigation, prevention and/or preparedness measures in usual times. However, there are common challenges in providing education of and for disasters, including flood risk actions. According to Kitagawa (2021), the shared challenges are centred on how to engage community members as well as individuals in disaster risk reduction, including in flood risk actions. Further challenges relate to how to sustain the community's interest in continued engagement in DRR activities, including those related to flood risk reduction. It should be recognised that not everyone has an appetite to engage in flood management activities. This is despite anecdotal evidence suggesting that individuals who have actually experienced a disaster at one point in their lives are more enthusiastic to engage in DRR actions compared to those who had never experienced one. Becker et al. (2017) have also pointed out that people are motivated to participate in DRR, including in flood risk management, through the experiences of unknown people. The media experiences of disasters equally play a critical role in driving people to get involved in DRR. This reinforces the emphasis on disaster education to invest in preparedness. Disasters should therefore be seen as historical processes. They involve continuous and persistent social construction and configuration of risk. As such, they are "detonators and amplifiers of pre-existing critical situations and a product of continuous and persistent construction" (Garcia-Acosta, 2017: 205). This is also true with flood disasters.

Role of education in building community resilience to flood risk

The increased occurrence of flooding and its related impacts are concerns globally. This creates many challenges in the management of floods, especially for developing countries where most communities are vulnerable to various hazards, including floods. In this context, reducing vulnerability as well as building resilience to flood risk, though challenging, remains a priority in most, if not in all, developing countries.

Literature shows that the word "resilience" is an ancient concept that originated in politics and natural history. Timmerman is credited for introducing the social review of resilience to the disaster discourse as the opposite of vulnerability (Timmerman, 1981). Having entered into the sciences in the 1960s, resilience has had varying interpretations.

The definitions of resilience that are popular tend to centre on the notion of "bouncing back" from a calamity such as flooding. It is the action or an act of rebounding or springing back or recoiling (Manyena, 2006; Wildavsky, 2012). The other definition of resilience is associated with elasticity. Various authors (e.g. OED, 2010; Leary, 2019) tend to agree that elasticity is the power or ability to resume an original shape or position after compression or bending. Such understanding implies a community returning to its original state prior to the disastrous occasion, regardless of whether it was at risk. For Pooley and Cohen (2010), resilience "defines the potential to exhibit resourcefulness by using available internal and external resources in response to different contextual and developmental challenges" (p. 34). The Paris Agreement and Framework Convention on Climate Change (UNFCCC, 2015), as well as the Sustainable Development Goals (UNGA, 2015), also define resilience as a preferred outcome of society or a community emerging from the clutches of adversity such as flooding. It bridges DRR and development goals. It also includes reduction in flood risk. From the previous discussion, it is clear that the term "resilience" has experienced a surge in popularity among various disciplines and scholars. However, this chapter adopts Pooley and Cohen's definition. As implied in their definition, resilience is dependent upon the capacity or ability of a system, community or society to adapt. Pooley and Cohen (2010) further equate resilience to a human characteristic that encompasses the ability to cope and adapt to new realities after experiencing a stressful occurrence such as flooding without much reliance on outside help. The targets for the Sendai Framework for Disaster Risk Reduction (SFFDRR) directly focus on improving resilience to risks including flooding at a global level by offering a practical tool for reducing vulnerability. In all these endeavours, the role of education is acknowledged.

Education plays a critical role in flood risk management and in building the resilience of communities to risks. For instance, it brings about desirable change in both culture and values for progress and response to and management of risks such as flooding. Its importance in promoting DRR, including flood risk and community resilience, has been acknowledged in three international agreements forged in 2015: the Sendai Framework for Disaster Risk Reduction (UNISDR, 2015), the Paris Agreement and Framework Convention on Climate Change (UNFCCC, 2015) and the Sustainable Development Goals (UNGA, 2015). The Hyogo Framework for Acton 2005–2015, the predecessor to the SFDRR, acknowledges the role of education in building communities resilient to hazards. Priority 1 of the Hyogo Framework specifically aims to:

- build knowledge of government officials at all levels and communities, among others, through experiences, lessons learned, good practices, training and education on disaster risk reduction, including the use of the existing training and education mechanisms and peer learning.
- promote the incorporation of disaster knowledge into informal and formal education as well as civic education at all levels as well as in professional education.
- promote national strategies that aim to strengthen public education and awareness in disaster risk reduction.

Priority 1 of the SFDRRR is "Understanding Disaster Risk Reduction". A critical interrogation of the focus of Priority 1 of the Sendai Framework for DRR (UNISDR, 2015), like in the case of the Hyogo Framework, tends to locate firmly the critical role of education in building community resilience to disaster risk, including flood risk. For any person or community to be classified as resilient, understanding the hazard and the risks it poses is imperative. Education is key in coming up with and implementing mitigating measures, which may include policy and practical activities as a response to the risk, including flood risk, thereby promoting community resilience.

Searching for resilience and vulnerability reduction

In recent years, resilience and vulnerability have received fair attention from academic disciplines. Efforts have also been made to extend resilience thinking and programming into policy at all levels of disaster risk reduction operations, including flood risk mitigation. This is reflected in the Sendai Framework for Disaster Risk Reduction (UNISDR, 2015), where the pre-occupation with improving resilience is noticeable. However, understanding the role of education in building communities resilient to adversity such as flooding cannot be fully understood in the absence of an understanding of the concept of "vulnerability".

Scholars in various disciplines have attempted to define "vulnerability" without reaching a consensus, and its meaning remains contested. After its identification as a concept by Rose and Killien in 1983, Havrilla (2017) argued that vulnerability was conceived as a descriptive term associated with risk. Drawing from the nursing discipline, the concept has become increasingly linked with health disparities. The idea of focusing on health-related vulnerability brings home

The role of education in flood risk management

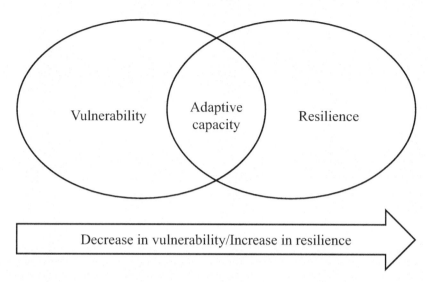

Figure 20.1 Vulnerability, education, and resilience relationship
Source: Adapted from the www.bing.com/images/search?q=diagrammatic

the need to hunt for resilience. Everyone is potentially at risk or vulnerable. Inequality, poverty and ensuing social or physical inequalities expose communities' vulnerabilities to disasters like flooding. However, the risk of vulnerability to disaster risk such as flooding is greater for those with the least social and human capital resources to prevent and/or ameliorate the origins and consequences of their proximity to risk such as flooding. The conditions for poverty, including inequality, are human-made or socially created. Therefore, it follows that poor people's vulnerabilities to different hazards, including flooding, are a result of their making or a result of other people's actions. Thus, the role of education in enhancing community resilience to flooding through being aware of existing vulnerabilities should be acknowledged. It enables graduates of various forms of education to apply their learning or learned skills, behaviours or competencies by mainstreaming DRR, including flooding risk reduction, into their daily lives (Luna, 2017). Through education, students and the community are equipped with knowledge and skills that help to enhance their capacities to adapt to new realities and reduce their vulnerability and exposure to risks. Awareness as an education activity would help communities at risk evade the impacts of a disaster such as flooding. Figure 20.1 is a diagrammatic representation of the existing relationship among vulnerability, education and resilience.

In this context, and as opined by Luna (2017), education acts as insurance for achieving and sustaining community resilience to disaster risk, including flood risk.

Chapter summary

The chapter examined the role of education in flood risk management. This was achieved through examining the forms of education in the context of building community resilience to flood risk. The increased occurrence of flooding in third world or developing countries has prompted the world to embrace resilience. The debate on what constitutes resilience expanded over the years, with the value of education in building community resilience acknowledged in various international agreements. The chapter argues that the value of education in managing flood risk cannot be underestimated. It is the source of skills, behaviours and capabilities required to build as well as enhance the resilience of communities to the risk of flooding plus the associated hazard impacts.

References

Adelekan, I. O. (2016). Flood risk management in the coastal city of Lagos, Nigeria. *Journal of Flood Risk Management*, 9, 255–264.

Arlikatti, S., Huang, S., Yu, C., and Hua, C. (2019). 'Drop, cover and hold on' or 'triangle of life' attributes of information sources influencing earthquake protective actions'. *International Journal of Safety and Security Engineering*, 9(3), 213–224.

Baker, J. L. (2012). *Climate change, disaster risk, and the urban poor: Cities building resilience for a changing world*. Washington, DC: World Bank Publications.

Bartlett, S., and Buron, D. (2020). *Introduction to education studies*, 5th ed. London: Sage Publications.

Becker, J. S., Paton, D., Johnston, D. M., Ronan, K. R., and McClure, J. (2017). The role of prior experience in informing and motivating earthquake preparedness. *International Journal of Disaster Risk Reduction*, 22, 179–193.

Centre for Research on the Epidemiology of Disasters. (2019). *Disasters in Africa: 20 year review (2000–2019)*. Issue 56, CRED crunch, November. https://cred.be/.

Chadderton, C. (2015). Civil defence pedagogies and narratives of democracy: Disaster education in Germany. *International Journal of Lifelong Education*, 34, 589–606.

Conway, D., Persechino, A., Ardoin-Bardin, S., Hamandawana, H., Dieulin, C., and Mahé, G. (2009). Rainfall and river flow variability in sub-Saharan Africa during the twentieth century. *Journal of Hydrometeorology*, 10, 41–59.

DasGupta, R., and Shaw, R. (2017). Disaster risk reduction: A critical approach. In Kelman, I., Mercer, J. and Gaillard, J. C. (eds.) *The Routledge handbook of disaster risk reduction including climate change adaption*. London: Routledge, 12–23.

De Houwer, J., Barnes-Holmes, D., and Moors, A. (2013). What is learning? On the nature and merits of a functional definition of learning. *Psychonomic Bulletin & Review*, 20, 631–642.

Dewey, J. (1916). *Democracy and education: An introduction to the philosophy of education*, 1966 ed. New York: Free Press.

Di Baldassarre, G., Montanari, A., Lins, H., Koutsoyiannis, D., Brandimarte, L., and Blöschl, G. (2010). Flood fatalities in Africa: From diagnosis to mitigation. *Geophysical Research Letters*, 37(L22402). https://doi.org/10.1029/2010GL045467

Domjan, M. (2010). *Principles of learning and behaviour*, 6th ed. Belmont, CA: Wadsworth/Cengage.

Egbinola, C. N., Olaniran, H. D., and Amanambu, A. C. (2017). Flood management in cities of developing countries: The example of Ibadan, Nigeria. *Journal of Flood Risk Management*, 10, 546–554.

Garcia-Acosta, V. (2017). Building on the past: Disaster risk reduction including climate change adaptation in the Longue Duree.' In Kelman, I., Mercer, J. and Gillard, J. C. (eds.) *The Routledge handbook for disaster risk reduction including climate change adaptation*. London: Routledge, 203–213.

Hamandawana, H. (2007). The desiccation of southern Africa's Okavango Delta: Periodic fluctuation or long-term trend? *Past Global Changes*, 15, 12–13.

Havrilla, E. (2017). Defining vulnerability. *Madridge Journal of Nursing*, 2(1), 63–68. doi: 10.18689/mjn-1000111.

Hegel, G. W. F. (1986). *Philosophical propaedeutic*. Oxford and New York: Basil Blackwell Ltd.

Hillard, L. (2019). *Cyclone Idai reveals Africa's vulnerabilities*. Council on Foreign Relations, 4 April. www.cfr.org/in-brief/cyclone-idai-reveals-africas-vulnerabilities (accessed on 12 March 2022).

Japan Bousaisi Organization. (2021). *What is a disaster prevention officer*. https://bousaisi.jp/aboutus/ (accessed on 24 March 2022).

Kitagawa, K. (2016). Disaster preparedness, adaptive politics and lifelong learning: A case of Japan. *International Journal of Lifelong Education*, 35, 629–647.

Kitagawa, K. (2021). Disaster risk reduction activities as learning. *Natural Hazards*, 105, 3099–3118.
Lachman, S. J. (1997). Learning is a process: Toward an improved definition of learning. *Journal of Psychology*, 131, 477–480.
Laraque, A., Mahé, G., Orange, D., and Marieu, B. (2001). Spatiotemporal variations in hydrological regimes within Central Africa during the XXth century. *Journal of Hydrology*, 245, 104–117.
Leary, J. P. (2019). *Keywords: The new language of capitalism*. Chicago: Haymarket Books.
Lumbroso, D., Brown, E., and Ranger, N. (2016). Stakeholders' perceptions of the overall effectiveness of early warning systems and risk assessments for weather-related hazards in Africa, the Caribbean and South Asia. *Natural Hazards*, 1– 24.
Lumbroso, D., Ramsbottom, D., and Spaliveiro, M. (2008). Sustainable flood risk management strategies to reduce rural communities' vulnerability to flooding in Mozambique. *Journal of Flood Risk Management*, 1(1), 34–42.
Luna, E. M. (2012). Education and disasters. In Wisner, B., Gaillard, J. C. and Kelman, I. (eds.) *The Routledge handbook of hazards and disaster risk reduction*. Abingdon and New York: Routledge, 750–760.
Luna, E. M. (2017). Education and training for disaster risk reduction including climate change adaptation. In Kelman, I., Mercer, J. and Gillard, J. C. (eds.) *Building on the past: Disaster risk reduction including climate change adaptation in the Longue Duree. The Routledge handbook for disaster risk reduction including climate change adaptation*. London: Routledge, 238–251.
Mahdavifar, M. R., Izadkhah, Y. O., and Heshmati, V. (2009). Appropriate and correct reactions during earthquakes: "Drop, cover and hold on" or "triangle of life". *JSEE*, 11, 41–48.
Manyena, S. B. (2006). The concept of resilience revisited. *Disasters*, 30(4), 434.
Masters, J. (2019). Africa's hurricane Katrina: Tropical cyclone Idai causes an extreme catastrophe. *Weather Underground*. www.wunderground.com/cat6/Africas-Hurricane-Katrina-Tropical-Cyclone-Idai-Causes-Extreme-Catastrophe.
OED (2010). *Oxford English dictionary: The definitive record of the English language*, 3rd ed. Oxford: Oxford University Press.
Ormrod, J. E. (2008). *Human learning*, 5th ed. Upper Saddle River, NJ: Merrill/Prentice Hall.
Plate, E. J. (1997). Dams and safety management at downstream valleys. In Betamio de Almeida, A. and Viseu, T. (eds.) *Dams and safety management at downstream valleys*. Balkema, Rotterdam: Sense Publishers, 27–43.
Pooley, J., and Cohen, L. (2010). Resilience: A definition in context. *Melbourne: Australian Community Psychologist*, 22, 30–37.
Sharpe, J., and Kelman, I. (2011). Improving the disaster-related component of secondary school geography education in England. *International Research in Geographical and Environmental Education*, 20(4), 327–343.
Sifuna, D. N., and Otiende, J. E. (2006). *An introductory history of education*. Nairobi: University of Nairobi Press.
Tazen, F., Diarra, A., Kabore, F. W. R., Ibrahim, B., Bologo, M., Traoré, K., and Karambiri, H. (2019). Trends in flood events and their relationship to extreme rainfall in an urban area of Sahelian West Africa: The case study of Ouagadougou, Burkina Faso. *Journal of Flood Risk Management*, 12, 1–11.
Timmerman, P. (1981). *Vulnerability, resilience and the collapse of society*. Environmental Monopoly 1. Toronto: Institute for Environmental Studies, University of Toronto.
UNDRO. (1991). *Office of the united nations disaster relief coordinator: Mitigating natural disasters: Phenomena, effects and options, a manual for policy makers and planners*. New York: United Nations.
UNESCO Institute for Statistics International Standard Classification of Education: ISCED 2011. (2012). http://uis.unesco.org/en/topic/international-standard-classification-education-isced (accessed on 10 April 2021).
United Nations Framework Convention on Climate Change (UNFCCC). (2015). *Paris agreement*. New York: United Nations.
United Nations General Assembly (UNGA). (2015). *Resolution Adopted by the General Assembly on 25 September 2015, A/RES/70/1*. New York: United Nations.

UNISDR. (2005). *Hyogo framework for action 2005–2015: Building the resilience of nations and communities to disasters*. Geneva: UNISDR.
UNISDR. (2015). *Sendai framework for disaster risk reduction 2015–2030*. Geneva: UNISDR.
Wildavsky, A. B. (2012). *Searching for safety*, 5th ed. New Brunswick: Transaction Publishers.
World Bank. (2019). Mozambique: Cyclone Idai and Kenneth emergency recovery and resilience project, July 2019
Zhu, T. T., and Zhang, Y. J. (2017). An investigation of disaster education in elementary and secondary schools: Evidence from China. *Natural Hazards*, 89, 1009–1029.

21
Emerging resilience to urban flooding in low-income communities
A socio-cultural perspective from Ghana

Clifford Amoako and Irene-Nora Dinye

Introduction – urban flooding in the developing world

Urban flooding is a critical challenge in cities of developing countries (Poku-Boansi et al. 2020). Over the past five decades, the frequency, manifestations and impacts of urban flooding, particularly in cities in the global south, have become increasingly known and widely documented by researchers, international organisations and development institutions alike. It has been described as "one of the major natural disasters which disrupts the prosperity, safety and amenity of residents of human settlements" (Jha et al., 2011a: 3) and the phenomenon with the most impact on the human population worldwide (CRED, 2015). It has also been seen as "economic risk because of the material damage caused" (Kovacs et al., 2017: 1) when floods occur in cities. The interest in urban flooding in the developing world is shaped by two main factors. First, the uncontrolled urban development that allows or pushes poor residents to live in hazardous areas, in precarious conditions, exposes many of them to flood hazards (Amoako and Inkoom, 2017; Gencer, 2013). Second, floods in cities of the developing world affect the poorest of their residents and widen the inequality gap (Kovacs et al., 2017; Abass, 2020). Slum dwellers and marginalised groups are the most affected by urban floods in the developing world (Adelekan, 2015a, 2015b; Adelekan, 2010; Alam and Rabbani, 2007; Revi, 2008).

The interest in urban flooding in the developing world has generated equal interest in flood risk management, especially in Eastern Asia, South America and Africa. Three main approaches, removal of houses and structures in flood zones, relocation of affected households and distribution of relief items by city governments, have been adopted for urban flood management in the developing world (Poku-Boansi et al., 2020). These approaches appear to pay less attention to the contributions of flood-affected residents. They rather tend to focus on the weaknesses and limitations of flood-affected residents and portray them as passive and incapable of responding adequately. These approaches treat flood victims as *clients* or *nuisances*, who should either be *served* flood interventions, manipulated or evicted (Amoako, 2016). Thus, the design of interventions in developing countries hardly considers the capacity of residents to cope with and respond to flood hazards (Amoako, 2017). Over two decades ago, Hewitt (1997: 167) referred to this framework as the "victims' approach".

In the recent past, there has been emerging attention on grassroots resilience and adaptive capacities of victims, especially of those in low-income urban communities, to respond to flood events. In spite of their increasing exposure and vulnerability to perennial floods and threats of forced eviction from city authorities, residents in flood-affected low-income communities continue to live precariously in these hazardous areas. They appear to have survived decades of flood events, resisted evictions by city authorities and rather increased in population and housing. For most of these flood-affected residents, responding to floods is becoming *learned, embodied and embedded in everyday living*. Are these communities *building grassroots resilience* to flooding? What has produced and continues to shape communities' capacities in *responding to* and *surviving* flood events? Are there any lessons from the emerging cases of adaptive capacities that could guide the design of urban flood interventions in the developing world? This chapter draws on examples from the developing world and empirical cases from Accra and Kumasi in Ghana to provide answers to these questions.

Drawing on seven years of empirical research conducted in five flood-affected low-income communities in the two cities (three in Accra and two in Kumasi), the chapter explores the evolving and enduring resilience exhibited through community mobilisation capacities, social capital and place attachment among flood-affected households. Decades of experiences and incremental responses to floods and how they influence communities' adaptive capacities and resilience are highlighted. The chapter also discusses the social, communal and structural responses to flooding resulting from this incremental learning of flood events. Urban flood management lessons from these emerging grassroots resilience and adaptive capacities have been purveyed.

The chapter's methodology could be categorised into three main qualitative approaches. First, the study is based on community interactions through household and key informant interviews, focus group discussions and field observations over a period of seven years in low-income communities in Ghana. The second approach involves consultations of relevant city and state institutions in flood management, social welfare and community management. Officials of these institutions were interviewed on their roles in building community resilience against urban floods. Third, the study drew from review of several secondary documents on urban flood vulnerability and resilience in the global south.

Emerging grassroots resilience and community adaptive capacities

The adaptive capacities of flood-affected residents in low-income communities have not received enough attention in research, policy and design of flood intervention programmes (Poku-Boansi et al., 2020). Earlier works on urban flooding have ignored the social, economic and political processes that make people capable of responding to hazards. Instead, they focused more on people's weaknesses and limitations and portrayed them as passive and incapable of responding to flood events (Hewitt, 1997: 167; Wisner et al., 2004: 13). In addressing the particular needs of affected residents, emerging works have stressed the need to identify households and community capacities in the design of sustainable interventions (Few, 2003: 51; Smit and Wandel, 2006: 286–287; Amoako, 2017; Poku-Boansi et al., 2020).

The previous argument points to the value of community "grassroots resilience" in urban flood management in low-income communities, as suggested by Amoako (2017). The debate on the definition of "community or grassroots resilience" has been ongoing for over two decades

(Furedi, 2008; Patel et al., 2017). The expression has also been given various names by different authors. For instance, communities' ability to prepare, plan and respond to hazards has been described with concepts such as coping ability, adaptability, robustness, flexibility, resistance and resilience (Smit and Wandel, 2006: 286–287). Many researchers connect "coping ability or adjustments" to "shorter term community capacity to adjust and survive", while "adaptive capacity" connotes long-term adaptation and recovery from the impacts of hazards (Parker, 2000: 34; Smit and Wandel, 2006).

Responding to the search for the most appropriate definition for "community or grassroots resilience", Patel et al. (2017: 10–11) encouraged academics, researchers and experts to clearly define the elements of resilience they are interested in since a comprehensive definition will be "too complex to apply at the local level". For this study, grassroots resilience is conceptualised as "a process linking a set of adaptive capacities to a positive trajectory of functioning and adaptation after a disturbance" (Norris et al., 2008: 130). This definition frames "grassroots resilience" as "a process and not an outcome" or "adaptability and not stability" (Norris et al., 2008: 130). In adopting this definition of resilience, this chapter explores the process communities go through and local socio-cultural and political resources they depend on in adapting to hazards (Ntontis, 2018: 25).

As a "process dominated" activity (White and O'Hare, 2014: 4), grassroots resilience is perceived to be participatory action and practice in which community capacities are built through socio-cultural and institutional changes along with recovery. In this chapter, we conceptualise grassroots resilience to include four inter-related stages that communities go through to respond to and/or cope with hazards (Folke et al., 2003). According to Folke et al. (2003: 354–363), these four stages are: (a) developing capacities to live with hazards through learning and experience; (b) consolidating the social learning process through which resilience is built; (c) nurturing diversity, flexibility and sustaining social memory for re-organisation; and (d) creating opportunities for self-organisation. These factors are operationalised through a number of elements and pre-conditions such as available local knowledge, social connections and inter-relationships, framework for communication, leadership structures, available resources and economic investments (Patel et al., 2017; Ntontis, 2018). We discuss how low-income communities build grassroots resilience through the use of available local resources, social capital, networks and relationships.

Characteristics and flood vulnerabilities in low-income urban communities

Most cities in the global south have long histories of flood hazards that have caused substantial impacts to lives and property (ILGS and IWMI, 2012). In these cities, flood hazards are more devastating in low-income communities due to the vulnerability of residents (Amoako and Inkoom, 2017). In a study of five African cities in Ghana, Uganda, Kenya, Nigeria and Mozambique, Douglas et al. (2008) revealed the range of physical, economic and socio-cultural impacts of flood events on low-income communities. These include loss of employment, outbreak of water-borne diseases, displacement of households and communities, mobility challenges and, in extreme cases, loss of lives. These flood impacts are exacerbated by unchecked urban growth, emergence of slums and poor housing environments that increase flood vulnerabilities in low-income communities. This makes the link between low-income communities and human vulnerability to flood hazards in cities of the developing world apparent and worthy of investigation.

Slum and low-income communities can be categorised into three main types (Amoako, 2016):

a. State-recognised *indigenous* communities that have grown into slums due to uncontrolled development. These were initially planned and occupied by original residents of the communities that grew into the city. They suffer uncontrolled and informal growth due the influx of new entrants into the city.
b. Unplanned, unrecognised squatter settlements on unapproved lands, facing forced eviction from city authorities. These are usually located on unused state lands, such as those acquired for community parks, dump sites, and ecological protection areas. They are illegally occupied, mostly by rural-urban migrants and other marginalised groups in the city.
c. Communities with unclear state recognition. They usually evolve organically as a result of rapid city expansion and uncontrolled growth. They are usually located at the fringes of the city and sometimes get planned and recognised later with urban planning interventions.

The empirical study selected low-income communities that fall within these categorisations, as presented in Table 21.1. The status of legitimacy and right to tenure of these communities shape their internal structure and political economy for responding to floods. These internal community socio-cultural dynamics and how they shape flood responses have been discussed variously across the developing world by many researchers (Pelling, 1998, 1999; Zoleta-Nantes, 2000, 2002; López-Marrero and Tschakert, 2011). This chapter also adds to the discourse on

Table 21.1 Study communities and nature of flood vulnerabilities

City	Selected communities	Type	Nature of flood vulnerabilities
Accra	Agbogbloshie	Initially planned and state-recognised indigenous community that has grown into a slum	Exposure to riverine and local floods from Korle Lagoon and Odaw River on heavy rainy days and during rainfall seasons
	Old Fadama	Squatter settlement on state-acquired land; unapproved and treated as illegal	The community forms part of the bigger communities at the banks of Korle Lagoon and Odaw River and experiences floods during rainy seasons
	Glefe	Coastal community with unclear state recognition; prone to flood hazards	The community is exposed to two lagoons, Gbugbe and Gyatakpo, and the Atlantic Ocean
Kumasi	Asawase	Indigenous community, grown into a slum and occupied by migrants from northern Ghana	The community is exposed to the overflow of the Aboabo River during the rainy season
	Aboabo	Grown organically with only a few houses approved by the municipal authority	Located along the banks of Aboabo River, exposed to the overflow of the river during the rainy season

Source: Summarised from Amoako, 2016; Amoako and Inkoom, 2017; Amoako et al., 2021; Abass, 2020

Emerging resilience to flooding in low-income communities

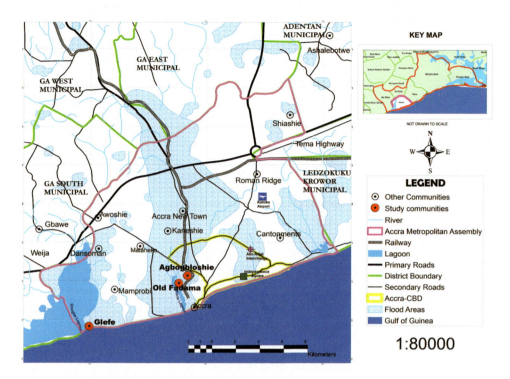

Figure 21.1 Locations of Agbogbloshie, Old Fadama and Glefe in Accra
Source: Amoako and Inkoom, 2017

how flood-affected communities draw on their internal strengths and resources to build adaptive capacities in response to flood vulnerabilities.

Location is a key factor of flood vulnerability and exposure in all five communities. Residents of the communities are exposed to floods due to their locations along river banks, draining the respective cities (See Table 21.1, Figures 21.1 and 21.2). In Figure 21.1, the study communities in Accra are within the catchments of the Odaw River and Korle, Gbugbe and Gyatakpo Lagoons. In the case of Kumasi, the two communities are in the catchment of the Aboabo River, as shown in Figure 21.2.

All five communities have a long history of riverine and local floods (Douglas et al., 2008; Abass, 2020) that occur annually during the rainy seasons in April to August and October to November. These communities have been affected by perennial floods for decades. The next sections present how residents are building their adaptive capacities and grassroots resilience through social and cultural networks and processes.

Local knowledge, social networks and responses to flood in low-income communities

Local knowledge on floods is produced through years of living in flood-prone communities and experiencing flood events annually for many years (Amoako, 2017). This knowledge is *experiential*, *embedded* and may be *transferred through social networks*. Thus, knowledge about trends and intensity of floods is produced by experience and lived through various changes to households'

337

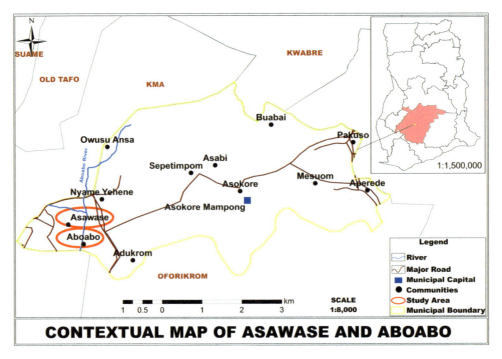

Figure 21.2 Location of Asawase and Aboabo in Kumasi
Source: Amoako et al., 2021

lifestyle and housing structure, which is learned by other households and community members through observation.

Community resilience is shaped by this local knowledge, frameworks for social connections and structures for community leadership in building adaptive capacities against flood events (Furedi, 2008; Patel et al., 2017; Ntontis, 2018). Thus, the many years of studies revealed that responses to flood events in the five communities are shaped by three main factors: depth of local knowledge on floods, previous experiences with floods and ability to mobilise social and physical resources. Local knowledge about the frequency and impacts of floods shapes communities' and households' responses; structural and non-structural. Every decision taken at the community or household levels is influenced by knowledge gathered from past flood events, while new lessons are learnt for future actions.

Most low-income urban communities do not have weather stations for recording and measuring the frequencies and intensities of floods. This reflects the situation of many low-income communities affected by floods in the cities of the global south, particularly sub-Saharan Africa (Jha et al., 2011a, 2011b; Adelekan, 2015a, 2015b; Adelekan and Asiyanbi, 2016). In the absence of official rainfall data, there is virtually no information on flooding in the study communities from city authorities. Hence, it is difficult to collect and objectively analyse data, except those experientially observed by flood-affected households. As a result, information on the frequency, impacts and community responses to flood events is based on community-acquired knowledge and stories of flood experiences of affected residents and key informants. Thus, flood-affected low-income communities build what can be termed as "collective social memories" of the trends, frequency and intensity of floods. Through these community flood experiences, affected households build their adaptive capacities and learn ways of responding to flood events.

Aside from the community experiences, adaptive capacities and knowledge built from responding to previous floods, communities' grassroots resilience is also built through existing social networks and protection systems. For instance, Pelling (1997, 1998, 1999) pointed to the role of local participation, social and political networks in building adaptive capacities against flooding in his work on flood vulnerability in Georgetown, Guyana. Similarly, in the study of flood affected low-income communities in Accra and Kumasi, social networks formed around ethnicity, religious groups and development associations were observed to be involved in community flood management (Amoako, 2016). These social networks are used as channels for mobilising residents in informal low-income communities for action against floods and other community hazards. For example, in parts of Agbogbloshie and Old Fadama, the continuous threats from city authorities to forcefully evict residents as a result of annual floods have generated various social groups to resist eviction and mobilise residents for flood prevention activities (also see Amoako, 2016). In Kumasi, Asawase and Aboabo have several local associations formed along ethnic lines, mobilising their members for communal actions against floods. These social groups evolve spontaneously in response to community needs, without any registration or recognition by the state (Amoako, 2016, 2018; Amoako et al., 2021). As a result, their activities are known after major flood events.

To gain the needed legitimacy for their spontaneous actions, some of the social groups form alliances with known national and international civil society organisations. For instance, flood victims and slum dwellers in informal communities in Accra have worked with a number of international civil society organisations in addressing their concerns on flood vulnerability, forced eviction and environmental management (COHRE, 2004; Grant, 2006; Afenah, 2009; Amoako, 2018). Another example of these trans-national alliances is when Slum Dwellers International (SDI), a network of urban poor residents in Africa, Asia and Latin America, partnered with a local civil society organisation in Ghana, Peoples' Dialogue on Human Settlements, to train community groups in mobilising social resources against floods, as well as creating alternative livelihoods for some flood victims. Their operations were reported in the three communities studied in Accra. Such trans-local and international alliances have become important in urban governance and environmental management in the developing world and have been used by many flood victims in low-income communities to develop their adaptive capacities to resist forced eviction and respond to floods and other hazards (Grant, 2006; UN Habitat, 2011).

At the household level, residents in low-income communities adopt various non-structural and socio-cultural strategies to respond and cope with flood events. These strategies include personal initiatives adopted through lived experiences as well as relying on social capital and community support. Inhabitants of these communities have come to conceive and believe that they only live in the city with the adopted initiatives and social capital built over the years of staying in those informal areas. In the five study communities in Accra and Kumasi, a key factor identified as important for building resilience at the household level is the knowledge and experience obtained from responses to previous floods. These experiences expose households to which flood response and coping strategies work and under which circumstances. Based on lessons gained from their responses to past flood events, many households select and adopt the most appropriate coping strategies for new occurrences. For instance, Amoako (2018: 962) points out that:

> Households who had parts of their houses destroyed repair the affected parts in anticipation of flood events in future.

The foregoing section situates emerging grassroots resilience within increasing experiential knowledge on floods and growing social networks in low-income communities in the developing world. In the next sections, we explore the experiential narratives of flood occurrences in the low-income communities studied and how those experiences shape socio-cultural flood responses and local adaptive capacities.

Experiential narratives of flood occurrence and socio-cultural impacts

Without any scientific methods for predicting and measuring the occurrence of floods, residents in flood-prone low-income communities grow in experiential knowledge of floods and their impacts (Amoako, 2018; Poku-Boansi et al., 2020; Amoako and Frimpong Boamah, 2020). In all communities studied, flood hazards are experienced annually and affect almost all residents in different ways. For instance, a recent study by Amoako et al. (2021: 6) in Asawase and Aboabo, Kumasi, revealed:

> 38% of households had been victims of flood events for more than ten years, 44% had experienced flood events between 2 and 5 years; while 18% had experienced flood events between 6 and 10 years in Asawase. In Aboabo, 76% had experienced floods for over 10 years; 18% between 2 to 5 years; and 6% had been victims for 6 to 10 years.

A woman affected by flood at Agbogbloshie in Accra showed her experience with the statement:

> We know the periods of intensive floods, and plan ahead to prevent devasting impacts on our families.

This is a snapshot of the case of many residents in flood-affected low-income communities in cities of the developing world. Most flood victims build experiences and adaptive capacities as they live longer in flood-prone areas and continuously assess the impacts of previous floods. Flood experiences are also transferred through social connections and shared knowledge. Through the sharing of previous flood experiences, residents perfect their understanding of the trends, intensity and impacts of floods. For example, at Agbogbloshie and Old Fadama, a respondent exhibited his experience and understanding of locational disparities in flood impacts as follows:

> There has been changes in the intensity and impacts of floods in this community in the last ten years or so. . . . It was very intensive in the southern parts of the community, close to the lagoon and less-intensive at the northern parts where the community begun. . . . This may be due to the construction of the storm drain.

In addition to this statement, another participant of the focus group discussions (FDGs) at Old Fadama also shared his experience on the nature and amount of rainfall that causes floods in the community. He argued from experience that the nature of floods in the community is not influenced by the duration and intensity of rainfall. Instead he stated that:

> when there is even moderate rainfall, the areas along the banks of the lagoon or Odaw River are always left with traces of floods and chunks of solid waste in the wake of the floodwater.

Like Old Fadama, Glefe lacks internal planning, lies within the lowest elevations of the Accra plains (Amoani et al., 2012: 3; Appeaning-Addo and Adeyemi, 2013: 3) and lacks storm drains to channel out flood waters back into the Gbugbe and Gyatakpo lagoons. As a result, the community experiences flood hazards more than once every year. A remark by a 56-year-old male participant at one of the mini-workshops explains this:

> As far as I can remember, this community floods every year. I have been here since I was born and there is flooding every year. Some parts of the western side of the community are constantly flooded. . . . I mean every day of the year . . . so those people live in water. In the last 20 years, the areas along the shore have been washed away . . . the earlier settlements are in the sea.

These experiential narratives show the depth and embeddedness of flood experiences at individual and household levels through over 35 years of social learning. After years of experiencing, responding to and learning from flood occurrences, the knowledge on the characterisation, frequency and intensity of floods increases. With this knowledge of floods, households and individuals design their interventions to reflect the knowledge acquired through their personalised experience with floods and inherit evaluation of previous coping strategies. Through these informal narratives of flood experiences, victims build their understanding and resilience to hazards, while non-victims get informed and prepare for future flood events.

These experiences of flood events also come with huge socio-cultural impacts. Families and friends are separated, common community spaces are destroyed and the communities are cut off for days and sometimes weeks. These experiences have shaped local flood responses, as well as generating further social learning, which in turn builds grassroots adaptive capacities.

Flood responses through experience and incremental social learning

Maskrey (1999: 85) states that the "capacity to absorb the impact of a hazard event and recover from it is determined by . . . its levels of social cohesion and organisation, cultural vision of disasters and many factors". The sub-sections that follow present the five main socio-cultural flood responses and practices observed across the low-income urban communities studied. Some of these social approaches for building grassroots resilience have also been observed in other parts of the developing world.

Local practices of flood prediction and preparations

The neglect of city authorities to intervene in flood management in many low-income communities in Africa (Amoako, 2016; Poku-Boansi et al., 2020) generates situations where flood victims and residents have gradually gained knowledge in predicting hazards and preparing for their occurrence. Victims' in-depth observation of trends and forms of floods, which are usually neglected by state institutions and research, are critical in their ability to predict not only the occurrence but the expected intensity. At Old Fadama, a respondent stated that:

> We experience the impacts of floods in different ways. . . . Sometimes the floods come into the compound without entering into the rooms . . . that takes just about an hour to recede. . . . But when floods enter our rooms. . . . This means it will have greater impacts. When that happens, the household involved will have to vacate their rooms for the entire day or even more.

Another respondent at Aboabo described his experience of flooding:

> It [level of flood water] could be anything between the ankle and waist level. When it is at the waist level it means you will pick nothing from your room – everything in the room will be destroyed. The floods rush so quickly into the room such that within a matter of 10–15 minutes you need to leave and run for your safety.

These quotations show residents' and flood victims' understanding of the *timing, character, intensity* and *expected impacts* of floods in the study communities and the corresponding actions or preparatory activities required. This in-depth knowledge of flood victims has been ignored in studies commissioned by both local governments and community-based and international development organisations. However, Codjoe et al. (2014) argue that local knowledge may provide important lessons for policy making in the developing world.

Community flood prediction and household preparation through past experiences and social learning, as argued by Amoako (2018: 962), evolve over time and become better as the community continues to engage with annual flood events. As affected residents get more rooted in flood experiences, social learning of responses and coping strategies, they are able to accurately predict rainfall patterns and flood events without measuring instruments and weather stations (Codjoe et al., 2014; Amoako, 2018). The longer residents live in the community, the easier it is for them to understand rainfall patterns and predict floods accurately. Community-level flood predictions are based on natural and geophysical occurrences that have been studied over years such as the intensity of the sun, direction of the movement of the moon and cry of some local birds (Codjoe et al., 2014; Amoako, 2018). Sometimes these community-level predictions fail, leading to disastrous consequences. In spite of the occasional failures in community-level flood predictions, they are still useful for community and household preparation and response.

Improvisation and investments in flood preparedness

In the study communities, preparation for floods is mainstreamed into everyday life and living. Households are always making investments in their housing units: changing the locations of valuable items or building extensions to their houses. For instance, in all five study communities, structural changes are made to houses in preparation for flood disasters. These physical changes to housing structures are aimed at improving their strength in anticipation of floods in the future. Structural changes include gradual alterations from wooden to concrete housing and changes made to parts of the house mostly affected by previous floods. In addition to this, some households raise the foundations of their houses or construct stairs to give access into such buildings during floods. These structural alterations are determined by the impacts of previous flood events. Flood victims make these incremental changes to their houses not only to make them resilient to future floods but to gradually improve them from temporary structures to permanent housing as their socio-economic conditions improve.

In Asawase and Aboabo, some households interviewed changed their affected walls and reconstructed their toilet and kitchen facilities as well as affected parts of their homes, while many others used sandbags to prevent flooding. Most of the affected houses were shared by more than one Households. Consequently, the structural changes are funded through financial contributions from the households occupying the affected houses. Beyond the structural changes made to houses, residents in the three communities studied in Accra mobilised themselves for communal works towards the prevention of floods. These communal efforts include cleaning and desilting major drains and reclaiming flooded areas with available materials such as sandbags, stones and gravel. Again, residents mobilise to clear public dump sites. Finally, there are emerging community engagements through the use of local information centres[1] to educate residents on flood prevention activities.

The community and households' flood responses discussed previously can be described as "traditional adjustment", as argued by Parker (2000: 34). These adjustments are made to reduce the full consequences of hazards but not to eliminate them. As pointed out earlier, these adjustments are learnt over time through experiences; social learning and knowledge transfer from flood-affected households. Again, it is important to note that these community flood responses depend mainly on local ideas, resources and materials which have been tried over many years (Amoako, 2018). Flood-affected residents make do with local resources and materials as a result of neglect or delays in state interventions. As a result, these local responses and traditional adjustments not only represent emerging grassroots resilience to flood but also depict community initiative and independence in responding to hazards. Thus, contrary to dominant narratives, flood-affected low-income communities rather build *social strength* and *independence of action* through their experiences with flood events.

Grassroots resilience in the previous narrative is simply the processes of regularly making changes to dwellings; taking precautions against floods; and preparing for their occurrence at individual, household and community levels. Hence, for flood-affected households in the study communities, living with and responding to floods are framed within the context of everyday life experience. From current and previous studies on flooding (Amoako, 2016, 2018; Poku-Boansi et al., 2020; Amoako et al., 2021), none of the community and household responses are by chance or randomly planned but result from many years of experiencing flood events, living with and responding to their different impacts using available socio-material resources (Amoako, 2016, 2018).

Spontaneous local mobilisation and international networking

Zoleta-Nantes (2000: 69) suggests that the "direct involvement and collective action" of low-income urban dwellers towards flood disasters is the first stage in reducing their impacts and residents' vulnerability. Residents in the study communities perceive flooding risk, occurrence and adaptation not as an individual issue but a collective and communal concern. This was shown in all five communities through reports of spontaneous community mobilisation during flood events.

Residents of the three communities in Accra were found to form action groups based on their occupations, ethnic background and neighbourhood. Leaders of these groups are selected based on existing power structures, such as individuals who settled in the area first or those who own land or properties in the respective neighbourhoods. In other circumstances leaders of the various ethnic groups or representatives of the communities at the local government level represent the various groups during their engagements with city authorities or international development agencies. The operational authority and mandates of the leaders of community groups are derived from the processes of their selection and knowledge in the issues the group intends to address. Through their leaders, the various community groups undertake various activities aimed at addressing flood hazards and building grassroots resilience. On their specific activities, a respondent at Agbogbloshie said this:

> As community leaders, we check people from indiscriminate dumping of rubbish into storm drains. We also prevent people from developing their properties in water ways. . . . In some cases we organise residents to embark on community cleaning and desilting major drains.

To mobilise flood-affected residents for action, community groups and their leaders adopt various communication channels. These include announcement through information centres, text messages on various phone-based platforms, information sent through religious leaders to their members and other social networks and channels through which residents are connected. Beyond these recorded group activities, there are cases where there was spontaneous mobilisation during floods. These spontaneous groupings are usually based on the areas of flood impacts or the parts of the community being threatened with forced eviction by the city authorities. In all the communities studied, there are various social groups and networks that have been harnessed to respond to floods, undertake flood prevention activities and/or resist forced eviction.

Similarly, at Aboabo in Kumasi, an opinion leader explained that residents support each other before, during and after flood events, stating that:

> We are all one family, so we help each other pack items into this structure [an old warehouse] which is not affected by floods, when someone is not there, we don't leave their belongings to get flooded but help pack them safely in the store.

In addition, community members mobilise to support the aged and other vulnerable persons in the communities to cope with flood events. Another community leader revealed that:

> We sometimes go and help the grandmothers who live in flood-prone areas to pack their items to our house [respondent's house].

At the Aboabo community market, it was identified that social capital and communal mobilisation played a role in structural flood response. A market woman explained that:

> we contributed to construct the main gutter at the entrance [entrance of community market] to enhance the flow of water, we are willing to contribute and construct other drains but the Assembly [local government unit] has promised to reconstruct the market.

The foregoing quotations and narrative point to increasing capacities for self-mobilisation and social networking against flood in the study communities. Derived from their separate and collective flood experiences, these capacities for community mobilisation in response to floods are an important source of grassroots resilience. Spontaneous community mobilisation for action against floods results from the absence of clear state flood intervention plans for low-income communities (Amoako, 2016; Poku-Boansi et al., 2020). Thus, social networking and protection through self-help and neighbourhood associations evolve to fill the state-support gap created by the neglect of city authorities. As can be seen from the study cases, the community associations are built around religious organisations and other social and ethnic groupings. Community mobilisation and organisation during hazards and disasters have been described by many researchers (Pelling, 1998; Few, 2003; Amoako, 2018) to be a central factor in building community resilience.

In addition to the social mobilisation discussed previously, community groups in the study communities also partner with local and international development organisations to enhance their community resilience against floods (Amoako, 2018). Through these *transnational alliances*, the capacities of community social groups to respond to floods, selecting appropriate methods and resisting the threats of forced eviction are built, sustained or improved (Amoako, 2016). Examples of these alliances were reported during field engagements, where community-based NGOs such as the People's Dialogue on Human Settlements (PD) and Ghana Federation of the Urban Poor (GHAFUP) operated in Agbogbloshie and Old Fadama and partnered with SDI to provide housing development support for flood-affected households (Afenah, 2009). Again, partnerships with international associations have been harnessed to demand justice in infrastructure provision and maintenance in the study communities. Thus, grassroots resilience is emerging as a transnational social resource shaping community flood response.

Resistance and emerging political agency

The previous narrative on the various social-cultural and political responses to flood in the study communities points not only to improved knowledge and incremental social learning on flooding but also an emerging political agency in flood-affected low-income communities to resist threats of forced eviction and social marginalisation. Here, references can be made to the alliances between residents of Agbogbloshie and Old Fadama on the one hand and the community-based NGOs and international institutions on the other to engage city authorities on forced relocation, as mentioned in the earlier sections of this chapter.

This is an emerging phenomenon through which residents of flood-affected low-income communities in developing countries are transforming their proclaimed status of poor and marginal slums to active urban citizens through cross-border activism (Appadurai, 2001; Grant, 2009) and thereby build strong local institutions for flood management and engagements with city authorities. In spite of their exposure and vulnerabilities to annual floods, all the low-income communities studied in urban Ghana have strategic partnerships with national and international development agencies. Through these international alliances, residents of these communities learn to understand their exposure and vulnerability to floods, design appropriate responses based on local socio-materials and negotiate their marginalisation while building political agency.

The emerging political agencies in the communities are shaping the trends of urban governance and flood management in three main ways. First, city authorities' hostilities towards flood victims occupying so-called hazardous zones are gradually giving way to negotiation and dialogue (Poku-Boansi et al., 2020). Second, community and household efforts towards flood management are gradually being recognised and discussed within research and policy (Amoako, 2018; Ntontis et al., 2018). Third, following from the first two points, socio-cultural and socio-material responses to floods are gradually gaining attention in flood research.

Key contributions, reflections and conclusions

This chapter presented the existing and emerging socio-cultural capacities and grassroots resilience within low-income urban communities to respond to flood hazards. Existing literature has explained why flood-affected residents continue to live in hazardous areas in cities where they are exposed and vulnerable (Pelling, Few, 2003; Wisner et al., 2004; Satterthwaite et al., 2007; Gencer, 2013; Amoako et al., 2021). The failure of city authorities and relevant state institutions to address flooding in low-income communities in the developing world has also been copiously discussed in the literature (Satterthwaite et al., 2007; Amoako, 2016; Poku-Boansi et al., 2020). It must be noted that the inability and/or negligence of state and city authorities in the developing world to design comprehensive flood interventions have gradually created low-income urban residents who have learned to live with and respond to floods as they occur.

Thus, the chapter argues that residents of flood-affected low-income communities build adaptive capacities over time through experiences and engagements with flood, and, contrary to the many arguments in the literature that they are only vulnerable and passive recipients of whatever happens to them, they are capable of responding to flood events. Rather than making them more vulnerable, residents in flood-prone low-income communities build their adaptive capacities instead through social learning, community networking and trans-national alliances. From empirical studies in Ghana, this chapter highlights the in-built social capital in low-income communities which has been harnessed at different levels for managing flood vulnerability. The chapter also discussed the accumulation of knowledge on floods, otherwise neglected in policy and research, and its possible role in shaping community and household flood responses. Another important contribution of the chapter is to point to non-state institutions and groups that spontaneously emerge to address floods and how they may shape urban governance and flood management in the developing world.

Rooted in the concept of grassroots resilience, the chapter draws attention to the enduring and complex socio-cultural and political economic responses to floods in low-income communities in cities of the developing world, which are usually ignored in mainstream research in urban flooding. Many useful and sustainable non-structural and socio-cultural interventions have escaped the attention of research. For example, all the study communities make use of their social cohesion and networking to mobilise their residents for community clean-up exercises and clearing of storm drains as pre-flood responses. Again, community engagements with city authorities are based on strong internal self-organisation. Spontaneous evacuations of flood victims in the study communities during flood events are also fuelled by local knowledge of areas usually affected and differentiated experiences of flood impacts. At the household level, local materials are being used for structural improvements to residential buildings to make them resilient to future flood events.

An understanding of these diverse community socio-cultural and political capacities has value in assessing grassroots resilience and incorporating local knowledge in the design of long-term and sustainable flood interventions. As its key contribution to the flood management discourse in cities of the global south, this chapter argues that, in the face of unclear and sometimes absent state flood management interventions, residents of flood-affected low-income communities are building socio-political agencies and grassroots resilience in responding to flood hazards. This grassroots resilience and adaptive capacities are yet to find the needed location in flood management policy documents in the global south.

Note

1 These are community media houses established by private persons and use loud-speakers placed at vantage locations in the neighbourhoods.

References

Abass, K. (2020). Rising incidence of urban floods: Understanding the causes for flood risk reduction in Kumasi, Ghana. *GeoJournal*, 87, 1367–1384. https://doi.org/10.1007/s10708-020-10319-9

Adelekan, I. O. (2010). Vulnerability of poor urban coastal communities to flooding in Lagos, Nigeria. *Environment and Urbanization*. https://doi.org/10.1177/0956247810380141

Adelekan, I. O. (2015a). Integrated global change research in West Africa: Flood vulnerability studies. In Werlen, B. (ed.), *Global Sustainability: Cultural Perspectives and Challenges for Transdisciplinary Integrated Research*, pp. 163–184. Springer, Zurich.

Adelekan, I. O. (2015b). Flood risk management in the coastal city of Lagos, Nigeria. *Journal of Flood Risk Management*, 10.

Adelekan, I. O., & Asiyanbi, A. P. (2016). Flood risk perception in flood-affected communities in Lagos, Nigeria. *Natural Hazards*. https://doi.org/10.1007/s11069-015-1977-2

Afenah, A. (2009). *An Analysis of the Attempted Unlawful Forced Eviction of an Informal Settlement in Accra, Ghana*, pp. 1–22. Development Planning Unit, London.

Alam, M., & Rabbani, M. G. (2007). Vulnerabilities and responses to climate change for Dhaka. *Environment & Urbanization*, 19(1), 81–97. https://doi.org/10.1177/0956247807076911

Amoako, C. (2016). Brutal presence or convenient absence: The role of the state in the politics of flooding in informal Accra, Ghana. *Geoforum*, 77, 5–16. https://doi.org/10.1016/j.geoforum.2016.10.003

Amoako, C. (2018). Emerging grassroots resilience and flood responses in informal settlements in Accra, Ghana. *GeoJournal*, 83, 949–962.https://doi.org/10.1007/s10708-017-9807-6

Amoako, C., Doe, B., & Adamtey, R. (2021). Flood responses and attachment to place within low-income neigbourhoods in Kumasi, Ghana. *African Geographical Review*. https://doi.org/10.1080/19376812.2021.1968445

Amoako, C., & Frimpong Boamah, E. (2020). Becoming vulnerable to flooding: An urban assemblage view of flooding in an African city. *Planning Theory and Practice*. https://doi.org/10.1080/14649357.2020.1776377

Amoako, C., & Inkoom, D. K. B. (2017). The production of flood vulnerability in Accra, Ghana: Re-thinking flooding and informal urbanisation. *Urban Studies*, 1–20. https://doi.org/10.1177/0042098016686526

Amoani, K. Y., Appeaning-Addo, K., & Laryea, W. S. (2012). Short-term shoreline evolution trend assessment: A case study in Glefe, Ghana. *Jàmbá: Journal of Disaster Risk Studies*. https://doi.org/10.4102/jamba.v4i1.45

Appadurai, A. (2001). Deep democracy: Urban governmentality and the horizon of politics. *Environment and Urbanization*. https://doi.org/10.1177/095624780101300203

Appeaning-Addo, K., & Adeyemi, M. (2013). Assessing the impact of sea-level rise on a vulnerable coastal community in Accra, Ghana. *Jàmbá: Journal of Disaster Risk Studies*, 5(8). https://doi.org/10.4102/jamba.v5i1.60

Centre for Resaerch on the Epidemiology of Disasters (CRED).(2015). *The Human Cost of Natural Disasters 2015. A Global Perspective*, Institute of Healt and Society (IRSS), USAID, UNISDR. http://repo.floodalliance.net/jspui/handle/44111/1165

Codjoe, S. N. A., Owusu, G., & Burkett, V. (2014). Perception, experience, and indigenous knowledge of climate change and variability: The case of Accra, a sub-Saharan African city. *Regional Environmental Change*. https://doi.org/10.1007/s10113-013-0500-0

Center on Housing Rights and Evictions (COHRE). (2004). *A Precarious Future: The Informal Settlement of Agbogbloshie*. COHRE, Accra, Ghana. www.mypsup.org/content/libraryfiles/60.pdf

Douglas, I., Alam, K., Maghenda, M., Mcdonnell, Y., Mclean, L., & Campbell, J. (2008). Unjust waters: Climate change, flooding and the urban poor in Africa. *Environment and Urbanization*, 20(1), 187–205.

Few, R. (2003). Flooding, vulnerability and coping strategies: Local responses to a global threat. *Progress in Development Studies*, 3(1), 43–58.

Furedi, F. (2008). Fear and security: A vulnerability-led policy response. *Social Policy and Administration*. https://doi.org/10.1111/j.1467-9515.2008.00629.x

Gencer, E. A. (2013). *The Interplay Between Urban Development, Vulnerability, and Risk Management*. Springer, New York.

Grant, R. (2006). Out of place? Global citizens in local spaces: A study of the informal settlements in the korle lagoon environs in Accra, Ghana. *Urban Forum*. https://doi.org/10.1007/BF02681256

Grant, R. (2009). *Globalizing City – The Urban and Economic Transformation of Accra*. Syracuse University Press, New York.

Folke, C., Colding, J., & Berkes, F. (2003). Synthesis: Building resilience and adaptive capacity in social-ecological systems. In Berkes, F., Colding, J., & Folke, C. (eds.), *Navigating Social-Ecological Systems: Building Resilience for Complexity and Change*. Cambridge University Press, Cambridge.

Hewitt, K. (1997). *Regions of Risk- A Geographical Introduction to Disasters*. Addison Wesley Longman, Essex, UK.

Institute of Local Government Studies (ILGS) and International Water Management Institute (IWMI). (2012). *Community Adaptation to Flooding Risk and Vulnerability.* ILGS & IWMI Collaboration, Accra, Ghana.

Jha, A., Bird, A., Jessica, L., Robin, B., Namrata, B., Lopez, A., Papachristodoulou, N., Proverbs, D., Davies, J., & Barker, R. (2011a). *Five Feet High and Rising: Cities and Flooding in the 21st Century.* Policy Research Working Paper. The World Bank, Washington, DC.

Jha, A., Bloch, R., & Lamond, J. (2011b). *Cities and Flooding: A Guide to Integrated Urban Flood Risk Management for the 21st Century.* The World Bank, Washington, DC.

Kovacs, Y., Doussin, N., & Gaussens, M. (2017). *Flood Risk and Cities in Developing Countries.* Technical Report No. 35, November. www.pseau.org/outils/ouvrages/afd_flood_risk_and_cities_in_developing_countries_en_2017.pdf.

López-Marrero, T., & Tschakert, P. (2011). From theory to practice: Enhancing resilience to floods in exposed communities. *Environment and Urbanization,* 23(1), 229–249.

Maskrey, A. (1999). Reducing global disasters. In Ingleton, J. (ed.), *Natural Disaster Management.* Tudor Rose, Leicester.

Norris, F. H., Stevens, S. P., Pfefferbaum, B., Wyche, K. F., & Pfefferbaum, R. L. (2008). Community resilience as a metaphor, theory, set of capacities, and strategy for disaster readiness. *American Journal of Community Psychology,* 41(1–2), 127–150. http://doi.org/10.1007/s10464-007-9156-6

Ntontis, E. (2018). *Group Processes in Community Responses to Flooding: Implications for Resilience and Wellbeing.* An Unpublished Thesis submitted to the School of Psychology, University of Sussex for the Doctor of Philosophy in Psychology.

Ntontis, E., Drury, J., Amlôt, R., Rubin, G. J., & Williams, R. (2018). Emergent social identities in a flood: Implications for community psychosocial resilience. *Journal of Community and Applied Social Psychology.* https://doi.org/10.1002/casp.2

Parker, D. J. (2000). Introduction to floods and flood management. In Parker, D. J. (ed.), *Floods,* vol. I, pp. 3–39. Routledge, London.

Patel, S., Rogers, M., Amlôt, R., & Rubin, G. (2017). What do we mean by 'community resilience'? A systematic literature review of how it is defined in the literature. *PLoS Currents Disasters,* 1, 1 February. http://doi.org/10.1371/currents.dis.db775aff25efc5ac4f0660ad9c9f7db2

Pelling, M. (1997). What determines vulnerability to floods? A case study in Georgetown, Guyana. *Environment and Urbanization,* 9(1), 203–226.

Pelling, M. (1998). Participation, social capital and vulnerability to urban flooding in Guyana. *Journal of International Development,* 10(1), 469–486.

Pelling, M. (1999). The political ecology of flood hazard in urban Guyana. *Geoforum,* 30(1), 249–261.

Poku-Boansi, M., Amoako, C., Owusu-Ansah, J. K., & Cobbinah, P. B. (2020). What the state does but fails: Exploring smart options for urban flood risk management in informal Accra, Ghana. *City and Environment Interactions,* 5, 100038. https://doi.org/10.1016/j.cacint.2020.100038

Revi, A. (2008). Climate change risk: An adaptation and mitigation agenda for Indian cities. *Environment & Urbanization,* 20(1), 207–229. https://doi.org/10.1177/0956247808089157

Satterthwaite, D., Huq, S., Reid, H., Pelling, M., & Romero Lankao, P. (2007). *Adapting to Climate Change in Urban Areas: The Possibilities and Constraints in Low- and Middle- Income Nations.* Human Settlements Discussion Paper. IIED, London.

Smit, B., & Wandel, J. (2006). Adaptation, adaptive capacity and vulnerability. *Global Environmental Change,* 16, 282–292.

UN Habitat (2011). *Global Report on Human Settlements 2011: Cities and Climate Change.* UN Habitat, Earthscan Publication Ltd., London.

White, I., & O'Hare, P. (2014). From rhetoric to reality: Which resilience, why resilience, and whose resilience in spatial planning? *Environment and Planning C: Government and Policy,* 32, 1–17. https://doi.org/10.1068/c12117

Wisner, B., Blackie, P., Cannon, T., & Davis, I. (2004). *At Risk: Natural Hazards, People's Vulnerability and Disaster.* Routledge, New York.

Zoleta-Nantes, D. (2000). Flood hazard vulnerabilities and coping strategies of residents of urban poor in metro Manila, the Philippines. In Parker, D. J. (ed.), *Floods*, Vol. I. Routledge, Taylor & Francis Group, London and New York.

Zoleta-Nantes, D. (2002). Differential impacts of flood hazards among the street children, the urban poor and residents of wealthy neighbourhoods in metro Manila, Philippines. *Mitigation and Adaptation Strategies for Global Change*, 7(1), 239–266. https://link.springer.com/article/10.1023/A:1024471412686

22
Towards a socially just flood risk management in developing countries

Lessons from serving the last mile in Malawi

Marc van den Homberg and Robert Šakić Trogrlić

Introduction

Importance of floods in developing countries

Hydrometeorological hazards such as floods and cyclones lead to most disaster losses and damages globally. Floods are the most frequently occurring type of disaster caused by a natural hazard (CRED and UNDRR, 2020) and affect more people globally than any other type of natural hazard. Climate change will increase the frequency and aggravate the severity of these hazards. In addition, Tellman *et al.* (2021) combined satellite images with population data to show that, globally, the number of people living in flood-prone areas is growing faster than the number living on higher ground and that these areas are concentrated in Asia and sub-Saharan Africa. In many cases, this has to do with growing population size (especially in coastal areas) and poorly planned socio-economic activities. Meanwhile, urban floods are increasing due to rapid urbanisation and more extreme rainfall events (di Baldassarre *et al.*, 2010a).

Introducing flood resilience, flood risk management, and flood risk governance

Although the exposure to floods has increased, the capacity of countries to manage and respond to floods has improved as well. Investments in flood protection, drainage infrastructure, early warning systems, improved building standards, schemes for supporting flood-affected people, and strengthened government policies enforcing risk-informed land planning can prevent floods and buffer the impacts when they occur (Browder *et al.*, 2021). *Flood resilience* refers to the ability of a system, community, or society exposed to floods to resist, absorb, accommodate, and recover from the effects of the floods in a timely and efficient manner, including through the preservation and restoration of its essential basic structures and functions (McClymont *et al.*,

2020). Flood resilience links directly to *flood risk management* (FRM) or is sometimes seen as an alternative to FRM. Cumiskey (2020) defined FRM as the continuous analysis and assessment of flood risk to generate knowledge, develop plans, and implement interventions to manage flood risk. A few decades ago, the term flood protection was mainly used to refer to the use of permanent or so-called grey infrastructure such as dikes or dams. But these flood protection measures are not sufficient to cope with dynamic flood risk. A grey infrastructure system is designed for a certain threshold or return period and has an inherent likelihood of failure, given that floods with higher return periods can occur, especially with climate change. This led to the paradigm shift from infrastructural flood protection to FRM, implying that protection from floods cannot be promised but that the risk can be managed (Begum *et al.*, 2010). Different FRM actions can be distinguished across the preparedness, response, recovery, and mitigation phases of the disaster risk management (DRM) cycle (Trogrlić *et al.*, 2019). Finally, it will also be useful to characterise *flood risk governance*. Alexander *et al.* (2016) define a flood risk governance arrangement (FRGA) as the actor networks, rules, resources, discourses, and multi-level coordination mechanisms through which FRM is pursued.

The role of communities

Both in developing and developed countries, local actors and communities have an important role in FRM. Kita (2017) explains that it is generally acknowledged that the actual implementation of global and national policies occurs at the local level where the actors are close to, or within, the policy issue. Mees *et al.* (2016) investigated the role of citizens in the coproduction of FRM in the EU. They defined coproduction as a conceptual umbrella that captures both the involvement of citizens within the decision-making and delivery phases of FRM as a public service. They found that the level of coproduction depends on the context in each of the EU countries studied. In developing countries, the role of citizens or communities is often determined by the resources and capacity available at the national and local governmental levels. If a state strongly relies on non-state actors for single or joint delivery of public services, it becomes a so-called 'hollow' state (Goldsmith and Eggers, 2004; Brinton Milward, 1996). Although they may retain monitoring and coordination functions (Kita, 2017), a parallel FRM system can come into existence based on interventions led by external actors, such as NGOs. While NGOs usually have a bottom-up, community-based approach, government actors have a top-down, national-level approach.

Social justice in FRM

Floods affect everyone regardless of whether they live in a developed or developing country. However, the degree to which people are affected depends on the flood resilience and FRM measures that have been implemented. A lower level of access to technology, funding, and capacity leads to communities being more at risk and less well protected from floods (van den Homberg and McQuistan (2019)). Thaler and Hartmann (2016) define justice, in the context of FRM, as the type of justice that concerns questions of allocation and distribution of resources and capital and wealth across different members of society. We focus on whether the FRM solutions are socially just for poor and vulnerable communities.

Assessment of whether FRM solutions are socially just is complicated and requires a multidisciplinary quantitative and qualitative approach (Vojinović and Abbott, 2012). Different principles can be used to evaluate social justice. We will focus on two forward-looking justice principles, distributive and procedural (as opposed to backward-looking principles such as compensatory or transitional justice) (van den Homberg and McQuistan, 2019). An example of procedural justice is if equitable procedures that engage all stakeholders in a non-discriminatory way are used to compare and evaluate different flood risk reduction measures. For example, grey infrastructure (such as levees, dikes, or dams) can decrease the frequency of inundation of flood-prone areas. But, as di Baldassarre et al. (2010b) showed for the Po river in Italy, it can also increase the flood discharge downstream, and it can lead to an increase in the use of the land close to the river due to a heightened sense of security. So, in this case, procedural justice would mean that the consequences for people living up- and downstream or close to the river and further away are taken equally into account. Distributive justice looks at how finance or capacity for FRM is divided among, for example, national and subnational government entities.

To assess whether FRM solutions are socially just, one can conduct cost-benefit or cost-effectiveness analyses. However, this is challenging. It is difficult to weigh the impact properly on poor and vulnerable people from a purely economic point of view; the monetary impact on them might be minimal compared to the impact on, for example, industry and infrastructure. For each phase of the DRM cycle, different actors are responsible and different funds are tapped into, making comparisons across phases complex. For example, how to compare permanent measures as part of the mitigation phase with forecast-based early actions as part of the early preparedness phase? Advances in flood forecasting have enabled a more prominent role for these forecast-based measures (van den Homberg et al., 2020). Bischiniotis et al. (2020) presented a methodology to compare permanent with temporary measures depending on, among other things, the forecast skill of the flood model. But this approach has not yet been used in practice.

Overall, FRM measures are often not designed to improve flood resilience coherently and holistically but rather – in a silo approach – to improve one dimension of flood resilience (often the dimension that can be more easily monetised). This can lead to conflicting outcomes for the aspects that cannot be easily monetised. For example, a dam can be built to protect an industrial area but can cause additional flood vulnerability to squatted areas downstream. In those cases, the risk for loss and damage of vulnerable communities is not adequately taken into account.

Objective

This chapter explores the challenges and opportunities in creating and maintaining a socially just FRM that adequately serves the last mile. With the last mile, we are referring to making sure that FRM actions reach and support the most vulnerable and poor communities. To this end, we have defined the following research questions that we will investigate for a case study in Malawi:

- Which generic public (government, NGO, or community-led) and private (household or individual) FRM actions exist across the disaster management cycle in developing countries?
- What are the challenges and opportunities of public and private FRM actions in reaching the last mile and being socially just?
- What are successful examples of integration between community-based (CB)-FRM and national-level FRM?

Our methodology combines conceptual ideas from literature with ground truth from fieldwork. We examine FRM interventions during the different DRM phases. For three DRM phases, we present examples from approaches coordinated and led by the national government, NGOs, and other development partners in collaboration with district governments or those organised by communities themselves based on their local knowledge.

Flood risk management in Malawi

Flood risk and impact

Natural hazards, most notably floods and droughts, lead to considerable impacts in Malawi (Nkomwa *et al.*, 2014). Figure 22.1 shows the occurrence of river floods from 1905 to 2021 and of flash floods from 1991 to 2021. Between 1967 and 2022, the country experienced 22 incidences of flooding. The most recent major events were in 2015, 2019, and 2022. Floods occur annually and can be attributed to the effects of El Niño and La Niña and the influence of tropical cyclones on the pattern and spatial distribution of rainfall (Government of Malawi, 2016b). These climatic conditions, in combination with increased exposure and vulnerability due to, for example, rapid population growth, deforestation, lack of proper urban planning, and high poverty levels, increase the flood risk. Pauw *et al.* (2010) calculated that floods and

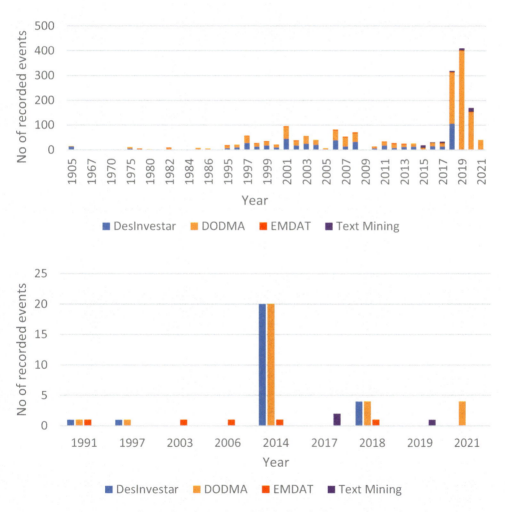

Figure 22.1 Overview of river floods (top) and flash floods (bottom) in Malawi from different data sources: Desinventar, EM-DAT, Department of Disaster Management Affairs (DoDMA), and text mining

droughts in Malawi decrease the gross domestic product (GDP) by 1.7% annually. The population in Malawi is predominantly rural (i.e., 85% of the total population) and dependent on agriculture, and approximately 70% live below the poverty line (Government of Malawi, 2016b). Hence, hydrometeorological disasters have considerable impacts, as they shatter livelihoods, create a reinforcing vicious disaster-poverty cycle, and lead to food insecurity. For example, the 2015 floods led to USD 70 million of losses in the agricultural sector (Government of Malawi, 2015a). Next to agriculture, floods impact sectors of energy, education, transportation, social protection, nutrition, housing, health, environment, water and sanitation, commerce, and trade. The recurrence of floods affects not only the local economy and its market prices of staple food but also local communities' cultural and psychological values (Mijoni and Izadkhah, 2009).

Flood risk management and governance

A range of private and public actions can be taken as part of FRM. Table 22.1, adapted from van den Homberg and McQuistan (2019), gives an overview of those actions for Malawi across the DRM phases, whereby the second column explains the objective of the actions in terms of risk reduction (to reduce the risk of impact as part of disaster risk reduction (DRR) or retention (to deal with the impact once it occurs during the response). Although DRR is mainstreamed into the legal and policy context in Malawi and several preparedness and mitigation activities can be observed, existing approaches still favour response and relief assistance (Šakić Trogrlić et al., 2018).

Malawi has a decentralised governance setup, with local governments (at the district level) having administrative and political power and a policy that stimulates community participation in development (Waylen and Martin-Ortega, 2013). Local government structures are at the district, area, and group village levels. At each of these levels, there are civil protection committees (CPCs) in charge of the coordination and implementation of DRR activities and policies. At the national level, the main agency for DRR is the Department of Disaster Management Affairs (DoDMA). The government's focus has been mainly on disaster response, where the cabinet can approve spending 2% of the total national budget for exceptional circumstances like disasters. Therefore, government institutions rely on NGOs and donors to fund the implementation of DRR-related plans (Waylen and Martin-Ortega, 2013; Kita, 2017). The scarce governmental resources available to implement FRM make the government highly dependent on international aid, resulting in a large number of international, national, and local NGOs implementing community-based projects in flood-prone areas through donor-funded packages (Nillson et al., 2010; Chiusiwa, 2015). Therefore, as also presented by Kita (2017), community-based approaches are the predominant way of dealing with disaster risks (including flood risks) in the country. Informal governance takes place via local leaders and community by-laws.

Flood preparedness: early warning systems

Early warning systems: global and national agreements

An essential component of flood preparedness is an early warning system (EWS). The National Disaster Risk Management Policy of Malawi (Government of Malawi, 2015b), agreed upon in 2015, has as a key policy priority the development and strengthening of a people-centred EWS. Also, key global agreements such as the Paris Agreement, the Sendai Framework for DRR, and the SDGs have objectives on EWS.

Table 22.1 Overview of public and private actions as part of flood risk management in Malawi across the disaster management phases

Disaster risk management phase		Flood risk management		
		Public action		Private action
Phase	Objective	Government	NGO	Community, household or individual level
Preparedness	Risk reduction; reducing existing risk	Install and maintain national early warning system, including dissemination and communication of weather forecasts	Locating relief items closer to the predicted to-be-affected area	Locally available early warning indicators based on local knowledge
		Local government structures prepare and use contingency and disaster risk management plans	Increase response capacity of communities through awareness-raising and training	A variety of early warning actions taken to prepare for the upcoming event
		Report on Sendai Framework for DRR indicators in relation to early warning	Create and train early warning teams at Village Civil Protection Committee level, train in search and rescue	Filling empty sacks with sand and placing them around the house
			Install community-based early warning system	Door-to-door communication of warning messages
Response and recovery	Risk retention	Lead damage and needs assessments	Contribute to damage and needs assessments or conduct specific ones tailored to the NGO's area of focus	People evacuate using locally available materials. Cattle can be relocated shortly before the floods arrive to the uplands
		First responders		
		Reconstruction of roads	Post-disaster public and donor assistance, such as relief items or cash transfers to households and money to governments for reconstruction of, for example, roads and embankments	
Mitigation	Risk reduction; preventing new risk or reducing existing risk	Government-led irrigation system	Capacity building of Village Civil Protection Committee	Decrease physical vulnerability of houses through measures such as improvement of foundation (houses of baked brick: raised foundation, houses of mud-brick: soil, plastered)
		Building of dikes at the community level	Training	Orienting houses so that entrance is opposite rainfall
		Afforestation and tree planting next to rivers		Planting palm trees
				Making a roof with a two-sided slope
			The construction of community-based flood protection infrastructure (i.e., dikes and levees), reafforestation of catchments and river banks, and other river training works (e.g., dredging)	Constructing check dams and dikes made out of local material
				Leaving a buffer zone between river and farm
				Planting of trees, grass, and shrubs along the river

Community-based, national, and local knowledge-based EWS

Figure 22.2 gives an overview of the different EWS present in Malawi, focusing on river floods. Before hydrometeorological scientific forecasts became available, communities in Malawi could only use their indigenous knowledge for early warning indicators that a flood would arrive. Nowadays, EWS in Malawi consists of an official EWS at the national level and a series of CB-EWS, in addition to the indigenous EWS. The Ministry of Agriculture, Irrigation and Water Development (MoAIWD) is responsible, at the policy level, for the national EWS, whereas the Department for Climate Change and Meteorological Services (DCCMS) provides the required forecasts. The Department of Water Resources (DWR) is in charge of monitoring the river catchments, and uses a flood routing procedure that gives forecasts with relatively low accuracy. Different NGOs and the Malawi Red Cross Society (MRCS) operate the CB-EWS. Most CB-EWS are in rural areas, but recently CB-EWS have also been rolled out in some cities. Due to the transitory nature of many residents in cities and the only recently increased focus of NGOs on urban DRR, urban CB-EWS are still less well developed. Ward civil protection committees (WCPCs) have been formed but have still limited functionality and capacity.

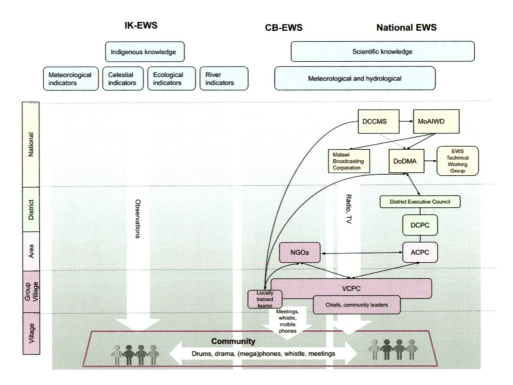

Figure 22.2 Diagram showing indigenous knowledge (IK), community-based (CB), and national EWS. DCCMS is the Department of Climate Change and Meteorological Services; MoAIWD the Ministry of Agriculture, Irrigation and Water Development; DoDMA the Department of Disaster Management Affairs; and DCPC, ACPC, and VCPC are respectively district, area, and village civil protection committees

Source: The authors

EWS typically consist of four components (UNDRR, 2022): risk knowledge, monitoring and warning, dissemination and communication, and preparedness and response. We will describe the last three components in more detail in the following and identify the relevant gaps.

The monitoring and warning component of the early warning system

Early warning systems in Malawi are primarily based on meteorological forecasts using either automated and/or community-based river gauges. The government has installed automated gauges mostly in the primary rivers and the NGOs' and MRCSs' manual gauges in tributaries or secondary rivers. The government-run gauges are quite well spread out over the country, whereas the CB-EWSs' gauges are mostly located in the most vulnerable and flood-prone areas.

Specifically, for the Shire River basin, the MoAIWD was responsible for the operation of the Flood Warning System for the Lower Shire Valley for several decades (Atkins, 2012). This system is a manually operated system that uses water level (at four gauges in the Ruo and Shire river) and rainfall readings to execute the flood routing procedure so that an initial alert can be given (Atkins, 2012; Nillson et al., 2010). The World Bank-supported Shire River Basin Management Programme (SRBMP), as will be further explained in a later section, has improved this system and developed an operational decision support system (ODSS) that can deliver a short-term forecast of three to five days. DCCMS, DWR, and DoDMA technically manage ODSS. Teule et al. (2020) assessed the forecast skill of the ODSS and also of the Global Flood Awareness System (GloFAS) for four locations in the Lower Shire river basin. This study shows that GloFAS does not predict absolute discharge values precisely but that GloFAS can be used to predict floods if the correct trigger levels are set per location for lead times of around seven to ten days. GloFAS can, in this way, complement the more accurate ODSS forecasts for shorter lead times.

The forecast skill of a CB-EWS is more difficult to quantify, given the unavailability of long-term data records. In general, one can say that a CB-EWS offers very local warning information and only for very short lead times. MRCS has also created WhatsApp groups of upstream and downstream communities, such as in the Karonga district, where the upstream communities can inform the downstream communities. Šakić Trogrlić (2020) found that, in some GVHs in the Lower Shire River basin, participants monitored the risk for flooding by using the CB-EWS installed by NGOs. Some community members still use reeds and triangulate the information between the reeds and readings from the CB-EWS. This shows that local knowledge is dynamic and continuously adapted through its interaction with external knowledge, resulting in hybrid knowledge (Hermans et al., 2022). Also, at the national level, the integration of multiple forecast sources offers the potential to contextualise the scientific forecasts that are more spatially aggregated.

The dissemination and communication component of the EWS

The official dissemination and communication component of the national EWS is structured as follows. The Department of Climate Change and Meteorological Service (DCCMS) generates different weather forecast products, such as short-term (within three to five days) probabilistic rainfall, temperature, and flood occurrence forecasts (only through cooperation with ODSS) and seasonal forecasts on rainfall and temperature. DCCMS shares these forecasts with the MoAIWD (Department of Disaster Management Affairs, 2019). MoAIWD subsequently shares the forecasts with the major national media house (Malawi Broadcasting Corporation) and DoDMA (Government of Malawi, 2016a). Via DoDMA, the forecasts go to the district levels; see Figure 22.2.

The dissemination and communication flow for a CBEWS can differ from one CBEWS to another. Often, a team is trained locally at the group village head (GVH) level, such as an early warning team or a disaster management committee to operate and use the CBEWS. The team consists of an upstream and downstream part, where the upstream team will inform the downstream one, such as via a WhatsApp group. The team often consists of members from the Village Civil Protection Committee (VCPC). This also means that capacity can get lost due to personnel changes that can take place, such as in periods of (re)election.

Dissemination and communication via local knowledge usually consist of door-to-door communication. Chavula (2015) reports that, at the village level, flood warnings will be conveyed through horns, shouting, and beating of drums.

No detailed statistics are available as to how well the national, community-based, and local knowledge EWS reach the last mile. Do the most vulnerable and poor people receive early warnings? Are the early warnings understood and acted upon? The reporting for the Hyogo Framework for Action and the Sendai Framework monitoring is often based on self-reporting at a highly aggregate level with insufficient ground-truthed data (van den Homberg and McQuistan, 2019). MRCS has more detailed information via base- and endline assessments as part of the EU ECHO projects, but only for a small area. From these assessments, it could be estimated that between 48% and a maximum of 75% of the population were covered by an EWS in Blantyre, Chikwawa, Nsanje, Thyolo, and Muzu. It was also shown that interpretation of the warnings required further strengthening and that there were gaps in the communication between the national, district, and community levels, with weak alliances between upstream and downstream VCPCs. These upstream GVHs are less flood prone and, therefore, not covered by other DRR programs. This disconnect reduces the time available to conduct early actions.

An improved reach of the EWS can be achieved by using additional or new communication channels. For example, MRCS is working on linking community-based radio stations to the DCPC to transmit forecasts and alerts and advise on responses recommended by the DCPC. There is also potential to make more use of digital and mobile-phone-based dissemination of early warning messaging, maybe even more so in urban than in rural areas. MRCS assessed that in Lilongwe, more than 30% of the households preferred social media, with the majority preferring SMS/phone-based communication.

The response component

MRCS conducted an assessment of the coverage and uptake of early warnings in the most disaster-prone traditional authorities (TAs) in the districts Chikwawa and Zomba (in 2020) as part of the EU ECHO project Increased Disaster Resilience through Early Action in Malawi (IDREAM). It turned out that the functionality of the EWS, the capacity of the TAs, the training followed, and the equipment available to respond to disasters are limited and dependent on intermittent project-based funding. Awareness of EWS proved limited among the surveyed households, ranging from 30% in Zomba to 100% of the surveyed respondents in Thyolo not knowing of their existence. Sustainability of EWS was also limited in some TAs, as the teams were primarily chaired by VCPC, leading to limited awareness of contingency plans and low technical capacity, as a reorganisation had just taken place in late 2020. Contingency plans at the district level were, except for Zomba and Chikwawa, not cascaded down or consolidated with TA or village-level contingency plans and were the result of a high-level analysis with limited involvement of communities. This resulted in a plan, with generic response scenarios, lacking specificity to make the plan actionable and the scenarios usable for simulation exercises.

Gaps in existing EWS

Although several stakeholders have been working on improving both the national and the CB-EWS over the last decades, several gaps in its functioning still need to be overcome. When we combine the project-based observations, the findings from a report by (Chavula, 2015), and the report from the First Stakeholder Workshop on Enhancing Early Warning Systems in Malawi (Chiotha et al., 2016), we can summarise the gaps as follows.

The monitoring and warning component

- *Lack of sufficient and well-functioning meteorological and hydrological stations.* The number of river gauges in primary and secondary rivers in Malawi is below the hydrological station density as recommended in the WMO guidelines (WMO, 2008). While the national meteorological policy (Department of Climate Change and Meteorological Services, 2019) recommends a target distance of 20 km between weather stations, the current obtainable distance is 80 km. For the existing automatic river gauges, DWR has several challenges in keeping them up and running. Vandalisation is an issue. Also, a lack of funding is hampering charging and changing the re-chargeable batteries in the gauges. Finally, the manual readers are often not being paid their 15.000 MWK monthly allowances. All these are causing river gauges that do not function to verify the estimated water levels with real-time observations. The so-called M-Climes program is supporting the installation of more gauges to narrow this gap.
- *There is limited and sporadic coordination between the government and NGOs* that run or have set up these gauges, and data are not consolidated into one easily accessible database. It is recognised that the inclusion of data from existing CB-EWS into a national-level database with the national EWS will contribute to their continued functionality and sustainability.
- *Difficulties accessing the available data.* Multiple CBEWS teams exist that monitor rivers using manual readers and disseminate early warnings to downstream communities once thresholds are met. However, this physical data collection is often not centralised or even digitised, meaning that valuable information is lost.

The dissemination and communication component

- *Insufficient multidisciplinary and multi-agency cooperation* exists within the national EWS (Chavula, 2015), as well as between actors in the national EWS and those of the CBEWS. There is clear potential for strengthening the linkage between the national EWS and the CB-EWS.
- *The content of the warning* is not sufficiently targeted to the end users (Chavula, 2015) and contains a limited translation of scientific and technical jargon (Chiotha et al., 2016). Similarly, for drought early warning in Malawi, Calvel et al. (2020) found that warning uptake among farmers will increase when they receive drought warning information provided as advice on agricultural practices rather than as weather-related information. In addition, both Chiotha et al. (2016) and Chavula (2015) stress the need to integrate indigenous knowledge with the scientific knowledge used in existing EWS. For flash floods in Karonga, Bucherie et al. (2022) showed that combining local and scientific knowledge provides improved understanding of flash flood processes within the local context. They also highlighted the potential of linking large-scale global data sets with local knowledge to improve the usability of flash flood warnings.

- *Measurement of the functioning of an EWS*, whether people receive, understand, and act upon the early warning, is a shortcoming. Only very high-level information is available on the coverage of both the national and CB-EWS. Chiotha et al. (2016) found, for instance, an accumulation of DRR programs (and hence CB-EWS) in a few districts. In addition, MRCS often faces turnover in the staffing of the CB-EWS teams, hampering the sustainability of CB-EWS that are started up.

In conclusion, the coverage and uptake of the scientific, national EWS at the community level is still limited due to the gaps and challenges described previously. The limitations in forecast skill (e.g., the spatial aggregation level) make it challenging to tailor forecasts to the local level. Also, the message content of early warnings can be too technical to be understandable or actionable. Communities suffer from the digital divide (lack of internet or mobile connection), as a result of which they often do not receive national early warning messages. An opportunity is to make sure a scientifically based, national EWS (Chiusiwa, 2015) benefits and learns from indigenous knowledge and CB-EWS.

Flood response in 2015

In this section, we explain and describe the response efforts from the government, NGOs, and communities to a flood event and explain the challenges faced in reaching the last mile. We will focus mainly on the 2015 floods, but some examples will be more general beyond this specific event.

Introduction to the 2015 floods

In January 2015, the southern region of Malawi received 400% more rainfall than usual, corresponding to a 1 in 500 years event (Government of Malawi, 2015a), creating floods that were among the most devastating in terms of geographical coverage and severity in terms of damage and losses over the last decades in Malawi. While 15 districts were directly affected, the whole country suffered from the effects. Water and electricity supplies were interrupted. Damage to roads and bridges disrupted business. An estimated 1,101,364 people were affected, 230,000 displaced, 106 killed, and 172 reported missing (Government of Malawi, 2015a). The floods were predominantly in the southern region, exacerbating an already precarious situation for rural households in this region. Houses and household properties were damaged. Standing and stored crops were also washed away, and animals were lost (Government of Malawi, 2015a). Several of these affected districts represent the poorest areas of the country. Based on the Integrated Malawi Household Survey of 2010–2013, the most highly affected districts – Nsanje, Chikwawa, Phalombe, and Zomba – have poverty incidences above the national average of 50.7%, ranging from 55 to 80%, as shown in Figure 22.3, which overlays a poverty map with the flood extent as derived from UNOSAT and the German Space Agency. When the poor are affected by floods, they lose a higher fraction of their wealth and income (Hallegatte, 2015). The poor usually live in houses with higher physical vulnerability (Wouters et al., 2021), such as houses built with traditional or semi-permanent housing construction material. Seventy-two percent of the households in the affected districts depend on agriculture and livestock-related income-generating activities, which are highly vulnerable to floods. Finally, poor households have less ability to deal with shocks given lower incomes and less asset diversification (Hallegatte, 2015).

Towards socially just flood risk management

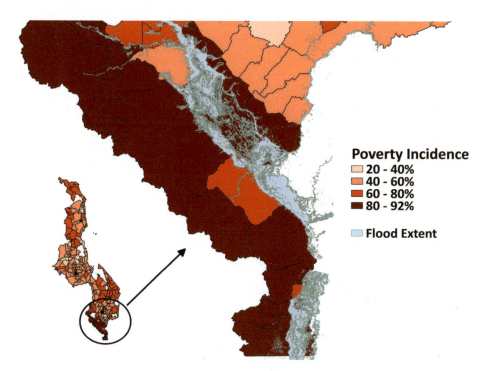

Figure 22.3 Overlay of the poverty headcount with the observed flood extent (Hallegatte, 2015)

Government and NGO response

On January 13, 2015, the president declared a state of disaster for 15 districts.[1] The government and, in particular, the Department of Disaster Management Affairs (DoDMA), realising that they did not have sufficient capacity to respond sufficiently, called for international assistance. The UN Cluster system was activated, and nine clusters became active. Hendriks and Boersma (2019) explained that the functioning of the UN cluster system was hampered by the lack of resources of DoDMA and by the fact that UN OCHA had no permanent presence in Malawi. Also, no valid preconceived response plans were available at the time of the 2015 floods (Hendriks and Boersma, 2019). The initial assessments were done by a United Nations Disaster Assessment and Coordination (UNDAC) team. The Malawi Inter-Agency Assessment Team, comprising government ministries and departments, UN Agencies, and the Malawi Red Cross Society, had conducted an initial assessment in Nsanje and Chikwawa, and the district civil protection committees had also conducted assessments in the other districts (Government of Malawi, 2015a). The diversity and multiplicity of organisations, some of them also new to Malawi, led to different ways of collecting data and made information management across these agencies more challenging (Hendriks and Boersma, 2019). Those organisations already active in Malawi for a longer time period and working close to the ground, near to communities, had, not surprisingly, the most trustworthy information. The diversity of organisations, in combination with limited coordination at the district level, led to a lack of impartiality in targeting those most in need. For example, some NGOs started distributing along roads so that those in camps deeper into rural areas got less, or churches could have a tendency to focus on their followers

(Hendriks and Boersma, 2019). Also, some districts were chronically understaffed, and even a year after the 2015 floods, some of Malawi's 28 districts still did not have DoDMA officers (Hendriks and Boersma, 2019).

Response of the communities based on their local knowledge

Šakić Trogrlić (2020) researched how some community members decided to evacuate during the flood event. The researcher's impression was that more people wait for the waters to arrive. People mainly used canoes carved from locally available trees for urgent evacuations, not carrying personal items or livestock. Canoes were also used to go to the closest maize mill and purchase food. In addition to canoes, research showed that people also use different ways of evacuation, such as walking, biking, and using the nearest trees to find shelter. Additionally, people might use locally available materials (e.g. banana trunks or grass doors) as a means of moving on the flowing water.

Some evacuate to existing shelters, while some will construct these upon arrival, and others will relocate to existing places such as schools, churches, tents, or permanent structures for evacuation. At times, trees and anthills are used for evacuation. In addition, people deploy relocation strategies for their livestock (Sakic Trogrlic and van den Homberg, 2018). Chickens and goats are kept in raised platforms made from locally available materials. Bigger animals, such as cattle, might be relocated to graze in the uplands and no longer close to the riverbanks.

Flood mitigation in the rural areas of the Lower Shire river basin

Government-led flood mitigation

Since the 1960s, the government built dikes in flood-prone areas (Nillson *et al.*, 2010). However, many of these suffer from a lack of maintenance (Botha *et al.*, 2018), decreasing their effectiveness over time. After the devastating flooding of 2015, several initiatives can be observed under the leadership of the Department of Disaster Management Affairs. For instance, the World Bank funded the Malawi Flood Emergency Recovery Project with components focusing on the rehabilitation and reconstruction of infrastructure like roads and dikes. This project can be classified as a mitigation activity. Similarly, the Shire River Basin Management Program has a component of planning and implementation of flood mitigation such as dikes at the community level, afforestation, and tree planting along the banks of flood-causing rivers. Figure 22.5 presents some examples of these government-led flood mitigation activities. A study by Šakić Trogrlić (2020) on the Lower Shire Valley suggested that given the severity of the flooding problem, more focus should be put on early warning systems than on building large-scale infrastructure.

Community-based flood risk management

With a lack of government resources, there is a high dependency on donor-funded community-based flood risk management (CB-FRM) projects (Nillson *et al.*, 2010; Chiusiwa, 2015). These include flood-related projects such as the creation of CB-EWS, capacity building of local civil protection committees, training in search and rescue, and so on. In terms of flood mitigation, these include, for instance, the construction of community-based flood protection infrastructure (i.e. dikes and levees), reafforestation of catchments and river banks, and other river training works (e.g. dredging). Examples are shown in Figure 22.5.

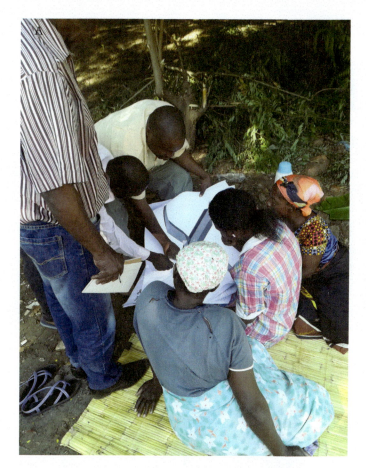

Figure 22.4 Images from the sites of government-led projects: a) villagers discussing dike design with consultants in group village head (GVH) Nafafa, b) replanting of vertiba grass GVH Monica, c) dike constructed as a part of the Malawi Flood Recovery Emergency Project, d) dike construction in GVH Tengani

Figure 22.4 (Continued)

There is an existing critique that flood mitigation infrastructure implemented by NGOs suffers from poor quality due to a lack of technical expertise (Nillson *et al.*, 2010). In addition, a big challenge for works implemented through CB-FRM, including flood mitigation works, is a lack of sustainability (Šakić Trogrlić *et al.*, 2018).

Flood mitigation through local knowledge

Local, indigenous, and traditional knowledge have significant potential in reducing disaster risks, including flood risks, as this knowledge has been developed based on experiences of phenomena and experimentation in finding solutions to complex problems (Mercer, 2010; Hiwasaki *et al.*, 2014; Dube and Munsaka, 2018). Previous research in Malawi has shown that communities have a rich local knowledge of flooding (e.g. Phiri and Saka, 2008; Chidanti-Malunga, 2011; Chawawa, 2018; Trogrlić *et al.*, 2019). Unfortunately, this knowledge is still largely not mainstreamed into Malawi's 'official' FRM (Šakić Trogrlić *et al.*, 2021) and thus remains untapped potential.

In terms of flood mitigation, local knowledge manifests in several aspects. For instance, the improvement of foundations is dependent on the housing type (Figure 22.6). Houses made of baked bricks generally have a raised foundation, whereas mud-brick houses have 'chiguwa' – soil plastered to strengthen the foundation. Houses with raised foundations are commonly found in the lowlands prone to flooding, raised to the levels based on previously experienced flood levels.

Figure 22.5 Examples of projects implemented as a part of community-based flood risk management: a) community-based dike, b) reafforestation

Source: Authors' own, April 2016

Figure 22.5 (Continued)

In addition, several approaches for flood risk mitigation are available locally, such as i) orientation of house entrances away from the direction of rainfall, ii) reinforcement of the walls using palm trees, iii) constructing a physical barrier for the water by placing bags filled with sand, iv) a roof with two-sided slope ('lamada'), v) constructing drains and ditches around the house, and vi) reinforcing foundations with stones.

Furthermore, there is often a 'buffer' zone between farms and a river (8–10 meters), with the idea of making space for the river and not presenting an obstacle to the flow of water. These examples serve as a local version of flood zoning.

Figure 22.6 Example of property level flood protection: a) reinforcement of foundations in Group Village Head (GVH) Mmodzi, and b) a house in GVH Misili with a raised foundation

Source: Authors' own, June 2017

Figure 22.7 Examples of river protection works from GVH Mmodzi: a) check dams, and b) a dike

Source: Authors' own, June 2017

Figure 22.7 (Continued)

Finally, several methods of preventing sedimentation of rivers can be seen through a flood mitigation lens: i) the construction of a physical barrier at the edge of the farms to prevent sediment from reaching the river and ii) planting of trees, Napier grass, elephant grass, reeds, and shrubs along the river to enforce the soil and prevent erosion. In some places, dams and dikes are made out of local materials to manage the water flow, and erosion processes were found, as shown in Figure 22.7.

Discussion and conclusion

Vulnerable communities are disproportionally affected by climate variability and climate change and consequently face higher risks and impacts of flooding. This chapter explores the challenges and opportunities in creating and maintaining a socially just FRM that adequately reaches and supports the most vulnerable and poor communities. We use the lens of distributive and procedural justice principles to identify the injustices in public and private FRM actions across the disaster management phases of preparedness, response, and mitigation. We discuss how the approach and results in our Malawi case study apply to other developing countries and give possible ways forward in tackling the injustices in FRM.

In terms of preparedness, the lack of sufficient and well-functioning meteorological and hydrological stations can be seen as a form of distributive injustice. We agree with the position of van den Homberg and McQuistan (2019) that the national meteorological and hydrological service in a developing country should not settle for an EWS with less lead time and spatial resolution than what is obtainable in a developed country. A form of procedural justice is that early warning messages are not adequately tailored to reach poor and vulnerable communities and to be understandable and actionable for them. More multidisciplinary and multi-agency cooperation is important in designing and running national and CB-EWS.

In terms of response, there are distributive and procedural justice principles at stake. Response efforts during the 2015 floods in Malawi did not always reach the people who were most affected. At the government level, there was an unfair distribution of resources and staff over districts, and NGOs focused on some districts more than on others. Although from a procedural point of view, most NGOs and governments have policies in place to ensure, for example, impartiality, in practice, distributive injustice can hamper compliance with these policies.

In terms of mitigation, Šakić Trogrlić (2020) found that most attention and resources go to response rather than mitigation. Distributive justice across the different disaster management phases is hence falling short. Given the limited resources available for mitigation, difficult decisions also have to be made as to investing in, for example, large-scale infrastructure, nature-based solutions (NBSs), or EWS. Public NGO and private actions consist primarily of NBS, whereas some multilateral donors such as the World Bank have funded grey infrastructure. No information has been found on overarching cost-benefit analyses that compare the different public and private actions or on initiatives that pool funds to jointly invest in a set of FRM actions that coherently lead to an optimal reduction of flood risks.

Across disaster management phases, we have seen that access to data of sufficient quality is essential to drive the equitable distribution of resources and capacity and to make informed choices on which public or private action to select and implement. As an example, there are very limited data on the performance of FRM measures, such as on the functioning of the different parts of the warning value chain of an EWS (Zhang et al., 2019). Subnational and disaggregated reporting is essential to monitor progress in delivering to the last mile. Also, it is difficult to get financial data on budgets for FRM across NGOs and government actors. Although Kita (2017) provided evidence that non-governmental organisations are delivering the majority of DRM services in Malawi, he pointed out that the government and NGOs have a symbiotic relationship, with the government endorsing NGO projects and giving NGOs the space to voice their views. Indeed, in developing countries with limited state capacity, a parallel donor-funded system often arises (Hendriks and Boersma, 2019). Kita (2017) explains that in Malawi, the central government is detached from community-level implementation, and the local government and politicians can position themselves to capture risk reduction finance.

Consequently, the integration between state-led FRM and CB-FRM is usually very minimal. This means that poor and vulnerable communities receive FRM support mostly via NGO-led public actions. However, this NGO support comes with challenges in terms of sustainability and spatial coverage. NGO interventions are often patchy (only at a few locations in a country) and dependent on temporary project-based funding (Šakić Trogrlić et al., 2018). Flood risk governance should therefore make sure a higher level of integration between these parallel systems takes place.

Our research is limited to river floods in rural areas, and we acknowledge that Malawi is one of the fastest urbanising countries globally, with an annual urban growth rate higher than 5% and an urban population of 20% of its entire population (UN-Habitat, 2022). Hence, more and more people will be exposed to surface and flash floods in urban areas (van Milligen de Wit, 2021). This means (future) flood risk governance needs to cover different types of floods, that is, flash, pluvial, and fluvial floods, in both rural and urban areas. These different types of floods tend to interact and can have compounding and consecutive impacts, especially in urban areas, also necessitating different FRM actions than in rural areas. Flood risk governance also needs to work across geographical levels, from regional to national to local. It requires coordination from national to community and across national boundaries. For example, early warning systems have to be cross-border in both Malawi and Mozambique to work effectively and to maximise efficiencies while delivering additional social and political benefits.

Although our findings are specific to Malawi, they are also applicable to other developing countries struggling to deal with the consequences of increased flood risk. We recommend using social justice principles such as distributive and procedural justice. These principles can drive the equitable allocation of means of implementation necessary for FRM, that is, technology, capacity, and funding, so that the last mile is served and no one is left behind.

Acknowledgements

Parts of this chapter have been extracted from the second author's doctoral research at Heriot-Watt University, resulting in a PhD thesis entitled 'The Role of Local Knowledge in Community-Based Flood Risk Management' (2020). Similarly, parts of the chapter are based on a consulting assignment on Indigenous knowledge and early warning system in the Lower Shire Valley in Malawi (2018) conducted by the first and second authors and funded by the EU ECHO action Enhancing resilience of vulnerable communities and building institutional disaster response capacity in Phalombe, Thyolo, Chikwawa, and Nsanje (Malawi).

Note

1 Nsanje, Chikwawa, Phalombe, Zomba, Blantyre, Chiradzulu, Thyolo, Mulanje, Balaka, Machinga, Mangochi, Ntcheu, Salima, Rumphi, and Karonga.

References

Alexander, M., Priest, S. and Mees, H. (2016) "A framework for evaluating flood risk governance," *Environmental Science and Policy*, 64, pp. 38–47. doi:10.1016/j.envsci.2016.06.004.

Atkins. (2012) *Shire Integrated Flood Risk Management Project Volume I – Final Report*. Available at: https://www.geonode-gfdrrlab.org/documents/470/download.

Begum, S., Stive, M.J.F. and Hall, J.W. (2010) *Flood Risk Management in Europe: Innovation in Policy and Practice*, p. 534. Available at: https://www.semanticscholar.org/paper/Flood-risk-management-in-Europe-%3A-innovation-in-and-Begum-Stive/6c413175999c0d0a91fa1eb7f4e0e72c224a7155.

Bischiniotis, K. *et al.* (2020) "A framework for comparing permanent and forecast-based flood risk-reduction strategies," *Science of The Total Environment*, 720, p. 137572. doi:10.1016/j.scitotenv.2020.137572.

Botha, B.N., Nkoka, F.S. and Mwumvaneza, V. (2018) *Hard Hit by El Nino: Experiences, Responses and Options for Malawi*. Available at: www.copyright.com/.

Brinton Milward, H. (1996) "Symposium on the hollow state: Capacity, control, and performance in interorganizational settings," *Journal of Public Administration Research and Theory*, 6(2), pp. 193–196. doi:10.1093/OXFORDJOURNALS.JPART.A024306.

Browder, G. *et al.* (2021) *An EPIC Response: Innovative Governance for Flood and Drought Risk Management Main Report*. Available at: https://www.worldbank.org/en/topic/water/publication/an-epic-response-innovative-governance-for-flood-and-drought-risk-management.

Bucherie, A. *et al.* (2022) "Flash flood warnings in context: Combining local knowledge and large-scale hydro-meteorological patterns," *Natural Hazards and Earth System Sciences*, 22(2), pp. 461–480. doi:10.5194/nhess-22-461-2022.

Calvel, A. et al. (2020) "Communication structures and decision making cues and criteria to support effective drought warning in central Malawi," *Frontiers in Climate*, 2(November). doi:10.3389/fclim.2020.578327.

Chavula, J. (2015) *Decentralized Early Warning Systems in Malawi*. Available at: https://www.mdpi.com/2071-1050/11/6/1681/htm.

Chawawa, N.E. (2018) *Why Do Smallholder Farmers Insist on Living in Flood Prone Areas? Understanding Self-Perceived Vulnerability and Dynamics of Local Adaptation in Malawi*. Available at: https://era.ed.ac.uk/handle/1842/31421.

Chidanti-Malunga, J. (2011) "Adaptive strategies to climate change in southern Malawi," *Physics and Chemistry of the Earth*, 36(14–15), pp. 1043–1046. doi:10.1016/j.pce.2011.08.012.

Chiotha, S. et al. (2016) *Strengthening Early Warning in Malawi: Proceedings of the First Stakeholder Workshop on Enhancing Early Warning System in Malawi: Golden Peacock Hotel, 10th and 11th August 2016*. Available at: https://www.worldcat.org/title/strengthening-early-warning-in-malawi-proceedings-of-the-first-stakeholder-workshop-on-enhancing-early-warning-system-in-malawi-golden-peacock-hotel-10th-and-11th-august-2016/oclc/1021064854

Chiusiwa, J. (2015) *Malawi National Progress Report on the Implementation of the Hyogo Framework for Action (2013–2015)*. Available at: www.preventionweb.net/english/hyogo/progress/reports/.

CRED and UNDRR (2020) *Human Costs of Disasters: An Overview of the Last 20 Years 2000–2019*. Available at: https://www.undrr.org/publication/human-cost-disasters-overview-last-20-years-2000-2019.

Cumiskey, L.R. (2020) *Joining the Dots: A Framework for Assessing Integration in Flood Risk Management with Applications to England and Serbia*. London: Middlesex University.

Department of Climate Change and Meteorological Services (2019) *National Meteorological Policy*. Available at: www.metmalawi.com.

Department of Disaster Management Affairs (2019) *National Disaster Risk Management Communication Strategy Voice from Primary Audience*. Available at: https://drmims.sadc.int/sites/default/files/document/2020-03/NDRMCS-CONSOLIDATED.pdf.

di Baldassarre, G. et al. (2010a) "Flood fatalities in Africa: From diagnosis to mitigation," *Geophysical Research Letters*, 37(22). doi:10.1029/2010GL045467.

di Baldassarre, G. et al. (2010b) "Analysis of the effects of levee heightening on flood propagation: Example of the River Po, Italy," *Hydrological Sciences Journal*, 54(6), pp. 1007–1017. doi:10.1623/HYSJ.54.6.1007.

Dube, E. and Munsaka, E. (2018) "The contribution of indigenous knowledge to disaster risk reduction activities in Zimbabwe: A big call to practitioners," *Jàmbá – Journal of Disaster Risk Studies* [Preprint]. doi:10.4102/jamba.

Goldsmith, S. and Eggers, W.D. (2004) *The New Shape of the Public Sector*. Washington, DC: Brookings Institution Press.

Government of Malawi (2015a) *Malawi 2015 Floods Post Disaster Needs Assessment Report*. Available at: https://www.ilo.org/wcmsp5/groups/public/-ed_emp/documents/publication/wcms_397683.pdf.

Government of Malawi (2015b) *National Disaster Risk Management Policy*. Available at: https://www.preventionweb.net/files/43755_malawidrmpolicy2015.pdf.

Government of Malawi (2016a) *Annex II-Feasibility Study I Scaling-up Early Warning Systems and Use of Climate Information in Malawi-Feasibility Assessment*. Available at: https://pims.undp.org/attachments/5710/214506/1697418/1697699/FP-UNDP-100915-5710-Annex%C2%A0II%20resized.pdf (Accessed: April 24, 2022).

Government of Malawi (2016b) *Malawi Drought 2015–2016 Post-Disaster Needs Assessment PDNA World Bank*. Available at: https://openknowledge.worldbank.org/handle/10986/25781.

Hallegatte, S. (2015) *Recent Floods in Malawi Hit the Poorest Areas: What This Implies*. Available at: https://blogs.worldbank.org/voices/recent-floods-malawi-hit-poorest-areas-what-implies (Accessed: February 25, 2022).

Hendriks, T.D. and Boersma, F.K. (2019) "Bringing the state back in to humanitarian crises response: Disaster governance and challenging collaborations in the 2015 Malawi flood response," *International Journal of Disaster Risk Reduction*, 40(May), p. 101262. doi:10.1016/j.ijdrr.2019.101262.

Hermans, T. *et al.* (2022) *Exploring the Integration of Local and Scientific Knowledge in Early Warning Systems for Disaster Risk Reduction – A Review*. Available at: https://link.springer.com/article/10.1007/s11069-022-05468-8.

Hiwasaki, L. *et al.* (2014) "Process for integrating local and indigenous knowledge with science for hydro-meteorological disaster risk reduction and climate change adaptation in coastal and small island communities," *International Journal of Disaster Risk Reduction*, 10, pp. 15–27. doi:10.1016/j.ijdrr.2014.07.007.

Kita, S.M. (2017) "'Government doesn't have the muscle': State, NGOs, local politics, and disaster risk governance in Malawi," *Risk, Hazards and Crisis in Public Policy*, 8(3), pp. 244–267. doi:10.1002/rhc3.12118.

McClymont, K. *et al.* (2020) "Flood resilience: A systematic review," *Journal of Environmental Planning and Management*, 63(7), pp. 1151–1176. doi:10.1080/09640568.2019.1641474.

Mees, H. *et al.* (2016) "Coproducing flood risk management through citizen involvement: Insights from cross-country comparison in Europe," *Ecology and Society*, 21(3). doi:10.5751/ES-08500-210307.

Mercer, J. (2010) "Disaster risk reduction or climate change adaptation: Are we reinventing the wheel?," *Journal of International Development*, 22(2), pp. 247–264. doi:10.1002/jid.1677.

Mijoni, P.L. and Izadkhah, Y.O. (2009) "Management of floods in Malawi: Case study of the Lower Shire River Valley," *Disaster Prevention and Management: An International Journal*, 18(5), pp. 490–503. doi:10.1108/09653560911003688.

Nillson, A., Shela, O.N. and Chavula, G. (2010) *Flood Risk Management Strategy: Mitigation, Preparedness, Response and Recovery*. Lilongwe: Department of Disaster Management Affairs.

Nkomwa, E.C. *et al.* (2014) "Assessing indigenous knowledge systems and climate change adaptation strategies in agriculture: A case study of Chagaka Village, Chikhwawa, Southern Malawi," *Physics and Chemistry of the Earth*, 67–69, pp. 164–172. doi:10.1016/j.pce.2013.10.002.

Pauw, K., Thurlow, J. and van Seventer, D. (2010) *Droughts and Floods in Malawi: Assessing the Economywide Effects*. Available at: www.ifpri.org/publications/results/taxonomy%3A468.

Phiri, I.M.G. and Saka, A.R. (2008) "The Impact of Changing Environmental Conditions on Vulnerable Communities in the Shire Valley, Southern Malawi," *The Future of Drylands*, pp. 545–559. doi:10.1007/978-1-4020-6970-3_49.

Šakić Trogrlić, R. (2020) *The Role of Local Knowledge in Community-Based Flood Risk Management in Malawi*. Available at: https://www.ros.hw.ac.uk/handle/10399/4241.

Sakic Trogrlic, R. and van den Homberg, M. (2018) *Indigenous Knowledge and Early Warning Systems in the Lower Shire Valley in Malawi*. Available at: www.researchgate.net/publication/327701675 (Accessed: February 25, 2022).

Šakić Trogrlić, R. *et al.* (2018) "Taking stock of community-based flood risk management in Malawi: Different stakeholders, different perspectives," *Environmental Hazards*, 17(2), pp. 107–127. doi:10.1080/17477891.2017.1381582.

Šakić Trogrlić, R. *et al.* (2021) "External stakeholders' attitudes towards and engagement with local knowledge in disaster risk reduction: Are we only paying lip service?," *International Journal of Disaster Risk Reduction*, 58(February). doi:10.1016/j.ijdrr.2021.102196.

Tellman, B. *et al.* (2021) "Satellite imaging reveals increased proportion of population exposed to floods," *Nature*, 596(7870), pp. 80–86. doi:10.1038/s41586-021-03695-w.

Teule, T. *et al.* (2020) *Towards Improving a National Flood Early Warning System with Global Ensemble Flood Predictions and Local Knowledge; a Case Study on the Lower Shire Valley in Malawi*. doi:10.5194/EGUSPHERE-EGU2020-507. Available at: https://ui.adsabs.harvard.edu/abs/2020EGUGA..22..507T/abstract.

Thaler, T. and Hartmann, T. (2016) "Justice and flood risk management: Reflecting on different approaches to distribute and allocate flood risk management in Europe," *Natural Hazards*, 83(1), pp. 129–147. doi:10.1007/s11069-016-2305-1.

Trogrlić, R.Š. *et al.* (2019) "Characterising local knowledge across the flood risk management cycle: A case study of Southern Malawi," *Sustainability (Switzerland)*, 11(6). doi:10.3390/su11061681.

UNDRR (2022) *Early Warning System*. Available at: www.undrr.org/terminology/early-warning-system (Accessed: February 24, 2022).

UN-Habitat (2022) *Malawi | UN-Habitat*. Available at: https://unhabitat.org/malawi (Accessed: March 2, 2022).

van den Homberg, M. and McQuistan, C. (2019) "Technology for climate justice: A reporting framework for loss and damage as part of key global agreements," in *Loss and Damage from Climate Change, Concepts, Methods and Policy Options*. Cham: Springer International Publishing, pp. 513–545. doi:10.1007/978-3-319-72026-5.

van den Homberg, M.J.C., Gevaert, C.M. and Georgiadou, Y. (2020) "The changing face of accountability in humanitarianism: Using artificial intelligence for anticipatory action," *Politics and Governance*, 8(4), pp. 456–467. doi:10.17645/pag.v8i4.3158.

van Milligen de Wit, M.N. (2021) *Urban Surface Water Flooding in Mzuzu*. Available at: https://d-nb.info/1234660121/34.

Vojinović, Z. and Abbott, M.B. (2012) *Flood Risk and Social Justice: From Quantitative to Qualitative Flood Risk Assessment and Mitigation, Water Intelligence Online*. Cambridge: IWA Publishing. doi:10.2166/9781780400822.

Waylen, K. and Martin-Ortega, J. (2013). *Report on knowledge exchange workshops on an ecosystem services approach*. Aberdeen: The James Hutton Institute.

WMO (2008) *Guide to Hydrological Practice Volume I Hydrology-From Measurement to Hydrological Information*. Available at: https://library.wmo.int/index.php?lvl=notice_display&id=541.

Wouters, L. *et al.* (2021) "Improving flood damage assessments in data-scarce areas by retrieval of building characteristics through UAV image segmentation and machine learning – A case study of the 2019 floods in southern Malawi," *Natural Hazards and Earth System Sciences*, 21(10), pp. 3199–3218. doi:10.5194/nhess-21-3199-2021.

Zhang, Q. *et al.* (2019) "Increasing the value of weather-related warnings," *Science Bulletin*, pp. 647–649. doi:10.1016/j.scib.2019.04.003.

23

Exploring the perspective of school children on flood risk management in developing countries

Lessons from Ghana

Henry Mensah, Grace Wanma, Divine Kwaku Ahadzie and Eric Kwame Simpeh

Introduction

Globally, the adverse impact of floods has been a major concern for schools and school children, who are particularly vulnerable to these natural disasters (Intergovernmental Panel on Climate Change (IPCC), 2012). School children account for about 50%–60% of flood victims, thus making them a highly vulnerable group (Seballos et al., 2011). For example, polluted floodwater causes sickness, while the destruction of school infrastructure contributes to school dropout or absenteeism, thus delaying academic progress and increasing children's exposure to various life stresses (Fothergill, 2017). In Nepal, about 57 schools were negatively affected by floods, and more than 23,000 children have died due to flooding (Schonfeld and Demaria, 2015). Similarly, in the United States of America, Hurricane Katrina in 2005 forced children to drop out of school while some schools were closed down (Frazier-Anderson, 2008). Getting children's perspective on flood disasters matters in cities' development process. Children are highly vulnerable to the effects of climate change, and those living in marginal environments and poverty are mostly at high risk (Peek, 2008; Hanna and Oliva, 2016). For school children, the occurrence of floods in the academic calendar can be a huge distraction and disruption to regular educational activities.

While there are vast studies on floods in the literature, the perspective of children has not been widely accounted for (Mort et al., 2018a; King et al., 2013), particularly on the African continent. However, as noted by Lopez et al., 2012) the perspective of children plays a major role in disaster risk reduction (DRR), as they form a large proportion of disaster-affected groups. Moreover, knowledge of children about floods is essential, as it allows for the definition of priority needs with respect to their safety both at home and in school (Mort et al., 2018b). The Sendai Framework for Disaster Risk Reduction (SFDRR) also recognizes that children can serve as agents of change for contributing to DRR (Muzenda-Mudavanhu et al., 2016).

DOI: 10.1201/9781003160823-28

Ghana has experienced severe floods affecting children, adults, and families in the past 20 years. Generally, flooding in Ghana is caused by heavy downpour, rainfall-run-off, and tidal waves. These are exacerbated by anthropogenic factors such as poor drainage systems, poor environmental attitude, and enforcement of urban planning policies. Flooding impacts several aspects of community life both in southern and northern Ghana. Indeed, in northern Ghana, apart from the impact of floods on agriculture (Musah and Akai, 2014), there are growing effects of flood risk on schools as well as on other infrastructure. For example, in 2009, over 121,000 people were displaced, about 5,104 houses were destroyed, 13 schools collapsed, and 30,000 acres of farmlands were destroyed due to flood occurrence (Kunateh, 2016). Sagnarigu Municipality is one of the rapidly urbanizing settlements in northern Ghana, and many of the areas are inaccessible, with infrastructure, properties, and livelihood destroyed during floods. Moreover, many residents and schools in Sagnarigu Municipality were submerged by flooding, which left people stranded. Hence Ghana is implementing a flood control programme to improve the resilience of settlements across the country, including in the Sagnarigu Municipality (Balgah et al., 2019). For example, drainage works, relocating flood victims, and construction of culverts are ongoing in many communities. Additionally, the government of Ghana has put measures in place to reduce vulnerabilities against climate change impact through policies. These policy provisions are contained in a myriad of documents such as the Ghana National Climate Change Policy (NCCP) (2014), National CC Adaptation Strategy (2010–2020), and NCCP Action Programme (Mensah et al., 2021). This is intended to support the resilience of vulnerable communities to climate-related risks.

However, as in many policy guidelines in developing countries such as Ghana, there is limited recognition given to children's welfare and needs, partly because of the limited knowledge and research information available (cf. Anderson, 2005). Here, it is argued that understanding children's knowledge and practices can help identify what is required and how children's participation can be facilitated to reduce their vulnerability to flood disasters. Furthermore, taking children's disaster knowledge and practices into account can support their families and communities before, during, and after floods and contribute to the DRR (Mitchell and Borchard, 2014). To fill this gap in the literature and contribute to the sustainable city development process, the authors examined the Sagnarigu Municipality of Ghana as a case study to generate some insights into the perspectives of school children with respect to flood risk management. This chapter addresses the following questions: 1. What are the causes of floods in the communities? 2. What are the effects of floods on the communities and schools? and 3. What are the coping strategies employed by school children during a flood?

Conceptual framework

Figure 23.1 helps to understand and analyse children's perspectives on the causes, effects, and coping strategies to reduce their vulnerability to flooding. Mainstreaming disaster education at the school level has been seen as an initiative in disaster risk mitigation to promote sustainable public awareness (UNDRR, 2020). Thus, when children acquire knowledge and are aware of flood risk, they are more likely to communicate their concerns and then prepare adequately for future flood occurrences, in turn reducing their vulnerability to flooding (Muzenda-Mudavanhu, Manyena and Collins, 2016). The awareness of flood risk allows school children to take necessary action to prepare and cope and even participate in DRR activities. The relationship between reduced flood vulnerability as elements of flood perception (causes, effects, and coping strategies) is influenced by flood education and awareness. Flooding harms

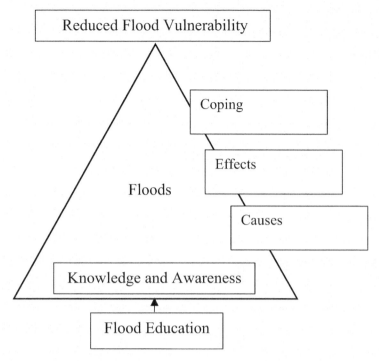

Figure 23.1 Conceptual framework
Source: Authors' construct

children's well-being, hence the need for flood education for children (Muzenda-Mudavanhu et al., 2016). Disaster education can provide appropriate knowledge, skills, and behavioural change for successful DRR.

With the Sendai Framework, the United Nations Educational, Scientific and Cultural Organization has developed the Comprehensive School Safety Framework that enables educational departments and associated partners to execute disaster management tasks effectively. The three pillars of this framework (UNESCO, 2013) are: (1) 'safe learning facilities: this requires that school site selection, design, construction and maintenance, incorporate disaster-resilient design, and construction to make every school a safe school is the core to safe learning facilities', (2) 'school disaster management; this demands, through collaboration between educational authorities, schools and local communities (including students and parents) and disaster management units, to conduct school-based mitigation and response assessment and planning, so that safe learning environment and continuous education are well maintained' and, (3) 'risk reduction and resilience education: this implies developing consensus-based key messages, such as teaching and learning materials in DRR towards engendering a culture of safety and resilient communities'. King (2002) proposes that flood risk awareness increases when a society is affected by a disaster and information about the hazard is more likely to be available. Therefore, assessing children's flood experiences will help determine the extent of flood awareness among children for insights on possible solutions that can lead to reduced vulnerability in the Sagnarigu Municipality and elsewhere.

Study area and methods

Study area

Ghana has 16 administrative regions, one of which is the Northern Region. The Sagnarigu Municipality, located in the northwest part of the Northern Region, has Sagnarigu as its capital city It covers a total land size of 200.4 km^2 and has a population of 341,711 (citypopulation.de/en/ghana/admin/northern/). It experiences one rainy season (a mean annual rainfall of 1100 mm within 95 days of intense rainfall) and one dry season in a year. The rainy season occurs from April/May to September/October, peaking in July/August.

Study methods

The study relied on both primary and secondary data sources. The primary data were collected from school children who have experienced flooding. Secondary sources of data include scholarly articles that focus on children-flood relationships in Ghana (e.g., Musah and Oloruntoba, 2013; Armah et al., 2010), flood reports from Ghana (e.g., the Environmental Protection Agency: Flood and Drought Risk Mapping in Ghana, 2012, the National Disaster Management Organization, Annual Report 2010, the National Disaster Management Organisation, Annual Report, 2011) and global perspective on flood risk (e.g., Hanna and Oliva, 2016; Martin, 2010; Muzenda-Mudavanhu et al., 2016; Shah et al., 2020; Kousky, 2016).

We argue that flooding is common in Sagnarigu and affects all school children directly and indirectly in the municipality, which suggests that all children are aware of and have experienced flooding. However, the impact on each child may vary. On this basis, the authors used a random sampling technique to get children's knowledge about flooding. The study focused on 14 upper primary pupils (9–11 years), 20 junior high school (JHS) students (years 12–14), and 26 senior high school (SHS) students (15–17 years). In terms of gender, there were 41 males and 19 females. The children's age group was selected based on Ghana's national education system (UNESCO Institute for Statistics [UIS], 2011). A total of 16 Focus group discussion (FGDs) were held from three schools, SDA SHS, Gumani Methodist JHS, and Higher Always International School. This resulted in selecting 60 respondents (two groups containing 10 respondents for each school).

The FGD was conducted in 2019 on weekdays from 12:00 pm to 4:00 pm. The whole discussion was audio recorded. The respondents were encouraged to ask questions, brainstorm, re-evaluate, and reconsider their understanding of the nature of the flood in Sagnarigu Municipality. The rationale of the FGD was explained to the children. They were informed about the scope of the study and the overall objective of the questions. They were also informed that they could leave the discussion at any point or request the termination of the interview. Written informed consent were obtained from the head of institutions and also the Parent Teachers Associations (PTA). In Ghana, all schools have a PTA which is actively involved in decision making for the running of the schools. Oral informed consent was also obtained from the school children for participation in the study and recording of interviews. They were also assured of the confidentiality of their information and responses. Drawing from the study's research questions, guiding questions for the FGD are illustrated in Table 23.1.

Using a content analytical approach, the transcribed data were analysed by Qualitative Data Analyses Software, NVivo desktop version 12.0. NVIVO 12 is professional software designed to support the researcher to present reliable coding in the shortest possible time. It also helps in the identification of patterns of ideas in compiling, disassembling, and reassembling phases.

Table 23.1 The guiding questions for the FGD

Themes	Key guiding questions
The causes of floods	What are the causes of floods in the communities?
	How do human settlement and land use contribute to flooding?
	How does the community drainage system contribute to flooding?
	How does waste contribute to flooding?
The effects of floods	What are the effects of floods in the communities and schools?
	What are the effects of floods on school infrastructure and learning materials?
	What are the effects of floods on school turnout and enrolments?
The coping strategies	What coping strategies have you adopted during floods and how do you use them?
	What do communities and schools need to know before, during, and after flooding?

Results

Situational analysis of floods in Sagnarigu Municipality

Causes of floods in communities

It was mentioned that the indiscriminate dumping of refuse into gutters has led to the choking of gutters. *Choked gutters* were repeatedly mentioned in the responses. Similarly, *rubbish, water, building, ways,* and *blocked* are some of the keywords that were prominent in describing the causes of floods in the municipality. Word context search query results (i.e. the world cloud) portray phrases like *choked gutters with rubbish, building in waterways,* and *blocked gutters.*

CHOKED GUTTERS

Regarding *choked gutters*, a respondent narrated that:

> In our area, the people dump refuse anyhow, so it blocks the gutter to prevent water from flowing. They sometimes wait until the evening when no one can notice them and eventually dump the refuse in the gutters.
>
> *(RP 1)*

> If there are rubbishes in the gutters, water will not be able to pass through freely. It will accumulate and cause a flood.
>
> *(RP 22)*

They expressed that the effects of flood could be minimized with a proper waste collection system. The city authorities cannot collect the volume of solid waste in the form of silt or rubbish. People dump solid waste directly into watercourses, drains, and culverts, reducing flow capacity and increasing overflows. The capacities of most drains have been greatly reduced by the common deposition of garbage and weeds along the drain banks.

In the opinion of RP 5:

> The gutters have overflown their paths. The gutters are filled with garbage because people throw rubbish into the gutters, especially when it is raining with the expectation that the running water will wash them away.

HUMAN SETTLEMENT

The respondents narrated that carelessness by the people and disrespect of urban law and regulations (e.g., building along waterways, unplanned settlements) contribute to flood risk. Respondents noted that some buildings in the suburbs of Sagnarigu Municipality are positioned close to drains, while others have been built a few meters away from the natural watercourse. This has increased the risk of these settlements to perennial flooding. It was revealed that *building*, *water*, and *ways* were prominent in the responses. This indicated that respondents described *building on or near water ways* as the major cause of flooding. This has been exemplified with the preceding narrations from the interviews:

> Building houses in marshy areas also caused flood, building your house on waterways can again cause flooding and affect people and buildings.
>
> *(RP 1)*

Similarly, R8 said:

> Flood occurs because some people build houses on waterways, so when it rains, all the houses get submerged, destroying properties.

> Many people are ignorant. They know that this place is a water-logged area, yet they build their houses on it.
>
> *(RP 32)*

POOR DRAINAGE SYSTEMS

The poor drainage system is one of the critical causes of floods. It was revealed that *drainage*, *lack*, and *gutters* featured prominently in the responses. Word context search query results portray phrases such as *poor gutters*, *narrow drains*, and *lack of drainage systems*. A respondent expressed that:

> I think our infrastructures are not properly built. Our gutters are too small for the water to pass through. When it rains heavily the water overflows its banks, causing havoc to the streets.
>
> *(RP 63)*

> The way the dormitories are built and how the gutters were constructed is the cause of the flood.
>
> *(RP 70)*

Current drains in the Gumani and Choggu, Yapalsi, are often choked with refuse and silt, which results in flooding. For instance, the major streams and gutters in Sagnarigu Municipality have become depots of wastewater, including faecal matter and solid waste. This has led to public health threats to people. In the Fuo, Gumani, and Choggu-Yapalsi communities, there are inadequately designed culverts for flood control, potentially resulting in flooding of the area. This challenge is compounded by continuous blockage of major drains resulting from lack of maintenance.

Effects of floods in communities and schools

Schools are exposed to various flood risks, and this has affected the quality of education of the school children. The results show that the word *destroys* featured most prominently in the responses. Similarly, *schools, learning materials,* and *life* are some of the prominent words in the description of the effects of floods in the communities and schools. Word context search query results portray phrases like *destroys properties, destroys life,* and *destroys infrastructure.*

DESTROYS PROPERTY, INFRASTRUCTURE, AND LEARNING MATERIALS

The respondents emphasized that infrastructural property is mostly affected by the flood. The analysis revealed that *property* was used to describe the effects of flooding. The destruction of residential properties and school buildings was also noted.

RP 24 noted:

> Yes, some apartments were destroyed. Others have the roofs of their houses removed. One of our neighbour's houses was destroyed to the extent that the ceiling ripped off and injured him. He came to our house for refuge.
>
> It destroys people's farms and their houses. It also removes the roofs of people's houses.
>
> *(RP 24)*

> but there were two residents that had their properties destroyed. One of them had his properties destroyed severely because his house was made of clay. The water washed the house and he got wounded. He came to my parents for first aid. The other person had his house built with cement, and escaped narrowly because his ceiling ripped off.
>
> *(RP 30)*

The material provided to teachers and children is adversely affected by floods. It was revealed that *learning materials* featured prominently in the responses. The respondents reported that floods washed away learning materials, and buildings developed cracks and holes. One respondent lamented:

> The flood destroyed my books when the floodwater entered our building through the holes and cracks. I have to borrow books from friends to enable me to study. It is not all the time my friends are willing to share their books with me.
>
> *(RP 64)*

Another respondent emphasised that school infrastructure was particularly affected.

> It destroys the school infrastructure such as the floor and the walls. The whole building is submerged during the flood, thereby soaking the learning materials.
>
> *(RP 63)*

DESTROYS HUMAN LIFE

It was reported that during the flood period, lives and properties were lost, and people were carried away due to the high volumes of water as they crossed the floodwaters. The word text query search in the data produced a pictorial of the context within which *loss of life* suffices as a key effect of flood. One of the respondents in Gumani had a major concern:

> During the flood, I heard that someone got drowned because the person was living in a flood-prone area. I saw how a woman was carried away in her car by floodwater in 2017.
> *(RP 50)*

Flood-related death in the municipality is attributed to the depth and velocity of the floodwater. According to the respondents, it's normal for children to lose their lives when it rains because children usually stay outside and play with friends in the rain, which is dangerous.

LOW TURNOUT AND LOW ENROLMENT

It was revealed that some children travel by road from their homes to school during floods. The respondents admitted that they cannot go to school when the roads leading to their schools are destroyed by flooding. Other respondents expressed that schools became flooded during the flood, preventing children from attending school. The inaccessibility of roads due to floods influences school attendance and the distribution of teaching and learning materials to schools.

For instance, a respondent lamented that:

> During heavy rains, roads in our area become flooded; it makes our movement difficult.
> *(RP 48)*

Another respondent recounted that:

> I will not be able to come to school, my colleagues will not come to school, and my teachers will not come to school.
> *(RP 1)*

> The school children will be in the house learning what they should have been learning at school.
> *(RP 8)*

> Some of us will stay in the house; we will not come to school.
> *(RP 17)*

> It prevents our access to learning materials like books. It also destroys buildings and learning materials.
> *(RP 51)*

HEALTH RISK

The qualitative data analysis revealed that floods destroyed health facilities, including all medical equipment and medicine, and disrupted healthcare delivery. It was also revealed that floods spread diseases such as malaria that affect the people. The water becomes polluted from pit latrines and open defecation, causing diarrhoea, respiratory tract infections, cholera, and other water-borne diseases. Another key point mentioned was that children usually play in the water during flooding and are constantly bitten by poisonous snakes.

> The floodwaters sometimes come with reptiles and bite us. The flood destroyed our school wall and water entered the classrooms washing away the learning materials.
>
> *(RP 75)*

> The floods lead to food insecurity because foodstuffs are lost during the floods.
>
> *(RP 73)*

> It displaces people and spreads disease, causing mortality.
>
> *(RP 58)*

Coping strategies

School children seek help, stay with other people until it is safe to go home, wait until floodwater subsides, stay at home or school when it is raining, and borrow learning materials from friends. *Help* was the most prominent word in the responses. Similarly, *home, stay, call,* and *school* were some prominent words in the description of coping strategies employed by school children. Word context search query results portray phrases like *help from neighbours, stay at home,* and *stay in school*. All the respondents narrated that they will seek neighbours' help and ask for support from NGOs or benevolent organizations. During a flood event, school children transport their valuable items, including textbooks and chalk, to a safe place. Generally, children depend on their family members for food and shelter.

WAITING/STAYING IN ONE PLACE

The respondents revealed that they either stayed at home or school or at a safer place for the rain to stop. The analysis shows that *stay, wait at home,* and *school* featured prominently throughout the responses. Some school children received support from their family members, while others resorted to parks and open spaces to seek refuge until the floodwater level reduced. A respondent indicated that:

> I stay in class if I am in school. And if I am at home, I stay in the room. I sometimes go to my auntie's house closer to the school.
>
> *(RP 2)*

BORROWING LEARNING MATERIALS

Respondents mentioned that they usually borrow books from friends and the school library. It emerged that borrowing learning materials from friends is a common strategy. A respondent indicated that:

> We depended on the notes our teachers gave us and borrowed them from friends who have the books.
>
> *(RP 72)*

However, some also face challenges in getting books from friends. A respondent reported that:

> We borrow books from friends and sometimes they refuse to give us. We studied with our friends who were not affected. We also copied the notes from our masters.
>
> *(RP 75)*

Towards flood management in Sagnarigu Municipality

The findings show that school children expect the authorities to resolve flood challenges in the municipality. It has also emerged that even when such support is provided, it is delayed and not enough. A respondent mentioned that:

> Some NGOs, Assemblyman, Parent Teachers Association, etc., can assist flood victims. It isn't easy to study without books. However, the support from benevolent organisations and NADMO are not enough for flood victims, particularly school children. But it is better than nothing.
>
> *(RP 62)*

Through communal effort, the school cleans drains and gutters. It is expected that people can support desilting or filling holes created by the flood. One respondent mentioned:

> We sought the assistance of some people in the community to help us cover holes with sand. We borrow sand to fill the eroded places.
>
> *(RP 11)*

while RP 12 recounted that:

> We should stop throwing rubbish in gutters and build our houses in the right places.

Another respondent expressed:

> Some days after the rain, we asked some people from the neighbouring community to assist us to fill the eroded areas to prevent any flooding in the future.
>
> *(RP 24)*

A respondent suggested that:

> Before the flood, we cover the potholes, and clean up the gutters so that the rainwater can easily flow through.
>
> *(RP 26)*

Moreover, some suggested that communal labour could help reduce the flood risk if properly enforced. A respondent expressed that:

> They could also organise communal labour. They should remove the rubbish from the gutters. They should also clean the environment to prevent diseases.
>
> *(RP 31)*

The school children understand that proper siting of buildings and structural retrofitting could reduce their vulnerability to floods. They confirmed that building houses with clay and/or mud cannot withstand the impact of flood and therefore expressed the need to build houses with flood-resistant materials. Moreover, building on water pathways must be avoided to avoid possible loss of life and property. RP 60 indicated that:

> The school wall can be built to retain the floodwater. Sacks containing sand can also be used to retain the flood from entering the school compound.

Similarly, the respondents expressed opinions on how to make school resilient to flood:

> They should construct gutters and bigger culverts in the school so that it can control the flow of water in the school.

while RP 67 adds that:

> I think they should rehabilitate the infrastructure that was not properly constructed to withstand a flood.

Discussion

Children's perceptions and interpretations of floods were overall well informed and largely grounded in scientific knowledge. The analysis indicates that the volume of solid waste in the form of silt or rubbish cannot be collected by existing systems in Sagnarigu Municipality. Some gutters are used for dumping solid waste, thereby causing a blockage; others have been greatly reduced by growing weeds. This implies a lack of enforcing local sanctions on indiscriminate waste disposal. The city authority needs to educate the public on the consequences of building in an undesignated space and enforce land-use planning in Sagnarigu Municipality (Tasantab, 2019). The results point out that flooding causes havoc to home, properties, and schools. Furthermore, flooding causes death through illnesses, injuries, snake bites, and drowning. Under these conditions, it is challenging for children to attend school and obtain a quality education (Mort et al., 2018a).

Interventions are needed from the government, NGOs, and the private sector to resolve the flood situation in the community. Moreover, teachers, parents, and traditional leaders play roles in flood disaster management (Santoro et al., 2019). Support from these stakeholders will enable the problem to be resolved holistically, leaving no one behind (Takeuchi et al., 2011). The coping mechanisms by school children are inadequate, so the government and NGOs would support the children's adaptation efforts. Following disasters, children are further disadvantaged in the help-seeking process and lack experience and knowledge about how to seek help (Silverman and La Greca, 2002). The flood often impacts an entire community, undermining children's sense of security and normalcy. Generally, most school children will obtain family support during a flood, while some children may be at risk of more extreme reactions. Therefore, strategies to reduce risks to school children would require providing early warning information and advice to households on how to respond to floods (Cools et al., 2016). This enables school children to adequately prepare for an impending disaster, reduce vulnerability, and improve resilience. Disaster education programming for children is critical to minimizing the vulnerabilities and impact of disasters. Education is regarded as one of the best media to prepare a community for DRR (Mensah and Ahadzie, 2020).

Indeed, the importance of disaster education is widely acknowledged globally (Azmi et al., 2020). For example, disaster lessons have been known to be integrated into geography and science classes in Australia, New Zealand, and Iran for children to recognize hazards to the climate, and environment. More so, the Caribbean conducted extracurricular programmes as preliminary testing specifically in disaster awareness promotion among children by using an interactive way of teaching and learning (Disaster Awareness Game; DAG). Japan took proactive steps by making crucial practices compulsory at school, such as emergency planning, safety inspections, and evacuation drills. The Philippines also has launched a children's book on disaster preparedness entitled *Handa Ako*; the book includes information on disaster awareness and preparedness. However, there is currently little evidence of any such emphasis in encouraging curriculum design for the uptake of children's education in appreciating the risk associated with flood disasters and effective management approaches in sub-Saharan African countries such as Ghana (Mudavanhu, 2014). As noted by Izadkhah and Hosseini (2005), education is one of the key ways of preparing a community for disasters. However, in most developing countries, education at all levels for disaster preparedness is lacking due to the lack of expertise and educational materials. In particular, Izadkhah and Hosseini (2005) opine that 'one of the best ways of publicising awareness programmes is the integration of these initiatives into children's programmes in both pre-school and school levels, as children are believed to be more receptive to new ideas than adults'. On the African continent, current best practice is in Kenya, where Mburu (2020) reports that the Education Ministry had integrated flood disaster management into schools' curricula towards mainstreaming flood disaster management into children's education. However, further evidence provided by Mburu (2020) reveals that there was a lack of instructional material to support the implementation of the programme. Mburu (2020) also observed that teachers had no training in disaster management for classroom instruction. In Ghana, even though flooding has become a perennial issue affecting school children in several negative ways, the Ministry of Education has yet to integrate flood disaster management into schools. Here, there are some lessons to learn from the Kenyan experience in this context. The missing link will be how to ensure that the appropriate instructional materials are developed for use in teaching and learning. Appropriate training for teachers and stakeholders in the educational sector is another missing link that needs to be addressed urgently. This chapter highlights opportunities for such engagement as a matter of need, given that children are amongst the most vulnerable groups confronted with flood hazards on the continent.

Conclusion and recommendations

This study explores the perspective of school children in Sagnarigu Municipality, Ghana, to understand the causes of floods, the effects of floods in communities and schools, and coping strategies that are employed by school children during flooding. Generally, the study revealed that the perception of school children agrees with much of the literature in flood research in the developing country context, including the causes, effects, and coping strategies. The present study argues that children are among the most vulnerable group affected disasters, and it is expected that their vulnerabilities should be given priority, which has not always been the case.

It is recommended that the government introduce flood risk management seminars/counselling sessions for school children and teachers to enhance their capacity for disaster preparedness and response. Also, disaster education and stress relief programming for affected school children should be mainstreamed. More so, the Education Service should consider incorporating flood risk management topics in basic and secondary school curricula to increase the preparedness level of school children. There is a need to strengthen educators' knowledge on disaster education to facilitate interaction with learners in delivering knowledge and trigger their curiosity about disaster learning. Also, municipal planning and works departments should enforce settlement planning and construction regulations to prevent people from building in flood-prone areas, including school infrastructure. Finally, there is a need to create socially inclusive disaster management committees at the school and community levels to respond to flood disasters. In terms of future research, apart from coping strategies, the various facets of the help-seeking process have been insufficiently studied in children and would be essential to further research.

Acknowledgements

The authors wish to thank the Head of Institutions and the Parent Teacher Associations (PTA) for permission to use their schools for the study. In this context, we would also like to express our appreciation to the school children who participated. We also wish to thank anonymous referees for their useful suggestions, which the authors believe have helped in improving the quality of the paper.

References

Anderson, K. G. (2005) 'Relatedness and investment in children in South Africa', *Human Nature*, 16(1), pp. 1–31, Springer. doi: 10.1007/s12110-005-1005-4.

Armah, F. A. et al. (2010) 'Impact of floods on livelihoods and vulnerability of natural resource dependent communities in northern Ghana', *Water*, 2(2), pp. 120–139. doi: 10.3390/w2020120.

Azmi, E. S., Rahman, H. A. and How, V. (2020) 'A two-way interactive teaching-learning process to implement flood disaster education in an early age: The role of learning materials', *Malaysian Journal of Medicine and Health Sciences*, 16(Supp 11), pp. 166–174.

Balgah, R. A., Bang, H. N. and Fondo, S. A. (2019) 'Drivers for coping with flood hazards: Beyond the analysis of single cases', *Jàmbá Journal of Disaster Risk Studies*, 11(1). doi: 10.4102/jamba.v11i1.678.

Cools, J., Innocenti, D. and O'Brien, S. (2016) 'Lessons from flood early warning systems', *Environmental Science & Policy*, 58, pp. 117–122. doi: 10.1016/j.envsci.2016.01.006.

Fothergill, A. (2017) 'Children, youth, and disaster', in *Oxford Research Encyclopedia of Natural Hazard Science*. Oxford University Press. doi: 10.1093/acrefore/9780199389407.013.23.

Frazier-Anderson, P. N. (2008) 'Public schooling in post-Hurricane Katrina New Orleans: Are charter schools the solution or part of the problem?', *The Journal of African American History*, 93(3), pp. 410–429. doi: 10.1086/JAAHv93n3p410.

Hanna, R. and Oliva, P. (2016) 'Implications of climate change for children in developing countries', *The Future of Children*, 26(1), pp. 115–132. doi: 10.1353/foc.2016.0006.

Intergovernmental Panel on Climate Change (IPCC) (2012) 'Managing the risks of extreme events and disasters to advance climate change adaptation', https://www.eea.europa.eu/data-and-maps/indicators/direct-losses-from-weather-disasters-1/ipcc-2012-managing-the-risks.

Izadkhah, Y. O. and Hosseini, M. (2005) 'Towards resilient communities in developing countries through education of children for disaster preparedness', *International Journal of Emergency Management*, 2(3), p. 138, Inderscience Publishers. doi: 10.1504/IJEM.2005.007355.

King, D. (2002) 'You' are on our own: Community vulnerability and the need for awareness and education for predictable natural disasters', *Journal of Contingencies and Crisis Management*, 8, pp. 223–228.

King, T. A., Tarrant, R. A. C. and Tchg, D. (2013) 'Children's knowledge, cognitions and emotions surrounding natural disasters: An investigation of year 5 students, Wellington, New Zealand', *Australasian Journal of Disaster and Trauma Studies*, 1, pp. 1–10.

Kousky, C. (2016) 'Impacts of natural disasters on children', *Future of Children*. doi: 10.1353/foc.2016.0004.

Kunateh (2016) 'Floods displace over 121,000 people in northern Ghana', www.ghanadot.com/news.ghanadot.kunateh.092809f.html (Accessed: 3 November 2020).

Lopez, Y. et al. (2012) 'Child participation and disaster risk reduction', *International Journal of Early Years Education*, 20(3), pp. 300–308. doi: 10.1080/09669760.2012.716712.

Martin, M. L. (2010) 'Child participation in disaster risk reduction: The case of flood-affected children in Bangladesh', *Third World Quarterly*, 31(8), pp. 1357–1375. doi: 10.1080/01436597.2010.541086.

Mburu, M. G. (2020) *Flood Disaster Risk Reduction Strategies and Participation Rates of Pupils in Primary Schools in Tana Delta Sub County, Tana River County Kenya* (Master Thesis). University of Nairobi.

Mensah, H. et al. (2021) 'Resilience to climate change in Ghanaian cities and its implications for urban policy and planning', *SN Social Sciences*, 1(5), p. 118. doi: 10.1007/s43545-021-00123-8.

Mensah, H. and Ahadzie, D. K. (2020) 'Causes, impacts and coping strategies of floods in Ghana: A systematic review', *SN Applied Sciences*, 2(5), p. 792. doi: 10.1007/s42452-020-2548-z.

Mitchell, P. and Borchard, C. (2014) 'Mainstreaming children's vulnerabilities and capacities into community-based adaptation to enhance impact', *Climate and Development*, 6(4), pp. 372–381. doi: 10.1080/17565529.2014.934775.

Mort, M. et al. (2018a) 'Displacement: Critical insights from flood-affected children', *Health & Place*, 52, pp. 148–154. doi: 10.1016/j.healthplace.2018.05.006.

Mort, M. et al. (2018b) 'From victims to actors: The role of children and young people in flood recovery and resilience', *Environment and Planning C: Politics and Space*, 36(3), pp. 423–442. doi: 10.1177/2399654417717987.

Mudavanhu, C. (2014) 'The impact of flood disasters on child education in Muzarabani District, Zimbabwe', *Jàmbá: Journal of Disaster Risk Studies*, 6(1). doi: 10.4102/jamba.v6i1.138.

Musah, B. A. N. and Akai, C. (2014) 'Effects of flood disasters on livelihood coping mechanism in Tolon/Kumbumgu district of northern region of Ghana', *International Journal of Agricultural Policy and Research*, 2(1), pp. 033–040.

Musah, B. A. N. and Oloruntoba, A. (2013) 'Effects of seasonal floods on households' livelihoods and food security in Tolon/Kumbumgu District of the Northern Region, Ghana', *American Journal of Research Communication*, 1(8), p. 12.

Muzenda-Mudavanhu, C., Manyena, B. and Collins, A. E. (2016) 'Disaster risk reduction knowledge among children in Muzarabani District, Zimbabwe', *Natural Hazards*, 84(2), pp. 911–931, Springer Netherlands. doi: 10.1007/s11069-016-2465-z.

Peek, L. (2008) 'Children and disasters: Understanding vulnerability, developing capacities, and promoting resilience – an introduction', *Children Youth and Environments*, 18(1), pp. 1–29.

Santoro, S. et al. (2019) 'Assessing stakeholders' risk perception to promote nature based solutions as flood protection strategies: The case of the Glinščica river (Slovenia)', *Science of The Total Environment*, 655, pp. 188–201. doi: 10.1016/j.scitotenv.2018.11.116.

Schonfeld, D. J. and Demaria, T. (2015) 'Providing psychosocial support to children and families in the aftermath of disasters and crises', *Pediatrics, American Academy of Pediatrics*, 136(4), pp. e1120–e1130. doi: 10.1542/peds.2015-2861.

Seballos, F. et al. (2011) 'Children and Disasters: Understanding Impact and Enabling Agency', *Children in a Changing Climate Research*, May, pp. 1–60.

Shah, A. A. et al. (2020) 'Looking through the lens of schools: Children perception, knowledge, and preparedness of flood disaster risk management in Pakistan', *International Journal of Disaster Risk Reduction*, 50, p. 101907, Elsevier Ltd. doi: 10.1016/j.ijdrr.2020.101907.

Silverman, W. K. and La Greca, A. M. (2002) 'Children experiencing disasters: Definitions, reactions, and predictors of outcomes', in *Helping children cope with disasters and terrorism*. American Psychological Association, pp. 11–33. doi: 10.1037/10454-001.

Takeuchi, Y., Mulyasari, F. and Shaw, R. (2011) 'Chapter 4 roles of family and community in disaster education', *Community, Environment and Disaster Risk Management*, pp. 77–94. doi: 10.1108/S2040-7262(2011)0000007010.

Tasantab, J. C. (2019) 'Beyond the plan: How land use control practices influence flood risk in Sekondi-Takoradi', *Jàmbá Journal of Disaster Risk Studies*, 11(1). doi: 10.4102/jamba.v11i1.638.

UNDRR (2020) 'Worldwide initiative for safe schools: For every new school to be safe from disaster', www.undrr.%0Aorg/publication/worldwide-initiative-safe-schoolsevery-new-school-be-safe-disaster (Accessed: 12 October 2020).

UNESCO (2013) 'Comprehensive school safety', www.unesco.org/new/%0Afileadmin/MULTIMEDIA/HQ/SC/pdf/Comprehensive_school_safety.pdf (Accessed: 15 February 2021).

UNESCO Institute for Statistics (UIS) (2011) 'Ghana: Age distribution and school attendance of girls aged 9–13 years', www.who.int/immunization/programmes_systems/policies_strategies/Ghana_country_report.pdf (Accessed: 11 July 2020).

24
Conclusions and final remarks

*Victor Oladokun, David Proverbs,
Oluseye Adebimpe and Taiwo Adedeji*

Introduction

This chapter summarises the key contributions from each of the chapters presented in this text, organised into the five major themes of: (i) impacts, challenges and particularities; (ii) preparedness, prevention, responses and recovery; (iii) risk assessments, flood mitigation and project management; (iv) infrastructure systems, urban systems and their management; and (v) community perspectives, resilience and adaptation.

Section I: impacts, challenges and particularities

In this section, the impacts and particular challenges faced by communities and governmental agencies in developing countries when confronting flooding were highlighted. The chapters brought together contributions from studies in Malaysia, Nigeria, Ghana and Sri Lanka and revealed some of the additional complexities involved, including some of the challenging circumstances found in developing countries.

Kabir et al. considered the effects of flooding on infrastructure in developing countries, focusing on Bangladesh, which faces regular disastrous flood events. Recent floods in 2017 and 2019 were highlighted as having negative impacts on critical roads and a huge population in coastal Bangladesh. The chapter reviewed the major impacts of flooding in Bangladesh, including a number of case studies from other developing countries. Geospatial techniques were used to identify affected road systems and population in Bangladesh. Based on the findings, relevant institutions were called upon to take necessary actions to mitigate the negative impacts of flooding in Bangladesh and beyond.

Akukwe et al. investigated the effects of flooding on agriculture and food security in southeastern Nigeria. They examined the relationships that exist between flooding and food security and offered ways to reduce flood impacts on agriculture and food security. The chapter adopted the framework of Food and Agricultural Organization–Food Insecurity and Vulnerability Information Management Systems (FAO-FIVIMS) to better describe the link that exists between the four pillars of food security and several influencing factors. Food security was conceptualised as a dependent variable with food availability, accessibility, utilisation and stability as the predictors.

Conclusions and final remarks

Data were obtained in two states in southeastern Nigeria that have suffered perennial flooding between 2012–2017. The findings revealed that flooding led to a significant increase in food insecurity in households. Also, the analysis revealed a negative relationship between food security and flooding. Flood-induced food insecurity which led to changes in food consumption patterns and affected the four pillars of food security was reported. For these reasons, they suggested ways to boost crop productivity and achieve food security in flood-affected areas, including the use of food safety nets, poverty alleviation for vulnerable households and techniques that boost crop productivity.

Simpeh et al. examined institutional and regulatory frameworks for flood resilience construction and adaptation in Ghana. They revealed that institutional and regulatory frameworks have a way of influencing the failures and successes of flood risk management in Ghana. They considered three main topics: (i) what aspects of the building codes/regulations relating to FRM are lacking and require improvement to regulate construction activities in flood-prone areas? (ii) What measures have been taken by the National Disaster Management Organisation (NADMO) to improve flood resilience in the construction industry? (iii) What are the effective approaches that can be adopted for the design and construction of flood-resilient buildings? They reviewed scientific literature and reports on institutional and regulatory frameworks for flood-resilient construction and adaptation and analysed these using Preferred Reporting Items for Systematic Reviews and Meta-Analysis (PRISMA) to extract the key findings. The findings revealed that flood risk management in Ghana is more of a reactive than a preventive approach. Also, there is no strong cooperation between and among necessary flood management organisations in Ghana or policies to support the development of resilient buildings. They highlighted the need for the development of a guide to adopt and promote flood resilience construction and adaptation.

Siriwardhanaa et al. discussed how resettlement has been adopted as a flood-preventive strategy to reduce the vulnerability of communities, especially those who live in flood-prone areas in Sri Lanka. While providing safe locations, resettlement needs to ensure long-term sustainability and satisfaction on social, economic and cultural aspects of the affected community. This chapter investigated the effectiveness of resettlement as a flood-preventive measure by probing into the lived experience of a community before and after resettlement. Accordingly, a case study was conducted for the Kalu River Bank Resettlement implemented following the 2017 floods in the Kalutara District in Sri Lanka. The findings disclosed that the resettlement of communities away from their native lands has both positive and negative consequences. Further, providing culturally and socially sensitive housing, infrastructure and land development was highlighted as being particularly important. The necessity of fulfilling community expectations through the development of a strategic resettlement policy framework rather than political favouritism was also reiterated.

Khaghanpour-Shahrezaee et al. considered the rising importance of being able to quantify resilience to improve risk management and how this has attracted remarkable attention due to the increasing number of natural and human-created disasters. They highlighted how the number of studies on resilient communities and recovery strategies related to developing countries has been rather limited. Generally, the main parameter used to estimate resilience has been *downtime*, that is, the time needed for calculating losses and planning the repair process. Their chapter investigated the downtime of buildings damaged by floods in developing countries using a three-storey building located in Nepal as a case study. In this work, the downtime was affected by three components: delays due to planning and finances, the time needed to repair building components and the time required for restoring infrastructure systems. Delays and utility disruption were calculated based on available data from previous events; finally, downtime due to repairs was obtained through a survey of property owners. Their proposed methodology is purported to support decision-makers in improved understanding of the state of their communities and towards helping future flood risk management responses.

Section II: preparedness, prevention, responses and recovery

This section of the text attempted to highlight some of the approaches used to prepare, prevent, respond to and recover from flooding in developing countries. Examples and cases are drawn from Nigeria, Indonesia, Mauritius, Nepal and Bangladesh.

Oluseyi described the efforts that are now being directed to the use of spatial planning instruments for the control of land use, building developments, drainage, public infrastructures and other components of urban space to enable free flow of excess storm water. Spatial planning is viewed as a rational and systematic process of guiding public and private actions and influencing the future by identifying and analysing alternatives and outcomes. A review of the various approaches to special planning was presented and highlighted some of the specific tools for planning in preparation for flooding using Nigeria as an example. The review highlighted the use of spatial data infrastructure and geospatial technologies to help prepare for flooding and improve the communication between emergency management agencies and government departments. The conclusions provided useful commentary on the benefits that these approaches can bring about based on lessons learnt from their implementation in the state of Oyo, Nigeria.

Prastica et al. considered flooding in urban areas in Indonesia. They highlighted the use of grey infrastructure such as canalisation as a leading trend for flood risk management. Moreover, they described the emergence of green infrastructure (GI) as a promising future urban flood mitigation strategy that needs further analysis and integration. Their chapter proposed an engineering approach for urban flood management by employing numerical programs as supporting tools, such as HEC-RAS and Personal Computer Stormwater Management Model (PC-SWMM), which align with the case studies in Jakarta and Ambon, respectively. HEC-RAS can provide policy scenarios regarding canalisation, whereas PC-SWMM delivers GI strategies. Both supporting tools produce quantitative results in terms of flood reduction and various scenario alternatives in the water-related disaster management field. The use of these tools is recommended to policy makers to assist in identifying whether projected scenarios can have a promising impact on the environment and urban flooding.

Chacowry considered the recovery process and defined it as the restoration of social, economic and environmental assets of flood-affected communities in developing countries and small island states. It is said that in developing countries, these processes present challenging issues for sustainable recovery in flood-prone communities, as they are influenced by a host of complex factors at local and national levels. Issues of social vulnerability, resilience and environmental justice were highlighted in the conceptual framework of recovery disparities. The importance of strengthening social capital was addressed and exemplified by a case study in the suburban town of Port Louis, the capital of Mauritius. This illustrates how a vulnerable community in a small island developing state perceives flood risk and how it develops coping strategies and plans to build resilience for early and sustainable recovery. Key challenges were highlighted and recommendations suggested for enhancing sustainable and speedy recovery.

Lageard and Bhattacharya-Mis considered how, in the wake of a worldwide health crisis, such as the COVID-19 pandemic, preparedness for disasters can be even more challenging, especially for communities in developing nations, who are already more vulnerable. Their evidence showed how investigating the dynamic balance between preparedness, communication and response helps in understanding the management systems of two very different forms of

Conclusions and final remarks

disasters happening together. They investigated a case in Bangladesh, one of the countries with the highest level of vulnerability to changing climate and large areas prone to frequent cyclones, with more than 80% of the population potentially at risk of flooding. While the country is renowned for its Cyclone Preparedness Programme, they highlighted the challenges faced in dealing with Cyclone Amphan during May 2020 and how this was hampered by the global pandemic. They examined the preparedness of the coastal population for dealing with the 'dual disasters' of Amphan and COVID-19. They interviewed community members and experts to elucidate the adequacy of their preparedness levels. The outcome of the exercise revealed that, due to inefficient institutional support and limited personal capacity, participants were not effectively prepared for the disasters. Their responses echoed some fundamental flaws in disaster management in the context of preparedness and response in tackling dual disasters for developing nations. Policy makers in Bangladesh and elsewhere are recommended to respond to these findings and to adopt new means of addressing and supporting vulnerable communities.

de Oliveira et al. reviewed flooding events in Brazil and highlighted the frequency of intense events each year, which are worsened by most cities' lack of infrastructure to face such events. The Santa Catarina state, located in the southern region of Brazil, suffers annually from flood disasters, mainly in urban centres. Due to these flood disasters and the consequential need for constant planning, protection and response actions, the Santa Catarina Civil Defence has become a national reference for good practice. The pattern of disaster events is the result of the interplay of climatic factors and characteristics of territorial occupation in Brazil. An urbanisation process marked by conflicting urban and environmental regulations further exacerbates the harmful effects of flood events. In practical terms, the removal of natural ground cover and its waterproofing effect has been recurrent, in addition to the occupation of flooded plain areas. The Santa Catarina Civil Defence's proactive stance in responding to flood events was consolidated in 2008, a year marked by an extensive disaster episodes in the state. This chapter highlighted some of the technical considerations and historical research of the state, with emphasis on the years of larger disasters. Other highlights presented were the administrative organisation, the actions of the Santa Catarina Civil Defence, the use of technological resources and future prospects.

Section III: risk assessments, flood mitigation and project management

This section of the book attempted to draw together ways in which policy makers and communities could take steps to mitigate the impact of flooding on vulnerable communities in developing countries. Systematic reviews were used to draw together and summarise previous research in this area. Two cases were drawn from Nigeria to highlight good practice in managing finances and policy making.

Karmaoui proposed a bibliometric approach to evaluating existing flood vulnerability information, with a focus on developing countries. The chapter demonstrated how Big Data taken from the most prominent sources can help researchers better understand and predict research trends in the subject of vulnerability. It concentrated on two important topics: (i) how scientific research on flood vulnerability was structured in these developing countries and (ii) the current research trends on this topic. The findings revealed five categories of flood vulnerability in developing nations, with ecological, social and economic study disciplines all being linked. A recent focus on techniques such as mapping hazards and climate change, modelling vulnerability, risk assessment, livelihood, sensitivity and exposure, regression analysis and urban flooding scale were underscored by trend analysis. These trends suggest that these variables and techniques could be integrated in flood vulnerability assessments. The findings outlined a way for quickly identifying the important parts of a flood vulnerability assessment, which aids in understanding the intricacies of flood effects.

Dutta and Kabir considered the use of flood risk assessments (FRAs) in formulating disaster management strategies and identified climatic and geographical factors together with spatial and temporal variability and impacts on society. While developing countries like Bangladesh, India, Nigeria and Indonesia are known to have high impacts, the FRA studies in these countries were found to be less noticeable before 2018. A systematic review was used to highlight models and techniques employed in FRA studies used in these countries. Additionally, required data and their sources, which were used for risk mapping, land use planning, flood mitigation and management, were also focused on with the approach. Most of the studies integrated statistical models; remote sensing; and GIS models like the analytical hierarchy process, hydrological and watershed models. Though most of the data were available locally for the mentioned developing countries, global disaster and spatial databases were also significantly used in the studies.

Ayorinde et al. highlighted how flood risk management projects in most developing countries suffered from poor financing and weak political will. They described how the low-resource economies in most of these countries are often a militating factor against the implementation of such projects. Unfortunately, the cost of implementing structural and non-structural measures for flood control can be daunting, especially in the type of low-resource economy operated by most developing countries. The governments of these countries do not usually prioritise the problems of flooding, as they have existential problems that far outweigh the problems of flood disaster. Most of the flood risk management projects in developing countries are usually funded by external donors through loans, grants and capacity development. This chapter documented the experiences of the Ibadan Urban Flood Management Project (IUFMP) in the area of project financing and implementation strategies and described the financing and implementation strategies used, including the successes and challenges.

Nkwunonwo considered the strengths, weaknesses and potential of flood risk management policies in developing nations, notably Nigeria. The chapter highlighted lessons from worldwide best practices and the opportunities that can contribute to realistic flood risk management in developing countries, drawing on previously published work. It investigated the strengths and weaknesses in light of global best practices, as well as the potential grey areas, obstacles and possibilities that come with applying fit-for-purpose procedures in developing countries. The importance of adopting policies based on indigenous approaches and solutions to flood risk was highlighted. The chapter closed by emphasising the need to adopt customised models and techniques that take advantage of free and open-source geospatial technology, as well as developing local technical capacity and building local infrastructure, as ways to help developing countries achieve their sustainable development goals.

Section IV: infrastructure systems, urban systems and their management

This section considered the key infrastructure needs of developing countries in responding to increased flooding. Cases were drawn from countries including India, Bangladesh, Indonesia and Nigeria to highlight some of the challenges and opportunities for improved management of these important resources.

Sugam et al. discussed the need and opportunities for integrated approaches to water resources management and specifically towards integrated flood risk management in developing countries. They presented a review of the existing literature to highlight some of the recent

Conclusions and final remarks

developments in this area and described a number of case studies from both developed and developing countries to highlight some of the challenges and opportunities. Specifically, for developing and under-developed countries, the transition from traditional structural measures is considered a challenging task, requiring financial as well as technical support from other countries and the private sector. The role of information, communication and education (ICE) activities across various departments, educational institutes and the local community level was highlighted as one of the main step towards the adoption of IWRM for FRM.

Wahab and Kasim described how the city is integral to the fulfilment of sustainable development. However, the rapid scale of urbanisation in developing countries comes with numerous environmental challenges such as meeting the accelerated demand for affordable housing, uncontrolled pollution, increased demand for transport systems, the upsurge of informal settlements and urban sprawl, poor waste management systems and uncoordinated land use planning and development. These environmental challenges have awakened the consciousness towards achieving resilient cities, especially flood-resilient cities within the developing countries of the world. In entrenching resilience in such cities, the approaches identified in this chapter included assisting communities in recognising, reducing and managing risk; strengthening urban land use planning and enforcing building standards; facilitating dialogue and collaboration to reduce risk; and supporting holistic risk reduction across sectors.

Pudyastuti and Isnugroho described some of the major flood challenges in Central Java, Indonesia, and how these have impacted infrastructures including transportation, housing, electricity and irrigation systems. They highlighted the need for further research regarding flood risk and towards improving the sustainability of the urban infrastructure in Central Java Province. These were found to include the development of appropriate sustainability indicators; the impact of flood risk management approaches; efforts required to improve the hydrology, hydraulic and other data availability for flood mitigation; the evaluation and assessment of drainage networks; appropriate actions to raise community awareness; and appropriate approaches to persuade political leaders to address environmental, sustainability and climate change issues in the province.

Chowdhury et al. considered the occurrence of disasters in the South Asia region, making this one of the most vulnerable regions in the world. They highlighted that countries in this region face different impacts from flooding and must take necessary actions to minimise the impacts on life and resources. Their chapter attempted to shed light on current practices and challenges regarding the use of natural flood management in Bangladesh, India and Nepal. These three countries were found to be adopting both structural and non-structural measures as flood mitigation approaches; however, there are still several significant challenges in these countries in various forms. To ensure sustainable natural flood management, proper formulation and implementation of national legal documents and international agreements should be established. This chapter aimed at helping the various stakeholders involved, including local and national policymakers, to take necessary actions to ensure sustainable natural floodwater management is encouraged in the region.

Section V: community perspectives, resilience and adaptation

This section considered some of the more modern social approaches to managing flood risk in developing countries, including the use of community-based approaches to improving resilience. Findings were drawn from cases in Nigeria, Ghana and Malawi.

Munsaka discussed the important role of education in supporting the development of resilient communities at risk of flooding in developing countries. The chapter highlighted the use of both formal and non-formal education and how this supports the concept of learning. The chapter highlighted the links between vulnerability, resilience and adaptive capacity and how education can play a key role towards increasing capacity. The author argued that the value of education in managing flood risk cannot be underestimated and highlighted the essential role this plays in developing the source of skills, behaviours and capabilities required to build the resilience of communities to risk of flooding as well as the associated impacts.

Amoako and Dinye considered how residents in low-income communities in cities of the developing world are gradually developing their adaptive capacity against floods. In spite of the increasing risks they are exposed to, residents in flood-affected low-income communities continue to live precariously in hazardous areas. The chapter highlighted how earlier works pointed to the limitations of flood-affected low-income households and portrayed them as passive and incapable of adequately responding to floods. Contrary to this 'victims' approach' to flood vulnerability, it was found that some low-income residents have developed grassroots resilience through decades of experience with flood events. For these residents, flood response was found to have been *learned, embodied and embedded in everyday life and living*. This chapter unpacked the emerging social networking and grassroots resilience to floods in urban low-income communities in developing countries. Empirical evidence drawn from two cities in Ghana – Accra and Kumasi – was used to establish a new framework for understanding community flood resilience.

Homberg and Trogrlić explored the challenges and opportunities in creating socially just flood risk management that adequately reaches and supports the most vulnerable and poor communities. Drawing on a case study in Malawi, they provided examples of both public and private flood risk management actions across the disaster management phases of preparedness, response and mitigation. They examined risks of having parallel systems of both NGO and state-led flood risk management with weak levels of integration, and recommendations were made for a more holistic flood risk governance approach across actors, geographic levels and flood types. Currently, developing countries are found to have lower levels of access to technology, funding and capacity required for effective flood risk management. Social justice principles such as distributive and normative justice are recommended to drive the equitable allocation of resources within a country, recognising the commitment to leave no one behind.

Mensah et al. explored the perspective of school children on the causes, effects and coping strategies for flood risk management in Ghana. The chapter revealed that school children perceived the main causes of flooding to be due to choked gutters and poor drainage systems. Additionally, flooding was found to have resulted in health risks for children and impacts included destroyed properties, damaged facilities and school buildings, which resulted in low attendance in schools. The coping strategies employed by school children during floods were to rely mainly on their instincts and to seek shelter in temporary havens. Pupils were found to be poorly educated about the risks of flooding, and recommendations were made towards integrating flood risk management into the curriculum, developing appropriate instructional materials and undertaking capacity building for teachers.

Final remarks

This final chapter has attempted to summarise the key findings from this unique collection of contributions, which have highlighted the array of challenges involved in managing flood risk in developing countries. The chapters have drawn on findings and experiences gained from a wide range of developing countries across the various continents and regions. The studies have highlighted the additional complexities and challenges, which encompass the economic,

Conclusions and final remarks

technical and political spheres. The chapters also highlighted the opportunities for sharing across developing countries as well as with other developed countries in the search for new innovative approaches to helping our communities become more resilient. In our efforts to become more resilient, it is essential that resilience becomes *learned, embodied and embedded in everyday life and living*. It is hoped that this text can play a contributing role in this endeavour and in helping communities in developing countries to bounce back from the effects of flooding.

Index

acceptable risk 233
adaptation 41
adaptive design 273
Adebimpe, Oluseye Adewale ii–iii, v, viii–ix, 1, 48, 390
Adedeji, Taiwo ii–iii, v, viii–ix, 1, 266, 267, 390
Adefioye, Abiodun vi, ix, 217
adoption pathway 41, 50–51
advance warning 140
agriculture and flood 24
Ahadzie, Divine Kwaku v, viii, ix, 41, 43, 48, 50, 375, 385
AHP analytic hierarchy process 189, 205
Ajibode, Ibadan Nigeria 217
Akukwe, Thecla Iheoma v, ix, 24–27, 29, 31, 34, 390
Alfath, Abdunnavi vi, ix, 109
Amoako, Clifford vii, ix, 44, 333–345, 396
ArcGIS 15, 189, 235
architectural 55, 62–64
audit 222
Ayorinde, Adedayo Ayodele vi, ix, 217, 394

Bangladesh 9–13, 21, 137–157, 198, 202, 252, 305, 310, 312, 314
beneficiary selection 61
best practices 172, 219, 223, 226, 231, 235–242, 394
Bhattacharya-Mis, Namrata vi, ix, 137, 392
bias 141
Braghirolli, Guilherme vi, ix, 161
Brazil 161–178, 393
building 12, 72; damageability 74; materials 57–67; regulations 42–52; resilience 4, 130, 132, 135, 270, 277, 327, 339; standards 2, 276–277, 350, 395

Canada 238
capacity to respond 138
carbon 271
catastrophic flood 14
Chacowry, Anoradha vi, ix, 123, 129, 392
China 183, 186, 236, 256, 323

chi-square test 80, 82
cholera 260, 383
Chowdhury, Md. Arif vii, ix, 199, 302, 304, 395
chronic: food 24; stress 129
Cimellaro, Gian Paolo v, ix, 70–71, 80
civil defence 162–177, 393
climate change 2, 9, 12, 21, 24, 26, 93, 178, 186, 189, 193, 229, 252, 266, 277, 350
Coastal Area Rehabilitation Project 140
community recovery processes 62
construction industry 48
COVID-19 137
COVID 19 standards 137, 148
critical infrastructure 273, 321
Cyclone Amphan 138
cyclone hazards 145

data: Copernicus Climate Data Store 210; Dartmouth Flood Observatory 210; global databases 210; NASA data portal 210; open-source data 238
Datta, Srijon vi, ix, 197
De Iuliis, Melissa v, ix, 70–72, 74, 78
density visualization 186
design standards 59
developing countries 1, 10
development control 96
Dinye, Irene-Nora vii, ix, 333, 396
disaster preparedness 138
downtime of buildings 70
downtime model 72
drainage 3, 4, 14, 42, 93, 94, 97, 102, 184, 198, 205, 224, 250–251, 285, 290, 293; analysis, 117–118; systems 3, 10, 12, 13, 48, 102, 124, 173, 183, 193, 232–233, 267

early warning systems 97
ecological networks 273
ecosystem services 236
engineering mobilisation process 76
environmental justice, 126, 132
environmental and social safeguard 225

Index

Fabiyi, Oluseyi. O vi, ix, 93, 94, 97, 217, 237
financing process 77
flash floods 203, 232, 353
flood: forecasting 256, 305, 307, 315; hazard 1, 2, 204, 257, 315; impact 4, 9, 14, 24, 35, 305; insurance 112, 173, 179, 259; management 1, 42, 217–228, 291, 303, 305; management practices 260; mitigation 110, 362, 364; models 202–203, 240; preparedness 102, 354; resilience 47, 350–351; responses 341, 360; risk 197; software 238
flood hydrograph 114
floodplain 259
flood-resilient construction 43
flood risk: analysis (FRA) 213; assessment 197, 203–205; communication 235; management 41, 46, 229, 235, 250, 251, 285, 290, 293, 326, 354; management cycle 124; management framework 98; management policy 229; management project 217–228
flow simulation 115
food safety net 37
food security 24, 25, 28, 36, 33, 35
fuzzy multi-criteria 207

geographic information system (GIS) 12, 105, 184, 205, 238, 326
George, Sherin Shiny vii, ix, 249
geospatial: analysis 13; impact 13; mapping 13; QGIS, 238
Germany 324
Ghana 41, 48
global: flood risk 198, 212; warming 55
greenhouse gases 9, 94
green roof 290
gross domestic product (GDP) 354
groundwater 13, 109

Habiba, Sayeda Umme v, vii, ix, 9, 10, 302
hazard risk maps, 41, 48, 326
health facilities 13, 55, 126, 382; hazards 132, 155–156, 225; risk 382
Henrique de Oliveira, Francisco vi, ix, 161
Hossain, Md. Lokman v–vii, ix, 9, 12, 197, 302
Hossen, Md. Nazmul v–vii, ix, 9, 197, 302, 315
hydrology 99

Ibadan, Nigeria, 217, 218, 274
Ibadan Urban Flood Management Project 217, 275
impacts: on buildings and infrastructure 12
India 311
Indonesia, 109, 205–207; Java Province 282
infrastructure 9
inspection process 76

integrated flood risk management 236
Isnugroho, Bingunath Ingirige vii, ix, 282, 395

Jakarta 110, 392
Japan 326
Java Province 207, 395
Juthy, Nure Tasnim vi, ix, 197

Kabir, Md. Humayain v–vii, ix, 9, 197, 249, 302, 390, 394
Karim, Mir Enamul vi–vii, ix, 197, 302
Karmaoui, Ahmed vi, x, 183, 184, 192, 393
Kasim, Oluwasinaayomi vii, x, 266–268, 270–273, 395
Kenya 14
knowledge: accumulated 322; gaps 303; indigenous 312; local 303, 315, 337–339; sharing 270
Kulatunga, Khaghanpour-Shahrezaee Udayangani v, x, 55, 144

Lageard, Sabiha vi, x, 137, 392
land use 2–4, 43–44, 49–50, 95–103, 118, 172–173, 192, 201–209, 221, 230, 258, 260–277, 290, 304, 392–395
level of risk, 276
Linheira, Guilherme vi, x, 161
living standards 242

malaria 125, 260, 382
Malaysia 285, 390
management plan 200, 209, 223, 227, 250, 259, 292–293, 304, 310
Master Plan for Urban Drainage 172
Mba, Lilian Chinedu v, x, 24
Mediterranean Sea 258
megaphone 140, 148
Mensah, Henry v, viii, x, 41, 45–46, 48, 375–376, 385, 396
mental health issues 126
mental strategic 129
meteorological 207; agencies 104; coverage 174; data 311; disasters 303; ensemble predictions 241; events 161; forecasts 260, 357; radars 174; satellite data 99; stations 305
Meteorological Services Department of Ghana 43, 50
Meteorology, Climatology and Geophysics Agency (BMKG – Badan Meteorologi, Klimatologi, Dan Geofi sika) 210
micro-insurance and appropriate grants 257
micro-level political engagement 56
micro-and meso-scale building 272
micronutrient 27
minimum river setback benchmark 221
Ministry of Environment and Forestry 210

Index

Ministry of Land (MoL) 209
Ministry of Lands and Survey, Abakaliki 210
mitigation strategies 183
mobile phone 148; accounts 153; alerts 147; based dissemination of early warning messaging 358; connection 360; cyclone-related information 148; floor barrier systems 51; network 146
modelling 118; software 111
MODIS 211–212
monetary impact 352; value of damage to property 218
monsoon season 10, 14, 59, 312, 315
mortality 56, 312, 383; rate 199
multicriteria index 207
multidisciplinary: concept 268; and multi-agency cooperation 359, 370; and multicultural perspectives 3; platform 233; quantitative and qualitative approach 351; technical and financial services 272
multipurpose: cyclone shelters 139; interventions 41; projects 176
multi-sectoral partnerships 140; approach 254; meetings 260; project 226
municipal spatial data infrastructure 96
Munsaka, Edson vii, x, 321, 364, 395

National Aeronautics and Space Administration (NASA) 210–212
national budget 354
National Commission on Floods 304
National Disaster Management Agency (BNPB –Badan Nasional Penanggulangan Bencana) 210
National Disaster Management Organisation (NADMO) 42–43, 378, 391
National Disaster Prevention and Reduction Day 323
National Disaster Risk Management Policy of Malawi 354
national EWS 311, 356–357, 359–360
national-level FRM 352
national and local governmental level 351
national meteorological policy 359
national and regional: laws 259; level 292–294; roadways and railways 307
National Strategy for Disaster Risk Management 305
national and subnational government entities 352
National Water Management Plan (NWMP) 254, 304
National Water Plan (NWP) 254
National Water Policy (NWPo) 254, 304, 311
Natural Disaster Relief/Calamities Act 305
Nawaz, Shahpara x
Nigerian Meteorological Agency (NiMet) 210
Nigerian Urban Development Decree 96
Nkwunonwo, Ugonna C. v–vii, x, 9, 197, 302

NOAA 212
numerical: computing 240; data reporting 78; mathematics research 240; modelling 213; models 110, 213; programs 392; schemes 241; simplifications 240
Nurfaida, Wakhidatik vi, x, 109
nursing discipline 328
NVIVO 379

Odaw River 336–337, 340
Office of the Surveyor General of the Federation (OSGOF) 210
Oladipo, Olakunle vi, x, 217
Oladokun, Victor Oluwasina i–iii, v, viii, x, 1, 2, 25, 233, 390
Oluoko-Odingo, Alice Atieno v, x, 24, 27
Ona River 224
operational 99, 140; activities 228; authority 343; costs 220; decision support system (ODSS) 357; disruptions 306; expenses 101; frameworks 233; inefficiency 14
optimise: flood simulation 241; prospects of IFRM 237
optimistic 'cathartic' idea 229
Ossai, Onyinyechi Gift v, x, 24
Our World in Data 212

Padilha, Victor Luis vii, x, 161
Padma River 10–11
Pakistan 10, 14, 21, 26, 183, 201, 256, 302
Pan-African flood forecasting system 241
Panceri, Regina vi, x, 161
pandemic 12, 126, 134, 135, 137–139, 141, 143, 149, 227, 392–393; see also COVID-19
Paris 252
Paris Agreement and Framework Convention on Climate Change 327–328
Pearl River Delta 183
perennial: flooding 29, 380; flooding problems 105; floods 334, 337; water 13
Performance Assessment Calculation Tool (PACT) 71
performance measures 252
performance standards 50–51; for resilient materials and technologies 50–51
permeable pavement 118–120
Personal Computer Stormwater Management Model (PC-SWMM) 110, 392
phases 127, 132
Phukan, Mayuri vii, x, 249
Physical Sciences Laboratory (PSL) 212
planning standards 94–95, 99, 102
pluvial flooding 232–233, 285, 288, 312, 370
political 65; actors 94; agency 345; alliances 44; aspect 56, 58, 64, 193, 291; benefits 371; capacity 242; connections 149; and economic conditions 236; economy 236;

Index

factors 236; favouritism 67, 391; interference 226; involvement 61–63; leaders 299, 395; party 154; party change 291; power 238; representatives 153; resources 335; responses 345–346; stability 276; undercurrents 226; views 113; willpower 233, 276, 394
pollutants 296
poor policy-operational linkages 140
population density 109–110, 117, 209, 258
population growth 2, 70, 125, 161, 209, 233–234, 242, 267, 308, 354
Po river 352
post-flood resettlement 56
Prastica, Rian Mantasa Salve vi, x, 109, 111, 113, 392
Precipitation Measurement Missions Data and Access 212
preliminary: flood risk assessment 238; flood water extent 13; testing 386
Prieto, Raidel Baez vi, x, 161
private sector 128, 261–263, 292, 323, 385
probabilistic approaches 71
procurement process 225
productive sector 171
project appraisal document (PAD) 220
project management 3, 181, 220, 226–227, 228, 393
project performance 255
project steering committee (PSC) 220
project technical committee (PTC) 220
property: damage 14, 183, 197, 209; destruction 60; losses 272; owners 392; ownership 236
Property Flood Resilience (PFR) *see* property-level flood resilience
property level flood protection 367
property-level flood resilience 50–51
Proverbs, David ii–iii, v, viii, x, 1, 2, 25, 41, 43, 47–50, 233, 390
psychology 268
public health 103; delivery infrastructure 140; and environmental preservation 47; threats 380
Pudyastuti, Purwanti Sri vii, x, 282, 284, 290, 395

qualified: personnel 106; staff 276
quality 202, 208, 226, 239; aspect 189; and availability 197, 199, 201; of life 126, 173; of shelters 143
quantification 70
quantifying downtime of residential buildings 71–72, 74
quantify resilience 70, 391; repair time 75; risks 173; water volume 113

radar images 175
rain: barrel 118–120; fed agriculture 24–25, 28; gardens 308; gauging seasons 291
rainfall: forecast 206; seasons 336

rainwater 290; harvesting 290, 324; infiltration rate 293
rainy season 336–337, 378
rational components 71
reconstruction: damaged caused by rains 171; infrastructure 362; of roads 355
recovery and reconstruction of essential services 166; disparities 126, 392; emergency 363
Red Cross 10, 43, 212, 292, 356, 361; Disaster Mapping 212
rehabilitation 134; of associated hydraulic structures 224; of associated and stand-alone structures 224; functions 313; measures 261; phase 132; programs 183; reconstruction of infrastructures 362; reconstruction phase 123–124, 127; recovery phases 132; resources 154; services 141; urban infrastructure 101
religions 291
remote area 140, 149, 154–156
remote sensing 26, 183, 205–207, 213, 240, 272, 394; approaches 183, 193; techniques 189; technologies 202
renovation and retrofitting 275
repair: of damaged road network 296; and reconstruction 12, 78; and retrofit 76; sequences 75
reservoir 256; flood control downstream 176; modeling 111; operating organizations 305
resettlement 55; programmes 56
residential 95–96, 221; areas 118; buildings 70–72, 83, 221, 267, 275, 346; items 270; and non-residential structures 44; properties 381
Resilience-based Earthquake Design Initiative (REDi TM) 71
resilience and emerging political agency 345
resilience function 71
resilient community 57, 70
resistance: and recovery 175; and resilience 335
restoration and recovery 123
retrofit: and adaptation of existing buildings 51; existing systems 2; process 77
riparian communities 127, 257, 262
risk-based framework 229
risk-based hazard 48
risk management plan 250
river: Basin Agency 114; canalization 110; capacity improvement 110; catchment 99, 356; channel capacity 99; channel depth 99; competence 99; confluence 209; data 203, 209; discharge 110, 202; embarkments 13; gauges 357, 359; modeling 110–111; morphological, geographical, socio-economic, disaster-related and climate model data 207; normalization 114–115; overflow 14; protection works 368; runoff 9; sedimentation 173; setbacks 95–96, 102; silting 291; training 94, 296, 355, 362; water level 204, 209

401

Index

river areas 292
riverbank 58–60, 63
river basin 258; authorities 257; management 257–258
river basins 2, 113, 205, 207, 241, 256, 258, 261, 282, 290, 304–305, 308, 311, 314
River Benue 232
river flood 111–112, 267, 353, 356, 370
river flow 114, 260
riverine floods 201–202, 256, 337
River Niger 232
River Ogbere 217
River Ogunpa 217
River Ona 217
River Soliette 183
River Watch Version 2 Satellite River Discharge Measurements 212
Rudorff, Frederico vi, x, 161

Sacramento-San Joaquin Rivers 259
sandbags 342
San Francisco Bay Area 259–260
sanitary 145; hygiene 155, 156; landfill sites 97
satellite 174, 208; data 99, 238; image 203–206, 208, 211, 213, 350; imagery 238; River Discharge 212
Scopus database 184
sea level: change 212; rise 205, 213, 252, 291, 304, 306
seasonal: floods 260; forecasts 358; monsoon rainfall 252
sea sustainability 258
sectoral: expertise 272; integration 255; needs 172
sector-specific development initiatives 277
Sendai Framework 377
septic tank 58
SERVIR 211–212
severe: disasters 70; economic losses 285; effects 11; environmental circumstances 272; flood disaster 123; flooding 9–10, 55, 252, 260, 302; floods 13, 60, 376; impact 55, 70; intensity 11; levels of food insecurity 25, 30; storms 9
sewage 3–4
sewerage 251, 308
sewers 117
Shire River Basin Management Programme (SRBMP) 357
Simpeh, Eric Kwame v, viii, x, 41, 48, 375, 391
sirens 31
Siriwardhana, Senuri v, x, 55–56
Siwaliks, Nepal 316
Slovic, Paul 129
slum dwellers 333, 339
Slum Dwellers International (SDI) 339
small and medium rivers 176
small and medium watershed 176
Soares, Maria Carolina vi, x, 161

social: activities 56; aspects 64–67, 291; capital 123, 127–129, 132, 135, 141, 151, 269, 334–335, 339, 344, 346, 392; communities 272; conflicts 64; cost 125; dimension of vulnerability 134; disparities and uncertainties 132–133; disruption and economic stress 129; distancing 126, 137, 155; ecology 268; economic impacts of resettlement 65; environment 61, 129; environmental safety 219; equity and environmental justice 126; events 59; exclusion 133; factors 120; groups 58, 64, 126, 141, 153, 339, 344; impacts through multi-sectoral partnerships 140; inequalities 125; infrastructure 56; injustice 126; interactions 58–59, 61–64; issues 67; justice in flood risk management 351; learning 252, 335, 341–342, 345–346; level 64; marginalisation 345; marginalisation and vulnerability 134; media 129, 147, 226, 358; networking for flood mitigation 131; networks 138, 337–340, 344; perceptions 58; power structures 139; problems 291–296, 302; protection 152–153; review of resilience 327; safeguards 219–220, 225; sector 166, 171; vulnerability 125; welfare 44, 249, 334
socialization 298
socially: inclusive disaster management 387; just flood risk management 4, 350–352, 396; sensitive housing 391
social sector 166
socioeconomic 189, 192, 204, 207; condition 125, 133, 305, 316, 342; damage 14; data 204, 209; demographic data 206; factors 36, 165; impacts 2, 4, 55–56, 65, 223, 305; inequalities 255; recovery programmes 153; recovery of vulnerable communities 126; sectors 9; status 139; variables 229; vulnerabilities 240
Socioeconomic Data and Applications Center (SEDAC) 211–212
sociology 197
socio-political: agencies 346; factors 255
Solo River 284, 290
Southeastern Nigeria 24–25, 29, 34–37, 205, 390–391
Southwest Area Integrated Water Resources Management Project (SWAIWRMP) 255
Southwestern Nigeria 220, 275
Southwest Indian Ocean cyclone season 325
space standards 96, 99, 101–102
spatial: analytical techniques 94; data 96, 103, 105, 175, 213, 224, 392; decision support systems 207; modelling 207; planning 4, 43–44, 93–96, 99, 101, 105, 107, 272, 291, 392; scale analysis 201
Sri Lanka 55–68
stairs 75, 84–85, 342
stakeholder and sector cooperation 290

402

Index

storm: damage 141; storm drain 340–341, 343, 346; surge 9, 93, 139, 147, 186, 203, 205, 207, 213; Warning Centre (SWC) 140; water 95–97, 99, 101–103, 107, 221, 259, 290, 296, 308, 392
strategic master-plans 223, 226
strategic partnerships 345
strategic planning 95, 266, 290
strategic resettlement policy framework 67, 391
structural: retrofitting 385; standards 57
Sulaiman, Muhammad vi, x, 109
supply chain for fund distribution 154
supply chains 271
supply channels 156
surface water 44
sustainable: cities 42; community livelihoods 230; development 47–48, 395; development initiatives 230; food management 201; natural flood management 314–315, 395; natural floodwater management 395; recovery 123, 126, 132–133, 392; urban development 172; urban drainage system 93; urban flood risk management 223; urban infrastructure 290
Sustainable Development Agenda 266
Sustainable Development Goal (SDG) 35, 37, 251, 394
Sweden 202

tangible assets 146
tangible and intangible impact 130
task team leader (TTL) 220
technical 250; audits 222; capacity 230, 238, 242, 358, 394; effectiveness 172; expertise 134, 364; guidance 62; knowhow 150; knowledge of flooding and its risk management 235; officers 59, 62–63; and political spheres 397; regulations 256; skills 242, 261; staff 57; support 57, 156, 395; Unit 219
Technical-Scientific Group 171
technology standards 103
temperature data 204, 211
temperature of the pacific ocean 162
temporary: havens 396; measures 352; project-based funding 370; shortage of drinking water 147
temporary shelter 57, 298; arrangements 310
tidal river management 304, 306, 310
toilet 342
traditional social structures 125
traffic congestion 234
transboundary river management 256
transboundary rivers 255, 282
trans-national alliances 339, 346
transnational social resource 345
transportation 273, 290, 354, 395; industry 10; infrastructure 284–285, 294; sector 14; and telecommunication 292

Trogrlic, Robert Šakic vii, x, 350–351, 354, 357, 362, 364, 370, 396
tropical: cyclones 312, 353; depression 315; zone 161
Tropical Africa south of the Sahara 220
Tropical Rainfall Measuring Mission (TRMM) 211–212
tropical wet-and-dry climate 29
tsunami 57, 125, 296; evacuation drills 324

UK charity 145
Ul Islam, Syed Labib vii, x, 302
uncertainty of climate 294
underprivileged and minority communities 126
UNESCO 323–324, 377–378
Unified National Program (UNP) for Floodplain Management 259
United Kingdom research into flood vulnerability 186
United Nations' agenda for disaster risk reduction 229
United Nations Centre for Regional Development [UNCRD] 48
United Nations Declaration 42
United Nations Disaster Assessment and Coordination (UNDAC) 361
United Nations Educational, Scientific and Cultural Organization 377
United Nations International Strategy for Disaster Reduction (UNISDR) 124–125, 127, 197, 315, 326, 328
United Nations (UN) Member States 266
United Nations Office for the Coordination of Humanitarian Affairs [OCHA] 27, 267, 271, 361
United Nations Office for Disaster Risk Reduction 326
United Nations Sustainable Development Goals (SDGs) 127, 233, 238, 266
United Nations urbanisation projections 270
United States 48, 171, 186, 238, 250, 259, 375
unplanned: development 93; economic development 252; urbanization 55
urban development 12, 68, 99, 103, 172, 201; application of barrier islands 207; boards 101; planning 205; plans 2; uncontrolled 333; uncoordinated 267
urban drainage: networks 97; project 173; system 117, 173; systems 232
urban flooding 234
urban governance 102
urbanisation 102
urban planning space standards 95
urban and regional planning standards 99
urban sewerage 251
urban waste management 97
utility disruptions 71–72, 74

Index

van den Homberg, Marc vii, x, 350–352, 354, 358, 362, 369
vapor transport 113
vector diseases 125
Vietnam 285
voluntary instruments 252
VOSViewer software 184–186, 189, 191–193
vulnerability assessment 70, 187, 193, 201, 233, 394
vulnerability index 189

Wahab, Bolanle vii, x, 266–268, 270–271, 275, 395
Wanma, Grace viii, x, 375
waste disposal 93, 97, 221, 227, 251, 282, 385; projects 251
waste management 103
water-borne diseases 125, 335, 383
water depth 241
water entry technology 50
water exclusion technology 50
Water Framework Directive (WFD) 257
water industry 10
Water management plan 259
Water Research Institute (WRI) 43

Water Resources Commission (WRC) 43
water resources management plan 292
water sector 304; sensitive urban design 117; services 262; stress 258
water supply 117, 139, 150, 186, 249, 251–252, 259, 290, 294; infrastructure, 294; systems 3–4
watertight toilet 58
weather radars 177
wet season 296
WHO 140, 268
Wijayanti, Amalia vi, x, 109
World Bank 24, 41, 47, 49, 105, 139–140, 150, 162, 166, 171, 198, 219–223, 225–227, 249, 266, 268–269, 272, 275, 325, 357, 362, 370; Data Catalogue 212
World Bank Data Catalogue 212
WorldClim 212
World Health Organisation 267
World Resources Institute (WRI) 198, 212

Zambezi river basin 260
Zika virus 137
Zimbabwe 260–263, 323, 325
zoning 57, 96, 118